WISDOM, KNOWLEDGE, AND MANAGEMENT:
A Critique and Analysis of Churchman's Systems Approach

C.West Churchman and Related Works Series
Series Editor - John P. Van Gigch
Professor Emeritus, California State University

Volume 1:
RESCUING THE ENLIGHTENMENT FROM ITSELF:
Critical and Systemic Implications for Democracy
Editor: Janet McIntyre-Mills

Volume 2:
WISDOM, KNOWLEDGE AND MANAGEMENT:
Editor: John P. Van Gigch, in collaboration with
Janet McIntyre-Mills

Cover design by Susanne Bagnato, a Graphic Designer in Adelaide, Australia.
Contact details are: sbagnato@picknowl.com.au

WISDOM, KNOWLEDGE, AND MANAGEMENT:
A Critique and Analysis of Churchman's Systems Approach

edited by
John P. van Gigch
in collaboration with
Janet McIntyre-Mills

 Springer

John P. van Gigch, Professor Emeritus
California State University
Sacramento, CA, USA

Library of Congress Control Number: 2006932258

ISBN-10: 0-387-35389-5 (HB)   ISBN-10: 0-387-36506-0 (e-book)
ISBN-13: 978-0387-35389-0 (HB) ISBN-13: 978-0387-36506-0 (e-book)

Printed on acid-free paper.

© 2006 by Springer Science+Business Media, LLC
All rights reserved. This work may not be translated or copied in whole or in part without the written permission of the publisher (Springer Science+Business Media, LLC, 233 Spring Street, New York, NY 10013, USA), except for brief excerpts in connection with reviews or scholarly analysis. Use in connection with any form of information storage and retrieval, electronic adaptation, computer software, or by similar or dissimilar methodology now know or hereafter developed is forbidden.
The use in this publication of trade names, trademarks, service marks and similar terms, even if the are not identified as such, is not to be taken as an expression of opinion as to whether or not they are subject to proprietary rights.

Printed in the United States of America.

9 8 7 6 5 4 3 2 1

springer.com

DEDICATION

Sadly, whilst this book was in press, John. P. van Gigch died peacefully on the 29 August, 2006.

His wife Ann and his daughter Monique De Santis were with him in his last hours.

To the end his life was characterised by energy, wisdom, work and concern about others and the world.

He will be greatly missed in the many roles he played in life; not least being his role as teacher.

In the years ahead we will benefit from reading and re-reading his lessons on ethical thinking and practice.

Thank you.

TABLE OF CONTENTS

Preface by van Gigch, J.P.   ix
Prologue by McIntyre-Mills, J.   xv
Summary of the Contents by van Gigch, J.P.   xxxi

A. Steps Toward a Reflexive, Complex and Evolving Approach: Wisdom, Knowledge and Management

1. Progress Achieving C. West Churchman's Epistemological Program: The Implementation of Science of Science and of Science of Ethics - van Gigch, J.P.   1

2. The Paradigm of the Science of Management and of Management Science - van Gigch, J.P.   15

3. Churchman's Contributions to the Advancement of Management Science - Agrell, P. S., Leonarz, B.   27

4. Expanding Churchman's Philosophical Discourse: New Perspective - Agrell, P. S.   33

B. Skepticism, Truth, Certitude and Wicked Problems

5. Certitude in a Post-Modern World? A Coherent Evolutionary Story - Bausch, K.   47

6. Addressing Complex Decision Problems in Distributed Environments: An Integrated Inquiring Systems and Sensemaking Perspective - Paul, D.   81

7. Epistemic Humility: A View from the Philosophy of Science - Matthews, D.   105

C. Setting The Tone: The Importance of Ethics

8. In Search of an Ethical Science - van Gigch, J.P., Koenigsberg, E., and Burton, D.   141

9. Ethics and Enlightened Personal Responsibility - François, C.   161

10. May the Whole Earth Be Happy: Loka Samastat Sukhino Bhavantu Festschrift in Honor of C. West Churchman - Beer, S.   169

D. Entrepreneurship

11. An Outline of a Descriptive Theory of the Enterprise - Eriksson, D.   183

12. Organizational Efficiency and Values - Parra-Luna, F.   205

E. New Paradigms: Applications of the Science of Management to Governance and Managerial Practice

13. Molar and Molecular Identity and Politics - McIntyre-Mills, J.   227

14. Education for Engages Citizenship - Hammond, D.   269

15. Dialogue For Conscious Evolution - Christakis, A. N.   279

F. The Use of New Metaphors to Refocus the Managerial Discourse

16. Making Friends of Enemies - Yu, J.E.   305

17. The Application of an Epistemological Inquiry to Increase Our Understanding of Complex Issues: Enlarging Churchman's Circle of Enemies - van Gigch, J.P.   331

18. *Epilogue*: Working The Boundaries of International Relations and Governance' Through Empathy, Listening and Questioning - McIntyre-Mills, J., and van Gigch, J.P.   347

VOLUME 2

C. WEST CHURCHMAN'S LEGACY AND RELATED WORKS

WISDOM, KNOWLEDGE AND MANAGEMENT

PREFACE

VAN GIGCH, J.P.

Series Editor and Volume Editor
in collaboration with McIntyre-Mills,J.

This is the second volume dedicated to C. W. Churchman. The volumes of the Book Series entitled *C. West Churchman's Legacy and Related Works* are edited for the purpose of promoting the cross-pollination among fields, areas of study and disciplines and people from all walks of life in order to enrich the foundations of Management as well as to provide solutions to its own theoretical, political and practical quandaries.

Each of the authors whose writing appears in this volume, make an original contribution to the Management discipline that needs infusions of novel ideas from several other disciplines if progress is ever to take place.

C. West Churchman (1913-2004) was a prolific author whose texts *The Systems Approach* (1968), *A Challenge to Reason* (1968), *The Design of Inquiring Systems*, 1971), *The Systems Approach and Its Enemies*, (1979) among others, encouraged us to study the philosophy and the epistemology of the management disciplines which had been neglected heretofore. As his texts attests, CWC (for C. West Churchman) was a strong believer that Management could never become a credible scientific discipline unless it re-discovered the source of its moral authority, based on healthy questioning within context.

Skepticism

Skepticism is the source of doubt. Skepticism (also spelled "scepticism), is the philosophical view that "we lack knowledge", (Dancy & Sosa, 1992), or in its more extreme formulation skepticism questions the premise that credible knowledge exists or that we can find justification(s) for it (Audi, 1999; Hecht, 2003).

C. West Churchman's Skepticism

C. West Churchman was a skeptic - let us say a "mild" skeptic. Let us explain why. There are all kinds of skeptics.

Descartes is the best known "modern" skeptic (as opposed to Ancient Skeptics) who in the seventeenth century laid the foundation for contemporary skepticism. In modern times to be a "skeptic" places you in good company, with the likes of Kuhn, Foucault, Derrida, Quine and Rorty (Audi, 1999:847). Modern Skepticism is applauded due to its articulated or may be non-articulated thrust in rejecting old forms of thinking and obsolete paradigms in favor of new approaches to the questions of coherence and truth.

Skepticism about Managers and Management

There is widespread skepticism about the ability of management and managers: The public has open doubts that they possess the ability and the will to manage the enterprise with which they are entrusted ethically.

Managers are hauled in court for malfeasance and overt dishonesty. It is not only a fact that the management discourse is flawed but that it smacks of outright deceit. Managers of large corporations are perceived as selfish and greedy. They do not promote the public good and are guilty of lying and of dishonesty toward the public, their shareholders, stakeholders and fellow employees. Accountability and responsibility are shunned and the reputation of high public officials as well as that of highly placed managers in the corporate world stands at a record low.

Skepticism is also exhibited in the lack of consensus that exists in solving our environmental, political problems, and economic problems and, in general, at a complete lack of rationality in the private and public discourse.

Restoring Confidence and Ethical Authority

In order to restore confidence and regain the high moral ground, management - be it management in the public or private domains - must modify its ways. This renewal can only be obtained in the framework of the hierarchy of inquiring systems which was pioneered by C. West Churchman who was a management scientist and a philosopher of management. He wrote and pioneered what we consider the source of invaluable knowledge which the authors of this volume are mining and enhancing to dispel the prevailing skepticism that nothing can be done about our state of the world problems- be they social, managerial, organizational or otherwise.

Multidiversity of Inquiring Systems

C. West Churchman (CWC - for short) invented the idea of the *hierarchy of inquiring systems* which is made up of several levels of inquiry which differ by the substance of their dialogue or discourse and by the rationality(ies) with which they specialize to resolve their respective problems.

Solving problems requires the knowledge of a multidiversity of inquiring systems and their respective rationality (ies).

Due to complexity, problems require the following levels of inquiry and of specialization. Starting from the bottom up:
- A *manager* (in the operational inquiring system) to implement a feasible and ethical solution
- An *engineer* (in the engineering inquiring system) to apply science to the real world
- A *scientist* (in the scientific inquiring system) to discover new ideas
- An *economist* (in the economics inquiring system) to weigh and balance the costs and solutions of the problems facing the enterprise
- A *lawyer* (in the legal inquiring system) to untangle conflicts and disputes inside and outside the organization
- A *politician* (in the political inquiring system) to obtain and broker consensus inside and outside the organization
- An *ethicist* (in the ethics inquiring system) to decide on the ethical standards, norms and morality that ought to pervade decision-making

Each of the inquiring systems listed above specialize according to:
- The *type of problem(s)* which they are asked to solve,
- The *type of rationality(ies)* and knowledge utilized for this task, and,
- The *level of logic* of the inquiring system in which the problem occurs.

As an example, the operational inquiring system is said to utilize its own rationality as well as that provided by the other inquiring systems. In terms of logic, it stands at the bottom of the hierarchy of inquiring systems due to its proximity to the real-world and to the relative low level of abstraction of its problems.

A distinction is also to be made between decisions and metadecisions, where decisions are made at lower levels of abstraction and logic, and metadecisions which correspond to those of higher levels of abstraction and logic.

For a more detailed description and explanation of the hierarchy and its properties, we refer the reader to *Metadecisions* (van Gigch, 2003).

The above listing does not include an inquiring system which is devoted to study the epistemology of the management process. For this purpose, we need epistemic rationality.

Epistemic or Epistemological Rationality

By *Epistemology* we mean that portion of the Philosophy of Science which studies the sources of knowledge of a discipline, guides how the discipline elaborates its theories and its methodology, determines the level of rigor of its scientific program and establishes the level of credibility that can be lent to the discipline in light of its approaches to marshall information and knowledge.

Systemic Rationality

We could not be C. West Churchman's admirers and disciples without advocating *Systemic Rationality or Systems Thinking*. *Systemic rationality* is a rationality which advocates the study of systems and problems as a comprehensive enterprise which encompasses all of the rationalities listed earlier. The latter cannot be studied separately. They must be studied in the context of the whole system.

Holism or Systems Thinking is the formal epistemology behind the Systems Approach and any theses that claim the superiority of the whole in relation to its parts or that the whole has properties that the parts may be lacking.

Epistemological Holism is the view that "that whole theories are the units of confirmation". It states that: "Whether a belief is justified depends upon the support of the whole structure of beliefs to which belongs" (Dancy and Sosa, 1992).

Two Kinds of Wisdom

Audi (1999) reminds us that Aristotle introduced a distinction between *theoretical wisdom* and *practical wisdom*.

Theoretical wisdom is related to the highest forms of rationality - such as those provided by the inquiring systems of high abstraction and logic, whereas practical wisdom is related to "the capacity for sound judgment in matters of conduct".

How to marry the two theoretical and practical wisdom is the subject of Kantian debate as well as of the Churchman's dialogue.

CWC's disciples have taken his message at heart and have extended his discourse to all of the Social Sciences, Political Science, Policy Sciences, Information Sciences and the like. This is what the volumes of this book series are all about. A sample of them are presented in the chapters of this book.

Steps Toward a Reflexive, Complex and Evolving approach

The papers presented in this volume are but small steps toward developing a reflexive approach to management that is different from a Taylorist Scientific Management and the new managerialism that has developed Frederick Taylor's legacy of control, without drawing on the creative potential of people within the organisation and its environment.

Our systemic approach is compassionate and considers the self, the other (including sentient beings) and the environment.

It addresses the points of view of all those who are to be at the receiving end of decisions and strives for sustainable options.

Having presented the need and the basis for such a discipline we now turn to the authors' contributions to this volume. Each of them in a small way indicates how progress could make management richer and more rigorous.

REFERENCES

Ackoff, and Strümpfer, 2003, "Terrorism: A systemic view", *Systems Research and Behavioral Science*, **20**:287-294.

Audi, Robert, 1999, *The Cambridge Dictionary of Philosophy*, 2nd.ed., Cambridge, New York. The label "Epistemic Modesty" is cited in the entry SKEPTICISM.

Churchman C. West, *The Systems Approach*, 1968, *A Challenge to Reason*, 1968, *The Design of Inquiring Systems*, 1971, *The Systems Approach and Its Enemies*, 1979.

Dancy, J. and Sosa, E., 1992, *A Companion to Epistemology*, Blackwell, Cambridge, MA.

Hecht Michael Jennifer, 2003, *Doubt: A History*, HarperSan Francisco, San Francisco

Lakoff, G., 1987, *Women, Fire and Dangerous Things: What Categories Reveal About The Mind*; 1980, *Metaphors We Live By*, with Mark Johnson; 2002, *Moral Politics: How Liberals and Conservatives Think*, and 2004, *Don't Think of an Elephant*.

van Gigch, J.P., 2003, *Metadecisions: Rehabilitating Epistemology*, Kluwer/Plenum, London/New York.

PROLOGUE

CONSCIOUSNESS, CARE TAKING AND COMPASSION, BASED ON LISTENING AND MAKING CONNECTIONS UNDERPINS SYSTEMIC GOVERNANCE

JANET MCINTYRE-MILLS

> "there is no universal good across the categories...the good cannot be some common [nature of good things] that is universal and single; for if it were, it would be spoken of in only one of the categories, not in them all. There is no single Idea across different sciences....(Aristotle in *Nicomachean Ethics*, chapter 6, p.1096 a.30 translated by Irwin 1985)

Consciousness is **not** the preserve of human beings (Greenfield 2002); however we do have the potential to make connections across sciences, experiential learning, species and cultural intelligences and the environment as care takers. We can construct our future or destroy it. Acknowledgement is the basis of accountability and risk management that 'sweeps in' social, cultural, political, economic and environmental factors and 'unfolds' the values or interests of stakeholders and sentient beings who are to be affected by decisions. A Socratic approach would argue that through dialogue we can test out ideas within context and that the principles of testing out ideas is the basis for better management within and across organizations and for operating in multiple arenas at the local, national and international level. Sometimes dialogue is not enough. We need empathy to work through scenarios of what matters across the boundaries of self-other and the environment. Being able to place ourselves at the receiving end of actions and imagine what pain or cruelty feels like to the other - irrespective of culture, age gender, ability or species.

> "the standpoint of perfection, which purports to survey all lives neutrally and coolly from a viewpoint outside of any particular life, stands accused already of failureof reference: for in removing itself from all worldly experience it appears to remove itself at the same time from the bases for discourse about the world. Our question about the good life must, like any question whaterver, be asked and answered within the appearances...When we ask about motion or time or place, we begin and

end within experience of these items: we say only what has , through experience, entered into the discouse of our group" (Nussbaum 1986:291).

Nussbaum (1978:263) makes distinction between humans and animals about consciousness and ability to think about the past and ability to make judgments. Current neurological research (Greenfield 2002) shows that this is incorrect .A laboratory rat and a human being have more neurological similarities than differences.

Consciousness is enhanced through communication and deliberation. All living things are conscious to a greater or lesser extent. Consciousness is a continuum. For the human animal and for other animals communication is essential and it can be used as *a means* and *an end in itself* where life is conscious of itself. At a point where life thinks about itself - thinking about our thinking and arriving at decisions based on a discursive process - is a means to achieving goodness. Eudaimonia "is a feeling of pleasure" associated with deliberation and it is *both an end and a means.* (see Nussbaum on 1978: 170). Participation in the decision making is essential for achieving resonance. This is important as an ideal and it has pragmatic consequences too.

" If Aristotle begins not with a priori first principles, but with a coherent articulation of shared reflections, the deductive enterprise will immediately have a different look, Aristotle will not , apparently , be emulating the Socratic effort to escape from the confusion of appearance to a static and stable truth .." (Nussbaum, 1978: 174).

West Churchman's Design of Inquiring Systems(1971) and Thought and Wisdom (1982) provides a way to address the problem , by argueing that by expanding mindfulness we will become more compassionate and able to make connections across self-other and the environment. In volume 3 I argue for developing the Design of Inquiring Systems as a means to enance systemic approaches to communication and governance. Working across conceptual and geographical boundaries is the basis for good governance. Fukuyama (2004) has stressed that although there is no universal formula for good governance, it remains an ideal to which we need to strive, whilst mindful of the many variables that need to be considered contextually and in terms of the consequences. He argues that attempts to impose solutions from Denmark or United States can lead to imposing models and processes that do not take local history, social, cultural, political and economic conditions into account. He argues that:
- building capacity in both the public and the private sector is important and it needs to be based on ongoing local programs that are mindful of specific sectoral needs.

- developing NGOs does not solve the problem of bad governance in the public sector and that building capacity in the public sector by holding up private sector models can also be inappropriate if we forget that in the public sector the public is the principal and government is the agent, but in the private sector the principals are the business owners and the agents are the managers.

Thus the argument that although similar 'off the shelf' approaches cannot work across sectors (including the NGO sector) is highly debatable, the principle of testing out ideas remains a core for good governance and good science.

Fukuyama (2003) stresses that so-called weak states need capacity building, without acknowledging (as stressed in Vol. 1) that capacity building is needed in both his so-called 'strong' and 'weak' states in the developed and less developed world. But how do we develop capacity when there are no formulas that work universally? I argue that the answer lies in accepting the need to ask questions and to apply participatory processes within context to ensure that local knowledge is taken into account. This point is also made by Fukuyama (2003) but he forgets the importance of humility and local contextual knowledge in his belief that managers and policy makers in strong states and strong economies need to build the capacity in the weak states. Many of the management and policy problems are created by globalization, colonization and pollution by the powerful nations and economies.

As policy makers and managers we need both the capacity and will to be accountable. What we need are processes for engaging others contextually. Aristotle talked of 3 wisdoms, namely: episteme (scientific knowledge), techne (craft/art) and phronesis, whereas C.West Churchman (1971) in his *Design of Inquiring Systems* talks of 5 ingredients for good research, policy and management, namely: logic, empiricism, idealism, the dialectic and pragmatism) and in *Thought and Wisdom* he develops compassionate praxis through 'unfolding' values and 'sweeping in' the many variables that could have a bearing on a decision. Aristotle underlines in *Nicomachean Ethics* the importance of being able to apply technical knowledge and scientific knowledge within specific contexts. He argues that this is the most important ability, the possession of which can be summarized as the kind of wise and prudent action that implies the possession of all the other kinds of knowledge. To sum up, it is an ability to make the most appropriate, wise or prudent decision. He called this concept phronesis[1]. This kind of knowledge

---

[1] Flyvbjerg, 2001: 57, summarises the three kinds of knowedge discussed by Aristotle in Nicomachean Ethics as follows:

is demonstrated by the ability to learn from experiences and case studies or examples. Aristotle stressed that phronesis cannot be summarized in some total system or definition. It is not a universal formula.

If we accept that there is no such thing as a total system (also one of the most important points made by West Churchman), then there can never be a formula for good management, even if we accept that questioning and testing out ideas helps us to understand issues better, in some instances listening and not questioning can be very important to the process. Acknowledgement of many voices, many experiences and many ways of seeing and doing are as essential for risk management as they are for accountable decision making.

Complexity management is only possible if the decision makers reflect the complexity of the issues they are dealing with. Working in teams and providing space and time for individual contemplation can be equally important for the testing out of ideas needs by those who have experience and those who are to be affected by the decisions (irrespective of age, gender, culture, education, income or other status indicators such as level of education).

Preston and Sampford (2002) drawing on a wide range of philosophers summarise three frameworks as a useful for policy makers and managers. The three frameworks are a useful starting point for enabling ethical literacy.[2] The frameworks include: idealism, pragmatism and virtue or values based approach. To develop this form of praxis (see Flyvbjerg, 2001) requires being mindful not only of some ideals (as per Habermas 1984 in a

---

- Episteme Scientific knowledge. Universal, invariable, context- independent. Based on general analytical rationality. The original concept is known today from the terms "epistemology" and "epistemic"
- Techne Craft/art. Pragmatic, variable, context –dependent. Oriented toward production. Based on practical instrumental rationality governed by a conscious goal. The original concept appears today in terms such as ' technique', 'technical" and "technology"
- Phronesis Ethics. Deliberation about values with reference to praxis. Pragmatic, variable, context–dependent. Oriented toward action. Based on practical value-rationality. The original concept has no analogous contemporary term". Flyvbjerg, (2001: 60) emphasized that the possession of prudence and the ability to make value judgment entails the wisdom to apply appropriate knowledge within context is essential , because it implies the possession of scientific knowledge, art and craft.

[2] Aristotle called this phronesis (or prudent decision making and action based on wisdom). West Churchman (1979, 1982) talked of using his design of inquiring system (based on questions to tap different kinds of knowledge) as a means to 'unfold' values and 'to sweep in' the social, cultural, political, economic and environmental factors within specific contexts. He did not explicitly focus on culture (see volume 1) , but his design of inquiring systems lends itself to building in culture and gender sensitivity to develop a post colonial approach that helps to rescue the enlightenment from itself, by embracing the serpent or the paradox.

Kantian sense) but also and the context of diverse stakeholders some with the power to decide what constitutes knowledge and others not (as per Foucault, 1967, 1970, 1980). Questioning and dialogue within some contexts can be impossible, but striving to question and to come up with more appropriate responses remains the goal for ethical and sustainable decisions.

We need to ask what are the implications of this decision and for whom? If we add a generational dimension and the notion that human beings are caretakers of our own species and all sentient beings then we are able to consider sustainable futures. Systemic approaches require considering:
- The big and small picture
- Logical relationships and an appreciation of the logic of many stakeholders, because understanding the shape of perceptions of the other helps to reduce risk and enhance appreciation for why people think the way they do;
- Idealism;
- Empirical data;
- Dialectical considerations within context based on rapport and resonance (Churchman 1982);
- Pragmatism based on the understanding of the 'boomerang affect' (Beck, 1992) of thought and actions that consider self and not the other or the environment;
- Compassion that flows from ethical literacy and working with ideals, contextual considerations and the implications these have for virtuous practice.

SYSTEMIC GOVERNANCE is the concept applied to ETHICAL AND ACCOUNTABLE management across conceptual and geographical categories . It is thus about considering where to draw boundaries. Managers and policy makers today need to be able to work wisely with many different kinds of knowledge. A globalised world is one where boundaries are challenged by the market and by communication of ideas and values (See volume 3). We can use dialogue underpinned by the wise management and application of knowledge to enable us to make socially and environmentally sustainable decisions. Different approaches to representation and knowledge are the subject of ongoing dialogue within and across the arts and sciences.

Debates are influenced by the power or powerlessness of the decision maker as every harried manager, policy maker, staff member or citizen or non citizen understands. Knowledge does not necessarily give power, it can lead to punishment, as detailed by Link (2005) in his article on journalists who expose issues in China.[3] Desires and emotions (Deleuze and Guattari in

---

3 Ranging from health issues such as AIDS outbreaks, SARS or food poisoning to criticisms of urban plans. Journalists soon learn that there are ways and means to report issues. Sometimes by publishing the material in overseas media or making oblique references.

Bogue 1989) also influence the way we represent and manage knowledge. Instead of emphasizing only the point of view of one person or one framework and labeling these as the objective facts (and personal experience as 'mere perceptions' or feelings) the social sciences have undergone a paradigm shift to recognize that there are many sources of meaning and also many sources of knowledge that can be useful for understanding the way in which we frame and reframe issues as policy makers and managers.

Churchman following Singer (1982) uses the key concepts of unfolding values and sweeping in the social, cultural, political, economic and environmental factors that shape our way of seeing and doing. Breadth and depth of vision can enable more prudent decision making within specific contexts.

Mindfulness can be aided by interpreting textuality or intertextuality as a means to enable us to trace connections across ideas so that many points of view can be expressed simultaneously, rather than hierarchically. A less prescriptive and more democratic approach is the appreciation of insights inspired by listening to many voices and exploring diverse meanings. Discordant disagreement is as important as harmonious co-creation. Enabling the discordant and harmonious themes to be expressed is about more systemic understanding. Where are the gaps? Where are the continuities? Technical knowledge, strategic knowledge and communicative knowledges have been identified by Habermas (1984) and different categories are given by Aristotle (Nicomachean Ethics). But these are just categories – neat ones at that. There are many voices, many experiences and many knowledges; all are expressive of meanings which can be viewed as more rational or more emotional, depending on the way they are interpreted.

Better decisions are made when we are mindful of more voices and more texts. So-called knowledge management (a misnomer in my opinion) has begun to do more than manage knowledge in terms of filing and retrieving it. There is also recognition that the nature of knowledge itself needs to be explored. Research these days needs to include ontology and epistemology in its design and the commentary on any framework or boundary around a research problem is itself of central interest (Gibbons and Images et al., 1994), Australian Research Council Guidelines for researchers). Not only the community, but also the public, private and non-government sectors have realized the value of tacit lived experience and also professional knowledge (State of the Regions Reports by Australian Local Government Association and National Economics Reports 2002, 2003). Making connections across different knowledges is good for democracy and good for quality of life and the environment that we live in. Greenfield (2002) stressed that

---

Knowledge is a matter of policy and management and who decides whether it should lead to apologies, retractions, counseling, abuse, imprisonment or death.

consciousness is a continuum. The more connections we make the more mindful we are. Being aware of our emotions and the role they play is very important.

My stance is to stress the value of theoretical and methodological literacy as an end in itself, because it supports mindfulness. But it **also** provides a means to respond to specific contexts, whilst mindful of logic, empiricism, idealism, the dialectic and ethical, systemic pragmatism (drawing on and adapting West Churchman's 1983 work on *Thought and Wisdom*), based on co-creation of ideas within context. We discuss striving for truth, valuing difference and working out what the so-called 'facts' mean to different stakeholders and how these understandings are constructed. Policy makers and managers need to think about values, viewpoints and how they could be represented and distorted and owned by them as the powerful decision maker. Policy makers and managers need to recognize that they need to think about theory, practice and representation. Whose voices will be heard? Whose will be loudest? As Nancy Hartsock stressed drawing on Foucault (in Allmendinger and Tewdr-Jones (2002: 148) 'the subjugated voice' is now being said to be just one of many – just as some unheard voices have obtained the right to be heard. We invite you, the reader to mine the rich veins of your own experiences[4] through unfolding your own values and considering the ways in which you have to act in context and the way in which your age, gender, societal status, your culture and experiences shape the way you see the world, your decisions and your actions.

If representation is the basis of truth then how can we know that what is represented is an exact match with reality? Who decides? This is where power, personality, values, perceptions and biological /chemical makeup come into play within a specific social, political, economic and environmental context. But the policy implications of postmodernism (that there are only *many truths* or worse *no truth*) makes society ungovernable. It also means that human rights have no basis for support and that the environment cannot be protected. So this approach is a 'dead end' from both an idealistic (or non consequentialist) ethical point of view and a pragmatic (consequentialist) point of view. Nevertheless the notion of one truth based on modernism runs the risk of trying to test out ideas by experts who do not have enough lived experience. The way forward for the enlightenment is to *expand the testing process, so that all those at the receiving end of a decision are part of the iterative and ongoing decision making process*. This is most important, because it enables testing out ideas in such a way that the complexity of the decision is matched by the complexity of the stakeholders and the context (This is the subject of the next volume 3).

---

[4] Consider doing auto-ethnography as explained by Ellis and Bochner (2000).

Based on a critical and systemic approach we can find or create points of connection. The testing out ideas in the interest of science and democracy needs to be expanded to include all those at the receiving end of a decision and the participants need to consider this generation of life and the next. But how can this be achieved? Volume 3 attempts to address this question.

Perceptions always filter our understanding and theoretical limitations can frame or reframe the way in which people have constructed their reality and so it is worth striving to explore ways to enable voices or texts or pictures to be shared in ways that enable the reader to see many interpretations. This is particularly useful for policy makers and managers. Representation can amount to nothing more that silencing or distorting the ideas of others. More than one interpretation of a story enables a richer or more complex representation. This is not such a difficult concept to grasp (see Allen 2000). Managers need to draw on many points of view if they are to make decisions that address risk and complexity. Decisions should be taken in such a way that they represent the complexity of the people who are to be impacted by the decisions. This makes sense in terms of risk management.

"Having lost the comfort of our geographical boundaries, we must in effect rediscover what creates the bond between humans that constitute a community" (Jean Marie Gu'ehenno 1995, 139 cited in Judt 2005: 41)[5].

REFERENCES

Allen, G., 2000, *Intertextuality: The New Critical Idiom*, Routledge, London.
Allmendinger, P. and Tewdwr-Jones, M., 2002, *Planning Futures. New Directions for Planning Theory*, Routledge, London.
Beck, U., 1992, *Risk Society Towards a New Modernity*, Sage, London.
Churchman, C. West, 1971, *The Design of Inquiring Systems. Basic concepts of systems and organization,* Basic Books, New York.
Churchman, C. West, 1982, *Thought and Wisdom*, Intersystems Publications, California.
Ellis, C. and Bocher, A., "Autoethnography, personal narrative, reflexivity: researcher as Subject", *Handbook of Qualitative Research*, N. Denzin, and Lincoln, eds., 2nd. ed.
Flyvbjerg, B., 2001, *Making Social Science Matter: Why social inquiry fails and how it can succeed again*, Cambridge University Press, Cambridge.
Foucault, M. and Gordon, C., eds., 1980, *Power / Knowledge: selected interviews and other writings 1972-1977*, Harvester, Brighton.
Foucault, M., 1967, *Madness and Civilization: A history of insanity in the age of reason*, Routlege, London.

---

[5] Judt, T., 2005, "Europe vs America", *The New York Review of Books*, pp. 37-41, he reviews Reid, T.R., "*The United States of Europe: The new superpower and the end of American supremacy*"; Rifkin, J., "*The European Dream: How Europe's vision of the future is quietly eclipsing the American dream*"; Garton Ash, T., "*Free world: America, Europe, and the surprising future of the west*".

Foucault, M., 1979, *Discipline and Punishment: The birth of the prison*, Vintage, New York.

Fukuyama, F., 2004, *State Building: Governance and world order in the 21st century*, Cornell University Press, New York, Ithaca.

Greenfield, S., 2002, *The Private life of the Brain: Emotions, Consciousness and the Secret of the Self,* John Wiley and Sons, New York

Habermas, J., 1984, *The Theory of Communicative Action: Reason and the rationalization of society*, Beacon, Boston.

Irwin, T., 1985, *Aristotle Nocomachean Ethics,* translated with Introduction and Notes, Hackett, Cambridge.

Link, P., 2005, "China: wiping out the truth", *The New York Review of Books,*. pp. 36-39

McIntyre-Mills, J. and van Gigch, J. 2004. *Rescuing the enlightenment from itself. Vol 1 of the C.West Churchman Series* . Kluwer-Springer. Boston

McIntyre-Mills, J. 2004. *Systemic Governance. Working and reworking the boundaries of international relations and governance.* Vol 3 of the C.West Churchman Series., forthcoming.

Nussbaum M, C. 1978 Aristotle's De Motu Animalium.Text with translation and commentary and Interpretive Essays Princeton University Press

Nussbaum 1986, M The fragility of goodness: luck and ethics in Greek tragedy and Philosophy. Revised edition Cambridge University Press.

Orr, M., 2003, *Intertextuality: Debates and contexts*, Polity Press, Oxford.

Preston, N., Sampford, C. and Connors, C., 2002, *Encouraging Ethics and Challenging Corruption*, The Federation Press, Sydney.

BIBLIOGRAPHIC DATA

Per Sirgurd Agrell lives in a suburb of Stockholm, Sweden where he is a consultant in defense and security systems in the firm Ekelöw Infosecurity AB. He is active in the European Systems and Operational Research societies. He contributes regularly in journals of the professional journals of both disciplines. He also interacts with his French counterparts and promotes interactions and exchanges with the French and the Swedish research communities.

Kenneth C. Bausch is affiliated with the Institute for 21st Century Agoras, Perimeter College (Atlanta). He works to promote The Emerging Consensus in Social Systems Theory. His latest publication (with Aleco Christakis who also contributed to this volume) is *Science: Pragmatic Design Dialogue, (2005), Information Age Press.*

Stafford Beer was one of the foremost cyberneticians of our age. His publications cover nearly two hundred items including eight books among which *Brain of the Firm*, *The Heart of Enterprise* and *Platform for Change* are his most renowned works. He is famous for introducing the concept of the "viable system" according to which viability of the enterprise is not a matter of economic solvency but follows cybernetic laws which are embodied in the "science of effective organization."

Charles West Churchman (1913-2004) to whom this volume is dedicated is honored throughout this Book Series. He was the author of *Prediction and Optimal Decision* (1961), The *Systems Approach* (1968), *Design of Inquiring Systems* (1971) and *The Systems Approach and Its Enemies* (1979) among others. As is evident in this volume, all contributors, one way or another, owe their inspiration to Professor Churchman's ideas.

Alexander N. Christakis is a principal of CWA Ltd. He claims to be Churchman's disciple, friend and admirer. His latest publication and major work is *Human Science: Pragmatic Design Dialogue*, (with Ken Bausch. See above), Information Age Press, 2005.

Burton Dean is Professor and Chair, Department of Organization and Management, and Director, Total Quality Management Certificate Program, San José State University, California. His publications include seven books and approximately 100 articles on manufacturing, technology, operations, systems and project management. He is involved in research and consulting assignments with over eighty regional, national and International organizations.

Darek M. Eriksson lives in Sweden where he combines his academic work with management consulting. He is affiliated with several universities and research organizations. He is currently managing editor of the journal *Cybernetics & Human Knowing* and consulting editor for *Information &*

Management Journal. His most recent publication is "An Identification of Normative Sources for Systems Thinking: an Inquiry into Religious Ground-Motives for Systems Thinking Paradigms," *Systems Research & Behavioral Science*, Vol.20, No. 6, pp. 475-487, 2003.

Charles O. François is Honorary President of the Argentine Division of the Int. Society for Systems Sciences. He has always been a Churchman's admirer. He is the author/editor of the *International Encyclopedia of Systems and Cybernetics*, Second Edition, published by Saur Verlag, München, Germany, 2004.

Debora Hammond is the author of the reading "Education for Engaged Citizenship." Churchman was one of her mentors in her Ph.D. Dissertation project at the University of California, Berkeley. She is presently an associate professor at the Hutchins School of Liberal Studies, Sonoma State University, Rohnert Park, California. She is the author of *The Science of Synthesis: Exploring the Social Implications of General Systems Theory*, University Press of Colorado, 2003.

Ernest Koenisgsberg is Professor Emeritus at the Walter A. Haas School of Business, University of California, Berkeley. He is proud to have been a student of West Churchman's Second Operations Research Course at the Case Institute in 1954. In addition to teaching, Professor Koenisgsberg has carried out and published numerous theoretical and applied studies in which he brought management science to industry, government and academia.

David Matthews is a Defence Scientist with the Defence Science and Technology Organisation (DSTO) of Australia. He has an Honours Degree in Applied Mathematics and has recently completed a Ph.D. in Systems Thinking, both from the University of Adelaide. His main research interests.

Janet McIntyre-Mills is a Senior Lecturer at Flinders University in the Flinders Institute of Public Policy and Management. Her recent publications include *Critical Systemic Praxis for Social and Environmental Justice*, 2003 and *Global Citizenship and Social Movements: Creating Transcultural Webs of Meaning for the New Millennium*" in 2000. She currently facilitates research with a transdiciplinary design team to address social inclusion with Aboriginal Australians.

David L. Paul received a B.S. from the Wharton School of the University of Pennsylvania, an M.B.A. from the Anderson School at UCLA, and a Ph.D. from the Graduate School of Business at the University of Texas at Austin. He has recently published in *MIS Quarterly*, and received the 2000 International Conference on Information Systems (*ICIS*) Best Dissertation Award.

Francisco Parra-Luna is Professor at the University of Madrid, Spain and has been the Director of the University Institute for Human Resources.

He is the author of eleven books on social system theory and its applications, among them *Elementos Para Una Teoría Del Sistema Social*, Universidad Complutense, Madrid, 1983 (Components of A Theory of the Social System) and *El Balance Integrado de la Gestion Estrategica* (Integrated Equilibrium of the Strategic Management Function) published by Deusto, Bilbao, Spain, 1993. Both in Spanish. At the moment he is involved in the celebration of 400th Anniversary of Don Quijote de la Mancha the masterpiece of Miguel Cervantes (1604-1614).

John P. van Gigch is the Series Editor for the Book Series Churchman's Legacy and Related Works which he is instrumental in getting started. van Gigch knew West Churchman as well as his family personally. Churchman wrote an Introduction for van Gigch's *Applied General System Theory* which was published in 1974 with a Second Edition in 1978. van Gigch remained close to West Churchman after the latter retired from the University of California Berkeley. Right to the end, they celebrated several of West's birthdays together.

Jae Eon Yu ("Making Friends of Enemies") is a Professor in the Department of Information Management, Seoul Information Technology University, Seoul, Korea. He claims to be a Poststructuralist/Postmodern systems practitioner and a Churchman's admirer. His most recent publication is: "Reconsidering participatory action research for organizational transformation and social change," *Journal of Organizational Transformation and Social Change*, Volume 1 (1.2 &1.3), 111-141 (2004).

SOME FOCUSING THOUGHTS

"What is in the nature of systems is a continuing re-viewing of the world, of the whole system, and of its components. The essence of the systems approach, therefore, is confusion as well as enlightenment. The two are inseparable aspects of human living. Finally, then, here are some principles of a deception-perception approach to systems:

1. The systems approach begins when first you see the world through the eyes of another. Another way to say the same thing is to say that the systems approach begins with philosophy....

2. The systems approach goes on to discovering that every world view is terribly restricted.

...For those who think in the large, the world is forever expanding; for those who think in the small, the inner world is forever contracting.

3. There are no experts in the systems approach....

And finally, my bias:

4. The systems approach is not a bad idea"[1]

---

[1] Churchman, C.W.C., 1979, *The Systems Approach*, Delta, New York, pages 231-232.

SUMMARIES OF VOLUME'S CONTENTS

A. Wisdom, Knowledge and Management

van Gigch, *"Progress Achieving C. West Churchman's Epistemological Program: The Implementation of Science of Science and of Science of Ethics"*, starts the volume by recalling one of C. West Churchman's earliest texts in which CWC articulates his epistemological program.

CWC did not want to oppose the purposes of Science and Ethics, but rather explore the relationship between science and values, and explore in depth how we could improve the understanding of the role of values in decision-making.

Science has made very few significant inroads to help management improve its methodology or its ethics - subjects which will taken up again in future volume of this Book Series.

CWC's called for a *Science of Management* which would take the form of an epistemological evaluation of the scientific methods of Management from the perspective of a metalevel, where an independent guarantor could assess their validity and truth content of *Management Science*.

In *"The Paradigm of the Science of Management and of Management Science"* **van Gigch** formalizes CWC's relationship between an inquiring system which is devoted to Management Science and a metasytem inquiring system which is devoted to the Science of Management. This hierarchy of inquiring systems embodies Churchman's reflexive loop the "X" of "X."

In this chapter, *The Science of Management* is described as a "metascience" because its "main problem" is to deal with epistemological questions and with the elaboration of the paradigm of the discipline. It also validates the scientific methodology which is employed by Management Science (a "normal science") at the lower inquiring level of logic where the everyday scientific problems of the real-world are solved.

The concepts of "metascience" and "normal science" were introduced by Kuhn.

Per Sigurd Agrell contributes two papers. In the first, *"Churchman's Contributions to the Advancement of Management Science"*, he lists Churchman's pioneering ideas after the revolutionary *Introduction to Operations Research (With* Ackoff and Arnoff*)*. In the second article, *"Expanding Churchman's Philosophical Discourse: New Perspectives"*, Agrell explains that management systems - and for that matter all systems - require a double control function, i.e. a mega-structure which "provides purposes and restrictions' and a meta-structure which provides inquiring systems with their rules of logic. It is in the interplay of mega-and meta-structures that organizational complexity can strive instead of destroy the enterprise.

B. Skepticism, Truth and Wicked Problems

Truth can only thrive in the context of discovery. Nowadays to be a skeptic is not a negative connotation. On the contrary: The skeptic asks questions, refuses the status-quo and looks for new approaches.

It is in this spirit that **Ken Bausch**, *"Certitude in a Post-Modern World? A Coherent Evolutionary Story"*, uses the related theories of self-reference and self-organization as guides to trace the story of how to establish certitude through an evolutionary history of cognition, representation, language information and epistemology.

Bausch's paper is sweeping in its scope and shows how truth "in the pragmatic sense, is an extension of the tests of survivability and productivity" two vital attributes of managerial systems. While Bausch's paper does not dwell with the details of how his theory can be applied to the organization, it shows how the spirit of discovery can dispel skepticism and "generate consensual grounds on how to construct knowledge" of a scientific discipline.

David Paul discusses wicked problems and makes the point that complex problems are the result of many interrelated socio-cultural factors and they can be perceived in different ways by the different stakeholders.

To conclude the section, **Mathews's** essay (*"Epistemic Humility"*) provides a time-line of the trend toward Skepticism from the perspective of the Philosophy of Science. He traces the journey of epistemology starting with intuitive vertificationism, continues with positivism, critical rationalism and social constructivism, before "suggesting a way forward towards a position of epistemic humility about the pronouncements of science".

C. Setting the Tone: The Importance of Ethics

In 1993, on the occasion of his 80th birthday, C. West Churchman consented to an interview with **van Gigch, J.P., Koenigsberg, E., and Burton, Dean**, *"In Search of an Ethical Science"*, which is re-published in its entirety. It sets the tone for later debates about the need for ethics in management practice.

Charles François, *"Ethics and Enlightened Personal Responsibility"*, claims that "only a genuine trans-cultural ethics can save us from a variety of possible global mega-catastrophe, mostly engineered by man himself."

François discusses the Why's, the Who's and the What's of what we believe to develop our worldviews and our mindscapes. Given that we are living a period in world affairs which is notorious by its lack to elementary ethical principles, François musing about Ethics comes at the right moment to infuse a dose of pessimism as well as one of optimism that we will ever reach wisdom and improvement of the human condition. Naturally these comments apply to the improvement of ethics in the private realm as well as the ethics of government officials, politicians and whoever is entrusted with the management of our lives and that of future generations.

This volume includes a paper written by another of the management "gurus" of this period, also deceased (Churchman passed away in 2004). We refer to **Stafford Beer**, *"May the Whole Earth Be Happy""*, where he recalls his friendship with CWC. They were obviously intimately related in heart and in thought. Beer describes the relationship between Eastern and Western mythological and religious traditions which provide a background to their mutual morality and their ethical kinship. Beer is known for several of the most insightful applications of Ashby's and Weiner's theories of control to the management of the firm through the use of information and the reduction of organizational entropy.

D. Entrepreneurship

Entrepreneurship is viewed as that portion of managerial practice where new theories and new ideas are tried and applied to the organization. Without endeavors of this sort a science such as a Science of Management would wither and disappear.

D. Eriksson, *"An Outline of a Descriptive Theory of the Enterprise"*, summarizes the thrust of his doctoral dissertation. He asks the question: How can we generate an ethically sound design" for the enterprise. He presents the proposed outline of a descriptive theory of an enterprise which is derived from the French systemicist of J.L. Le Moigne.

They provide a theory of modeling for the enterprise which when applied to real-life cases provides strikingly accurate and counter-intuitive conclusions of how the enterprise can be improved and modified to achieve its avowed purposes and functions.

Eriksson must be lauded for his efforts to bridge the language and cultural gap between French and Anglo-Saxons renditions of two distinct managerial styles.

Parra-Luna, *"Organizational Efficiency and Values"*, is a sociologist who studies the concept of organizational efficiency and provides and axiological (operational) definition of the concept of efficiency.

After sketching his Theory of Organizational Efficiency he postulates that his theory represents a progress toward an axiological (axiomatic) definition of efficiency which yields six equations by which efficiency can be defined and profiled. These definitions can be used in the context of small enterprises as well as in large organizations.

E. New Paradigms: Applications of the Science of Management to Governance and Managerial Practice

A Science is meaningless without a context in which new theories, new approaches and new ideas are applied to the real-world.

Janet McIntyre-Mills in her article *"Molar and Molecular Identity and Politics"*, discusses how "policy workers and managers need to work 'with' rather than 'within' frameworks to achieve accountable policy decisions with the stakeholders She presents conceptual tools which can be used to enhance systemic and managerial governance. She is an advocate of "compassionate dialogue that explores paradoxes and considers the rights and responsibilities of caretakers" and stakeholders. What could be more important objectives than "the promotion of two-way learning, socio-economic wellbeing and diversity, she asks. We could not agree more.

In *"Education for Engages Citizenship"*, **Debora Hammond** studies US federal mandates for meeting narrowly defined content and assessment standards in Education. She decries the fact that these mandates trump all values except market forces. Pedagogical practices which include interdisciplinary curriculum, dialogue and critical inquiry while cultivating ethical practices in management are left behind. Hammond advocates alternative approaches which foster more "interdisciplinary approaches to learning," "cultivation of skills and collaboration to develop participatory decision-making and the emergence of a truly democratic society." While education may appear remotely related to the corporate board room, Hammond shows that "cultivating democracy and social justice" are the cornerstone of an ethical enterprise - be it in education or in business.

Indeed The Science of Management should not harbor barriers among disciplines which foster similar purposes and objectives.

Aleco Christakis, *"Dialogue for Conscious Evolution"*, focuses on the critical role of dialogue in the evolutionary guidance of social systems. He describes the design of the Structured Design Process (SDP) which "is suitable for engaging stakeholders to design their social systems in the context of the escalating complexity of the Information Age."

Christakis shows how the SDP has been applied for the last twenty years to address national, international and inter-organizational design challenges. SDP is primarily based on three axioms: a) the escalating complexity of social systems designing, b) the cognitive limitations of the observer in processing too many observations simultaneously and c) the difficulty in determining the relative saliency of the observations. SDP is a perfect example of the importance of how new paradigms evolve in the midst of a Science of Management to contest the primacy of ordinary Management Science.

F. The Use of New Metaphors to Refocus the Managerial Discourse

Jae Yu, *"Making Friends of Enemies"*, evaluates the contributions of contemporary philosophies to develop approaches to facilitate communication and participation in managerial practice. Frankly philosophical in nature and in approach, this paper is the most abstract and theoretical of the articles of this volume. However, it shows the world wide scope of Churchman's influence. Jae Yu is a professor in Korea who obtained his doctoral degree in the UK under the sponsorship of Michael Jackson one of the most influential educators and practitioners of systems thinking extant.

Ken Bausch and van Gigch in Volume 1 of the Churchman's Legacy Book Series presented an article each on the so-called 'confrontation' between the Systems Approach and Its Enemies. See C. West Churchman's *Enemies of the Systems Approach*, 1979.

We recall that Churchman identified Politics, Morality, Religion and Aesthetics as "enemies." They were labeled 'enemies' of the enterprise not because they engaged in open warfare, but because in they hindered efforts of the planner to be comprehensive and more systemic.

van Gigch in *"Churchman's 'Circle of Enemies' Provides an Epistemological Tool to Enlarge the Scope of Available Knowledge"*, uses Lakoff's (2002; 2004) theories of frames to show that Churchman was mistaken to use the "enemy" label. By using Lakoff's study of metaphors, frames and framing, van Gigch shows that the four 'enemies' identified by Churchman are not enemies at all. They should be re-labeled or re-framed in order to dispel the erroneous perception that Churchman is AGAINST Politics, Morality, Religion and Aesthetics. As a matter of fact they are very much his 'friends' and the label 'enemies' should be erased from his texts.

Furthermore, van Gigch recalls the metasystem approach in the context of which conflicts are "dissolved,' "constrained" and "controlled" by realizing that some of these organizational disagreements only arise because the participants - managers etc. - operate at different levels of logic and of control. We would like to note that the idea to revisit the "circle of enemies" was gleaned from Ackoff and Strümpfer (2003) in which they discussed how to counter terrorism and its frightening consequences.

REFERENCES

Ackoff, and Strümpfer, 2003, "Terrorism: A Systemic View", *Systems Research and Behavioral Science*, 20: 287-294.

Audi, Robert, 1999, *The Cambridge Dictionary of Philosophy*, 2nd.ed., Cambridge, New York. The label "Epistemic Modesty" is cited in the entry SKEPTICISM.

C. West Churchman, *The Systems Approach*, 1968, *A Challenge to Reason*, 1968, *The Design of Inquiring Systems*, 1971, *The Systems Approach and Its Enemies*, 1979.

Dancy, J. and Sosa, E., 1992, *A Companion to Epistemology*, Blackwell, Cambridge, MA.

ACKNOWLEDGMENTS

All the volumes of the present Book Series are a labor of love and I thank the contributors warmly. Thanks to Janet McIntyre-Mills who contributed to this volume and to the series by remaining optimistic and working to make the volume and the series possible. Our thanks to the publishers who have made this series possible.

SECTION A

WISDOM, KNOWLEDGE AND MANAGEMENT

Chapter 1

PROGRESS ACHIEVING C. WEST CHURCHMAN'S EPISTEMOLOGICAL PROGRAM
The Implementation of Science of Science and of Science of Ethics

VAN GIGCH,J.P.

Abstract: We provide an overview of CWC's[1] epistemological program by referring to his multiple writings. We take a metasystem approach which means that we articulate CWC' vision of a hierarchy of inquiring systems in the context of whose different logic levels, his Dialectics between opposites, between arguments and counter arguments, between ethical extremes, between parties to a conflict are discussed and settled. We describe how CWC's viewed several of his favorite philosophical issues such as his quest for the sources of knowledge to improve the Science of Ethics, the factors inhibiting agreement in decision-making, the limits of rationality, and the call for a Science of Science.

Where relevant we cite from CWC's texts to give the reader the flavor of his inimitable prose.

Key words: epistemology, inquiring system, ethics and decisions

1.1 EPISTEMOLOGY AND EPISTEMOLOGICAL PROGRAM

By Epistemology we mean that portion of the Philosophy of Science which studies the sources of knowledge of a discipline, guides how the

---
[1] The abbreviation "CWC" used throughout the text stands for C. West Churchman.

discipline elaborates its theories and its methodology, determines the level of rigor of its scientific program and establishes the level of credibility that can be lent to the discipline in light of its approaches to marshal information and knowledge.

The Epistemological Program refers to an author's efforts to elaborate the Epistemology of the discipline in the context of which he/she makes his own philosophical and methodological pronouncements.

1.2 C. WEST CHURCHMAN'S KEY TEXTS DISCUSSED HEREIN

C. West Churchman laid out an epistemological program in two of his earlier books: TEI (Theory of Experimental Inference) and POD (*Prediction and Optimal Decisions*), published in 1948 and 1961, respectively.

We summarize for our readers what this epistemological program was all about and give an assessment of whether any progress was made to implement it.

1.2.1 Theory of Experimental Inference (Churchman, 1948, 1960) (to be known below as TEI)

In this early text which was published in 1948, CWC laid out the claim "that ethical judgments can be [should be] included in the scope of science". At the outset, we must translate CWC statement to clarify what it means. By ethical statements he does not necessarily refer to their moral content but rather to the necessity of evaluating their worth and establishing their value. Additionally, he established the requirement of an epistemological guarantee of their worth. His claim seeks to extend and apply scientific inquiry in realms or domains other than the traditional physical or natural sciences.

CWC already spoke in TEI of a "Science of Ethics" which, we repeat, is the claim that evaluative judgments should be subjected to scientific methods.

In so many words, CWC asserted that "the one-truth methodology" (his words) of the Scientific Method must apply to both hard and soft-variables domains. For CWC there cannot be one epistemology of the hard sciences and another for the soft sciences: CWC stated: "The attempt is made here [in TEI] to outline how non-scientific problems-solving can be studied with the Scientific Method" (p.238).

Finally he stated that the Science of Ethics must be subjected to all the methodological and epistemological controls of traditional discourses in other disciplines.

1. PROGRESS ACHIEVING C. WEST CHURCHMAN'S EPISTEMOLOGICAL PROGRAM

CWC believed in the progress of science, which for him meant, the improvement of the scientific validity of judgments and of all the variables used in decision-making.

1.2.2 Prediction and Optimal Decision: Philosophical Issues of a Science of Values. (Churchman, 1961) (to be abbreviated below as POD)

1.2.2.1 Three Key Words Easily Overlooked

At first glance POD's title seems to read: "Prediction and Optimal Decision": [A Science of Values]. For many years, the present reviewer took the abbreviated title to mean that CWC wanted to apply scientific rigor to determine - A Science of Values - i.e. the meaning of values and proceed to their measurement. Fortunately, a closer reading of the title adds three important key words ["Philosophical Issues of"] which can be easily be overlooked: the complete title is: *"Prediction and Optimal Decision: Philosophical Issues of A Science of Values"*.

Those three words easily overlooked add a completely new meaning to CWC' intent. CWC did not want to oppose the purposes of Science and Ethics, but rather explore the relationship between science and values, determine their commensurability and explore in depth how we could improve the understanding the role of values in decision-making.

1.2.2.2 Hierarchy of Inquiring System: The Metasystem Approach

While he did not articulate the metasystem paradigm which provides an approach to reconcile this split, he certainly inspired it. The metasystem approach provides a process by which some of logic gaps/splits can be resolved. We deal with the following three subjects:
1. The bridge between Fact AND Value,
2A. The bridge between Science as a methodological/ technological discipline AND Science as the guarantor of its own of Epistemology.
2B. The bridge between Science AND Ethics,

We will deal with these issues after introducing the Hierarchy of Inquiring Systems and the Metasystem Approach.

1.2.2.3 Hierarchy of Inquiring Systems: The Metasystem Approach

To understand C. West Churchman's writings it is advantageous to take what we call a metasystem approach. This approach is actually based on CWC's own writings although it has been formalized later after reading his work. The elaboration of the Metasystem Approach can be found in J.P. van Gigch, (2003a), where more details on this approach are available.

We will articulate the approach step by step and then proceed to use quotes and comment on C. West Churchman's work to elicit some of his most important lessons.

1.2.2.4 An Inquiring System

An inquiring system is a system which has three distinct characteristics:
1. A Main Problem to solve and,
2. A Logic. In turn, the Logic is characterized by:
3. A Level of Abstraction, and,
4. A Subject Matter.

Three Main Inquiring Systems

For our purposes, the hierarchy of inquiring systems consists of three Inquiring Systems:
  The Real-World Inquiring System,
  The Science Inquiring System, and
  The Epistemology Inquiring System.

Level 1. The Real World Inquiring System

In the real-world inquiring system we deal with reality and apply the information provided by the inquiring systems of higher logic and abstraction. In this case this would consist of specifications and norms stemming from the Science Inquiring System.

Level 2. The Science Inquiring System

In this inquiring system, the information originating in the Epistemology Inquiring System is incorporated in the methodologies of the Science Inquiring System so the Science can proceed to elaborate its theories with logic and rationality which has been vetted by a system of higher logic and abstraction.

Level 3. The Metalevel: The Epistemology Inquiring system

The Epistemological Inquiring System acts as a guarantor of the logic, rationality and veracity of lower level systems. It deals strictly with those issues of the Philosophy of Science which must apply in a system of thinking to be coherent and consistent.

1. PROGRESS ACHIEVING C. WEST CHURCHMAN'S EPISTEMOLOGICAL PROGRAM

We note that each of the inquiring systems mentioned above stand at different levels of abstraction and that its logic differs from that of one another inquiring system. These distinctions allow it to deal with different levels of discourse and explains as we shall see the beginning of a reconciliation and compatibility with "what is said" at one level and "what is meant by it" at another where the logic, language and abstraction may be radically different.

1.3 APPLICATION OF THE METASYSTEM APPROACH TO INTERPRET CHURCHMAN'S EPISTEMOLOGY:

1.3.1 The Fact/Value Split

On the surface, Science and Values do not intersect:
Scientists are strictly dealing with facts, Ethicists with values. Physicists and other hard systems scientists will confirm that they are not in the business of dealing with those "softie" variables called "values." The logical process to find the truth - they say - is not a psychological or social science game which involves making judgments about beliefs, morality or other soft considerations. Theirs is strictly a "hard," (from hard system variables) logical inquiry which leads to heretofore unknown facts and more facts about the universe. They are dealing with matter - they say, physical variables and strictly natural properties.

Thus they want us to believe that there is a philosophical split/gap between the world of facts and that of values.

In POD, CWC intended to explode the myth by proposing to explore the philosophical issues which seems to separate these those two universes of discourse.

The fact-value distinction takes many pages of philosophy to discuss. For our purposes and obviously for CWC, this distinction appeared bogus. Facts and values are blurred, and values play a large part in the hard sciences, in spite of the physicists' protests. In turn, facts are fundamental in soft system domains.

1.3.1.1 CWC's 'Grand' Intent

We are in 2004, and POD was published more than forty years ago. Reading it anew enables me in (retrospect) to understand what CWC meant by this most important treatise.

CWC's intent can be garnered from the following quote:

> "We conclude [he states] that a science of ethics is a necessary condition for a science of science".

What does this quotation mean to us nowadays?

Throughout the book, and of course throughout his life, CWC searched for a way to measure and apply scientific methods to study values - the bedrock of ethics. To him, "scientific," means a rational way of reaching conclusions which can be guaranteed.

Nowadays we look somewhat with suspicion at any author who would even attempt such chimerical objective. And indeed, throughout his career, CWC was not accepted by traditional academic disciplines. While chairman of a philosophy department he soon realized that he could not pursue his quest which is outlined in POD. He then started what became Management Science. At the beginning, this fledging discipline kept searching for normative theories to improve decision making. But somehow it was overtaken by an exaggerated emphasis of quantitative methods which was not what CWC intended. Certainly he looked for theories, methodologies by which the "soft side" of humanity i.e. beliefs, judgments, conflict feelings and other psychological and behavioral attributes could be measured and evaluated. In CWC parlance, "measurement" acquires a special meaning: If we ever want to improve our understanding of how human beings make decisions or behave, we must develop a more rigorous science by which they can evaluated. In POD, CWC studies the classical Scientific Method to determine whether starting from observations and explanations of soft variables, we can finally predict what the outcome of a decision will be. His assessment is that the Scientific Method and other so called scientific approaches to evaluate soft variables leaves a lot to be desired. He concludes that before we can start to measure values and come up with a Science Of Ethics we must study and formulate a Science Of Science.

1.3.2 A Science of Science

> "Our science is developed our ethics is not." (CWC: POD. p1).

1. PROGRESS ACHIEVING C. WEST CHURCHMAN'S EPISTEMOLOGICAL PROGRAM

"We can conclude [POD with the statement] that a science of ethics is a necessary condition for a science of science." (CWC in POD, 1961, p.380) [Bracketed words are those of the present author].

"We must seek the development [of] a scientific method for verifying ethical judgments." (CWC: POD, p.2)

We go back to CWC's earliest writings such as TEI and POD.

There, CWC indirectly referred to the effects of measuring and evaluating soft system variables erroneously. He studied how managers made decisions and came to the conclusion that in the real-world, soft systems variables such as judgments, beliefs, feelings and the like were not only badly understood but were poorly evaluated. He considered the application of the so-called Scientific Method to improve the evaluation of soft-systems variables and develop a so-called Science of Ethics. He compared the results of the Scientific Method in hard-system domains (such as found in the physical and natural sciences) to that of soft-system domains (such as found in the biological and behavioral sciences) and found the results wanting.

CWC's aim was to improve measurement and evaluations so that the decisions would become more "ethical," by which he meant that they would reflect more accurately the state of the world and would lead to improvements which he could not find. Time and time again, CWC analyzed management decisions to find that they led to unanticipated consequences and that the Scientific Method was not the appropriate approach for soft-systems. Thus, CWC called for a Science of Ethics where science would buttress ethical norms and guarantee evaluations and improve results.

CWC advocated one common epistemology for all sciences (hard and soft). CWC and some of his colleagues started on this path by applying the quantitative methods of Operations Research to soft-system variables and started Management Science. These methods certainly had a normative content in that they aimed at optimizing the variables which represented the real-world through mathematical models but they could only find solutions in the context of limited models of the real-world. This was far from the comprehensiveness which CWC wanted to achieve with the Systems Approach.

Later, when Management Science was well established as a scientific discipline, CWC continued to criticize its epistemological content which led to an overemphasis of mathematical models. This approach was not the Science of Ethics that he had originally advocated. He then called for a Science of Management which could provide the appropriate epistemology and methodology for the soft-systems which were found in the real-world.

At this point, we can better understand the duality between Management Science and a Science of Management by referring to the hierarchy of inquiring systems which was postulated earlier.

At logic level 1, the activities of the real-world take place.

At logic level 2, the scientific activities of the science (Management Science) take place. Finally, at logic level 3, the epistemology of the science is elaborated as a Science of Management: it represents the study of the knowledge provided to the Management Science at the lower level of logic as well as an inquiry in the validity of its methodologies. Furthermore, the Science of Management seeks to guarantee the results of the Management Science by evaluating its results. This characterization of the three levels of logic corresponds to the hierarchy of inquiring systems summarized above. It embodies CWC's vision of a Science of Values and of a Science of Science by which he concluded POD. For more details refer to a recent paper on the subject (van Gigch, 2003b).

1.3.2.1 Meaning of a Science of Science: "The 'X' of 'X'"

In *Challenge to Reason* (Churchman, 1968) CWC threw the gauntlet: he defied us to take a reflexive step and "climb" above the real-world level to the metalevel to change the logic of our inquiry. As we explained at the beginning of this essay, the logic by which Science is used in the real-world changes drastically when we take "The X of X" an assess the epistemological implications of the science. We recall that in this context to take the "The X of X" is equivalent to investigating the sources of knowledge of the Science and evaluating it "epistemologically" or from the perspective of a higher frame of logic and of abstraction.

1.3.3 A Science of Ethics

CWC was an optimist. While realizing that conflict was the essence of the relationships among human beings, he also reasoned that progress in the so-called Science of Ethics could be realized by the contribution of science.

Progress may have progressed in the realm of hard sciences but scant progress has been achieved in the realm of the soft sciences since CWC enunciated his epistemological program. In this he had many colleagues that agreed with him.

1. *PROGRESS ACHIEVING C. WEST CHURCHMAN'S EPISTEMOLOGICAL PROGRAM*

1.3.3.1 Progress of A Science of Ethics [or Lack Thereof] in Management

If anything the Science of Ethics and the practice of ethics has reached a very low level of implementation and acceptance. There is rampant dishonesty in management, lying and fraud are the order of the day. The "ENRONS" and other frauds make the daily news.

Whether it is for political reasons or greedy reasons for self-aggrandizement and enrichment, the area of the "spin", has arrived where every piece of news is manipulated for selfish reasons which cloud the truth and obliterate transparency from public view. Even in the US military where the principles of management started and where the earliest definition of accountability and responsibility started have been blurred. The US military does not even bother to use traditional budgeting when submitting its financial needs to Congress. In other words the budgets - a traditional albeit imperfect forecast of future needs has been made superfluous.

The meaning of ethics has vanished and the inner controls and motivation to do good are non-existent.

Science has made very few significant inroads to help management improve its record, its methodology or its ethics. If CWC had lived to witness the high record number of frauds and other misdeeds which are occurring daily as we speak (June 2004) he would have been appalled but not surprised.

Frauds and lack of ethics in business transactions are at a record high. The business page of any newspaper reveals infringements of regulations as well as illegal deeds on the part of corporate America. Greed of top executives is rampant.

The epistemological program outlined by CWC in his early writings whereby Science - and in particular the Scientific Approach - would help articulate a Science of Values has not come to pass.

What happened to CWC's good intentions?

Nowadays greed pervades and all the enemies of the Systems Approach are out in full force to prevent the implementation of CWC's epistemological program to be implemented.

1.4 THE "IS" AND THE "OUGHT"

CWC claimed that the "IS" and the "OUGHT" are very similar. They only differ in that the former is in the NOW whereas the OUGHT is in the FUTURE (CWC: POD, ch 15). Seen in this perspective they both require scientific justification so that prediction - for the future-- is as credible as declarations of what is happening now. Thus, the only differentiation is due to the addition of the concept of time.

In modern discourse, the OUGHT implies an obligation. It constitutes the embodiment of the categorical and conditional imperatives postulated by Kant.

CWC argued that Kant's imperatives are not effective to motivate humankind to achieve what is ethical - whereby "ethical" we now refer to achieving what is Good.

We can again use the metasystem approach to advantage to formalize the differentiation between the "IS" and the "OUGHT."

The "IS" is what is happening in the real-world in real time. Scientifically we use observations and descriptions to provide information about what "IS." This information can be found at the logic level 1.

The "OUGHT" is in the realm of speculation and prediction. It implies a normative concept which can only be found at the metalevel - i.e. logic level 3.

In the hierarchy of inquiring systems, the metalevel - level 3 - is the level of Metaethics (also labeled Moral Epistemology) where the meaning of Good/Bad is discussed and elaborated from an epistemological point of view. Epistemology and Metaethics are both involved in elaborating Kant's imperatives and defining what is meant by "OUGHT". The middle level 2 is the level of Normative Ethics where the definitions obtained at the metalevel are used to formulate norms i.e. rules by which the behavior of humankind will be controlled and assessed. We are referring here to a comparison between ideal and actual behavior. Finally, level 1 is the realm of Morality where the actual behavior of an individual is measured/evaluated in terms of the norms established at level 2. Obviously, what is taking place in the real-world, is the realm of the "IS".

If as CWC stated above, Kant's imperatives are not effective to motivate humankind to motivate it to behave ethically, the question remains "What will?" and "Where can we find the source of this motivation?"

1.5 THE MOTIVATION TO BEHAVE ETHICALLY

When CWC considered whether he could trust humankind to resolve conflicts ethically, or to eliminate calamities like poverty, delinquency and wars, he fluctuated "dialectically" between despair and hope.

On the one hand, CWC despaired (One side of the Despair-Hope pair) to see the poor and the wretched of the world abandoned to their plight, while developed nations benefited from wealth and opulence.

1. PROGRESS ACHIEVING C. WEST CHURCHMAN'S EPISTEMOLOGICAL PROGRAM

On the other hand, he was hopeful (The other side of the Despair-Hope duality) that ways could be found to improve the distribution of wealth and health, promote peace and in general strive for the Beauty, Love, and the Ultimate Good.

We asked the question above: "Where does humankind find the motivation to behave ethically without coercion, incentives, regulations or the force of the law? Several ethical theories have been advanced to answer this question:

The Social Contract. Schneewind (In Singer, 1991:137) argues that the autonomous individual will follow the precepts of what has been generally called the "social contract" which is an unwritten covenant between an individual and society according to which the individual behaves ethically for reasons of self-preservation, "natural sociability", and "self-interested aims". To do so is to follow human reason. Virtue is a two-way street which imposes obligations upon us which are in turn rewarded by reciprocal virtuous acts from others towards us.

Natural Law. The motivation to act ethically has been said to spring also from "natural law" (Buckle in Singer, 1991:162), "an unchanging normative order that is part of [society and] the natural world".

Egoism. Others have argued against the plausibility of natural law by siding with those who argue that humankind only behaves either driven by Egoism (To live in such a manner as to follow selfish interests) or as a means to the "Common Good" (a "practical ideal to follow - at least in the economic sphere," a view advocated by Adam Smith in 1776) (Baier in Singer, 1991:200).

Legal Rationality. It is also acceptable to believe that human beings only behave legally, i.e. follow the laws and circumvent them when overtaken by greed. This kind of morality is driven by what is called Legal Rationality). Another theory of why and how human beings behave morally is called Consequentialism and Utilitarianism. According to Consequentialism people behave by weighing the consequences of their acts and actions. In Utilitarianism they act by weighing the "utility" of their actions against the futility of inaction

Kant's Theory of Duty. CWC mentions Kant's imperatives according to which individuals act morally pushed by the belief in duties and obligations, without necessarily relying on ulterior motives or rewards.

Which one comes closer to CWC's inner beliefs? It is hard to say. We know from CWC's writings that he did not believe in the force of Kant's imperatives. We think that CWC felt that human beings are intrinsically good but that society corrupts the best intentions and leads individuals to be greedy, selfish and to act unethically. CWC believed in the existence of a higher Good, the most admirable Beauty and the warmest felt Love. Throughout his life, CWC was concerned with the problems of how these ideals could be achieved. Naturally these questions remain as unsolved today as they were a long time ago.

1.6 CONCLUSION

Has the epistemological program outlined in CWC's two early books been achieved?

Systems thinking has come a long way. It has evolved in the direction outlined by M. Jackson (2003) among many others. For a while CWC's influence was felt in the new realm of management science and Operations Research. However, these disciplines failed to heed the program charted by Churchman when they became purely quantitative and mathematical. That's exactly the opposite of what CWC advocated. In summary he wanted:

1. To improve Management Science by calling for a Science of Science - that is an epistemological evaluation of the scientific methods from the perspective of a metalevel where an independent guarantor could assess their validity and truth content of the Practice of Science.
2. To develop a Science of Ethics whereby scientific pronouncements would be supported by evaluative and ethical significance.
3. To develop one truth-epistemology, which would apply both to the hard sciences and to the soft sciences. Instead we still have a history of duality of epistemological rigor and of scientific achievement.
4. Unfortunately, the progress to establish a Science of Science and a Science of Ethics in the direction and meaning proposed by Churchman has been sporadic and not as vigorous as would have been desirable. Our scientific disciplines still neglect epistemology, and overlook the importance of suffusing decision making with a satisfactory ethical justification. Their neglect threatens the discipline survival.

REFERENCES

Churchman, C. West, 1948, *Theory of Experimental Inference,* MacMillan Co., New York, 4th ed., 1960.
Churchman, C. West, 1961, *Prediction and Optimal Decision: Philosophical Issues of a Science of Values*, Prentice Hall, Englewood Cliffs, N.J., USA.
Churchman, C. West, 1968, *Challenge to Reason*, McGraw-Hill, New York.
Jackson, Michael C., 2003, *Systems Thinking: Creative Holism for Managers*, Wiley.
Singer, Peter, 1991, *A Companion to Ethics*, Blackwell, Cambridge, MA, USA.
van Gigch, J.P., 2003a, *Metadecisions: Rehabilitating Epistemology*, Kluwer/Plenum, New York and London.

1. PROGRESS ACHIEVING C. WEST CHURCHMAN'S EPISTEMOLOGICAL PROGRAM

van Gigch, J.P., 2003b, "The paradigm of the science of management and of the management science disciplines," *Systems Research and Behavioral Science,* **20**(6):Nov-Dec. An Article among others celebrating Churchman's 90th Birthday.

Chapter 2

THE PARADIGM OF THE SCIENCE OF MANAGEMENT AND OF MANAGEMENT SCIENCE

VAN GIGCH, J.P.

Abstract: This essay explores C. West Churchman's contributions to the development of the paradigm of the science of management and of management science. No other contemporary thinker has contributed more than Churchman to the elaboration of the management paradigm. In his unique way, and throughout his professional life, he has raised our level of consciousness concerning the unfinished business of the discipline. Given the eclectic nature of Churchman's mind, it would be impossible to do him justice and make a complete list of the methodological and epistemological issues that he has studied. In this essay, we illustrate the ideas that, to our mind, will have the greatest impact on the future direction of the discipline.[1]

Key words: management science; science of management; C.W. Churchman, Festschrift, legacy

2.1 THE HIERARCHY OF INQUIRING SYSTEMS

In his *Design of Inquiring Systems*, Churchman (1971) introduced the concept of "inquiring system", a system designed to produce knowledge that is relevant to solve present-day problems. According to the metasystem

---

[1] A version of this paper appeared in *Interfaces: Journal of the Institute of Operations Research and Management Sciences*, on the occasion of Professor Churchmann's 80th birthday. See *Interfaces*, Vol. 25(2): 81-88, 1995. Used by permission of the Institute for Operations Research and the Management Sciences. See also *Ssystems Research and Behavioral Science*, 2003, 20: 499-506.

paradigm (van Gigch, 1991b, 2003), a scientific discipline can be characterized as a hierarchy of three inquiring systems, as follows:
1. An *implementation inquiring system*, or "practice level" of the discipline, where practitioners solve the everyday problems which take place in organizations and where the tools of the discipline are being applied to real-world problems.
2. A *science inquiring system*, or "science level" of the discipline, where the scientific methodology of the discipline is being developed and shaped. This level can also be characterized by Kuhn's (1970) level of "normal science", i.e. the level at which the scientists of the discipline formulate and solve the everyday scientific problems of the domain.
3. An *epistemology inquiring system*, or "epistemological level" of the discipline, which is the level at which the reasoning processes, the logic and the anomalies confronting the discipline are being discussed. In contradistinction with the level where the problems of "normal science" are solved (see above), the epistemology level is where "extraordinary science", i.e. innovation, creativity and the paradigm, are conceived.

2.2 THE X OF X.

The above hierarchy serves to point out the distinction between management science and the science of management. Churchman (1968a) is the pioneer who first emphasized this distinction. He created the "X of X", a self-reflective loop that posits the existence of:
1. The *science of management*, (X), a metascience which is carried out in the epistemology inquiring system. It is concerned with the design of the science of management as "a science" (qua-science). It is concerned with (a) the purposes of the science; (b) the definition and renewal of its paradigm, formulating the domains of application of the paradigm; (c) the investigation of the science's foundations; (d) the assessment of its methodology, its logic and methods of reasoning; as well as with (e) the generation of new problems. See also Chapter 1, herein.
2. *Management science*, (the X of X) devoted to the application of the *science of management* to the solution of organizational problems. Management science is mostly carried out in the so-called "science inquiring system" where management science experts and scientists study organizational problems and later implement the chosen solutions at "the practice level" or implementation inquiring system.

In this essay, we are mainly concerned with a discussion of the science of management and with Churchman's contributions to its formulation. One of Churchman's earliest publications, his *Introduction to Operations Research*

2. THE PARADIGM OF THE SCIENCE OF MANAGEMENT AND OF MANAGEMENT SCIENCE

(Churchman et al., 1957), while meant as a management science text, included many discussions concerning some of the scientific and epistemological problems of the budding discipline. Unfortunately, this material was summarily discarded in later management science texts which concentrated, instead, upon optimization methods, without ever considering their justification or consequences.

Churchman was the first to draw attention to the importance of the science of management, whose study takes place at the level of the epistemology inquiring system. As a cursory survey of the contemporary management literature easily reveals, most of the work and publications in the field deal with questions about management science or its practice. A very small percentage, may be less than 10%, treats the science of management and epistemological issues. This situation is very unfortunate because, when a discipline neglects its foundations and its epistemology, it is bound to flounder and even disappear (van Gigch and Pipino, 1986) (van Gigch, 1989, 2003). It is to Churchman's credit that most of his writings have been concerned with the epistemological issues of the discipline. Indeed, as we will try to illustrate in this essay, his main concern has always been the elaboration of the paradigm of the science of management and that of the management science disciplines.

2.3   JUSTIFICATIONS FOR A PARADIGM

We use the definition given by Kuhn (1970), according to whom a paradigm is "the complete constellation of beliefs, values, techniques, etc., shared by the members of a scientific community". We add Morgan's (1980) clarification that a paradigm "denotes an implicit or explicit view of reality" and that "it contains [a discipline's] core assumptions that characterize and define [its] worldview". As we know, a paradigm constitutes the essence of a discipline. Scientists in the discipline use the statement of the paradigm to establish the problems that it must solve. Usually, the scientific establishment works in the direction of the prevailing paradigm, i.e. its rules and procedures are directed by the currently accepted paradigm. A paradigm is essential to discover anomalies and dilemmas where, in the language of epistemology, anomalies are unresolved questions which run counter to the anticipated results predicted by the paradigm, and dilemmas are epistemological or methodological debates which pit different groups in the discipline. Dilemmas can also stem from unresolved questions that are presently discussed and researched by the members of the discipline.

Subsequent texts (Fuller, 1999; Kuhn, 2000) on Kuhn's work do not alter the above view.

2.4 QUESTIONS WHICH THE PARADIGM MUST ANSWER

A paradigm should help provide answers to at least the following six fundamental questions:
1. What are the *main sources of knowledge* of the discipline?
2. What constitutes the *object of study* of the discipline?
3. What are the *main schools of thought* underlying the discipline?
4. What are the *main purposes* of the discipline?
5. What are the *significant instrumentalities (methodologies)* used by the discipline and, by derivation, its *main activities*?
6. What are the *anomalies* and *unresolved problems* which are facing the discipline?

Questions 3 (main schools of thought) and 5 (significant instrumentalities or methodologies) are not relevant to the present essay and therefore will be omitted from our discussion.

2.5 CHURCHMAN'S CONTRIBUTIONS TO ANSWER THE PARADIGM'S FUNDAMENTAL QUESTIONS

Question 1. What constitutes the main sources of knowledge of the discipline?

In *Design of Inquiring Systems*, Churchman (1971) questions the philosophy of past and contemporary thinkers to discover how their logic and methods of reasoning can be used to design the modern inquiring system, a system designed to produce knowledge which is relevant to solve present-day system problems. He surveys the thinking of Leibniz, Locke, Kant, Hegel and Singer for ideas that could be incorporated in the management paradigm. Churchman's study inspired countless publications in which this approach was emulated. For the present essay, it will suffice to recall studies whose authors explore the work of philosophers, beyond the initial five studied by Churchman. Van Gigch (1988a, 1988b, 1988c, 1991b), Snell (1988), Pavesi and Pavesi (1989, 1991) and Herrscher (1989) review the work of Descartes, Destutt de Tracy, McCulloch, de Condillac, Hesse, and Cassirer. These studies show that many of the systems concepts which are used today can already be found in the work of thinkers as ancient as

Ramon Lull, who lived in the thirteenth century. In this vein, other studies inspired by Churchman could be planned in this series to study contemporary thinkers such as Schrodinger, Feynman, Simon, Newell and many others. Readers have an open invitation to participate in this open project.

It is interesting to speculate about the direction which management information systems will take in the future and whether some of the important features identified by Churchman will be incorporated in future inquiring systems. In an indirect way, Churchman's thinking has been incorporated in new sciences such as information sciences, artificial intelligence, cognitive science and knowledge engineering. The latter are noticeably bypassing management science and operations research and new disciplines may, in the long run, supersede the old ones altogether.

Question 2. What constitutes the main object of study of the discipline?

Churchman has always tried to steer the discipline to reconsider its domain and its subject matter (Mitroff and Churchman, 1992). Churchman separates problems into two classes: *ordinary* and *extraordinary* problems. To his dismay, the profession has concentrated on the former and neglected the latter. He calls problems "extraordinary" when "they are beyond the ordinary", or meaning that "they are astonishing, horrible or deadly". "Examples are the world military threat, world pollution threat, growth of population and the enormous extent of poverty" (Churchman, 1984). It is pertinent to select a quotation from Churchman's own writing and note that it was published as early as 1984:

> "In a world of food plenty, we starve millions of people, especially little kids. There is the image of a refugee camp in Somalia where at one time in a field were termite mounds, until the children found that termites soothed their thirst and hunger. Another image is of a man in a small house with one window, located in the Ganges Valley. He had just enough energy to move away from the hot sunlight pouring through the window."

> "How dare humans treat other humans so that the latter must live their lives in this manner? Where is the compassion that humans are supposed to have with respect to one another? If the city management of San Francisco daily selected six prisoners in the city jail to be tortured in Union Square, wouldn't the public rise in moral outrage? But starvation is a torture, carried out in the cruelest of ways." (Churchman, 1984).

It is tragic that almost twenty years after the above was written little has changed. If anything, conditions for the wretched of the world have grown steadily worse. In 1992, Mitroff and Churchman (1992) repeated their outrage by challenging the institution of the science of management for its lack of conscience for not tackling the important problems which beset the modern world. In a "manifesto", they urged the system sciences and, implicitly, the management science community to heed their call for urgent changes in the direction of these sciences. In other words, Mitroff and Churchman pleaded for a change in the *object of study* of the management discipline and of its paradigm. They called for a "different definition and concept of science" which will result in a "desperately needed inquiry into the management of world problems", and for "action research" of the kind that will "meaningfully improve the human condition".

They stated:

The key word [should be] to meaningfully improve the human condition.

By "improve", we mean actions which are ethically sound, not necessarily scientifically or theoretically.

How can we persuade some humans ... to conduct research into how the human condition can be persuaded to embrace a *different way of thinking and behaving about world problems*? [Emphasis is ours.]

We have no sound way to pose global problems such as "world starvation" except to know that it exists.

During all of his academic and professional life, Churchman has championed the intervention of science as an activist institution which has "fundamental moral and ethical responsibility to develop programs that serve humanity's broader needs ... to recognize and to take seriously the emotional complexity of humans and organizations" (Mitroff and Churchman, 1992).

Again, as a springboard for action and change, Churchman relies on "the X of X", the reflective quality of science, which allows it to become critical of its own epistemology and paradigm. Ackoff (1979a, 1979b) and van Gigch (1989, 2003) have also criticized management science ("the X") for failing to be reflective and for neglecting to investigate and evaluate the underlying justification of the discipline ("the X of X"). Echoing Churchman, and as stated earlier, van Gigch (1989, 2003) has even argued that, unless the discipline makes a concerted effort to re-study its foundations and its purposes, it will founder and may even disappear as a scientific discipline.

2. THE PARADIGM OF THE SCIENCE OF MANAGEMENT AND OF MANAGEMENT SCIENCE

Question 3. What are the main schools of thought underlying the discipline?

Answering this question is not relevant to the present study. For a partial answer to this question readers are referred to van Gigch and Pipino (1986), van Gigch and Le Moigne (1989) and van Gigch (2003).

Question 4. What are the main purposes of the discipline?

A determination of the purposes of the discipline is intimately tied to the definition of the management paradigm and to a discussion of its epistemological issues. An exhaustive study of the purposes of the management discipline would require several volumes. A more limited intent recalls Churchman's contributions to the subject, by referring again to some of his publications.

Churchman refers to management as a "purposive" and "purposeful" science. In *Prediction and Optimal Decisions* he argues for a science of values (Churchman, 1961). He is always concerned with the morality of systems and organizations, where by "morality" he means the ethical/moral justification for the results of actions undertaken in the name of science. He is concerned with the effects of the scientist's recommendations on recipients and clients. He never conceives science to be neutral and uninvolved. On the contrary, for Churchman, the institution of science must be conscious of the consequences of its designs. It must be alert that it does not solely dwell upon theoretical propositions and that it must also become actively involved in the ethical implementation of its recommended solutions (see Chapter 1, herein).

Churchman explores the formulation of a multilevel ethic which implements the three-level hierarchy of inquiring systems described earlier. The postulation of a multilevel ethic allows us to conceive the elaboration of different rationalities depending on the level at which they originate (van Gigch, 1991, 1992). Multilevel ethics has a direct application to production and environmental management where the question of abiding by environmental, pollution and conservation constraints must be justified. Obviously, rationalities at the implementation level call for making survivability and economic viability the main objectives of the organization. At the next of level of inquiry—the science inquiring system—the design of the production system must be technologically sound and economically feasible. Finally, at the higher level of inquiry, Churchman invokes the ultimate values of *"the good"* and *"the beautiful"* to imply that, even if our designs are economically and technologically minded, they will be worthless unless they are *"ethically sound"* and take into account the *"improvement of the human condition"*.

Van Gigch (1991, 2003) elaborates a similar hierarchy of ethical imperatives which depend on:

(a) the level of logic at which the problem at hand occurs; and

(b) the subject matter or domain to which the problem pertains.

The hierarchy advanced by van Gigch (1992, 1997, 2003) includes a management imperative as well as a scientific, a legal, a political, an epistemological, an ethical and, even an aesthetic imperative. It echoes the earlier concept advanced by Churchman. See above.

Question 5. What are the significant instrumentalities (methodologies) used by the discipline and, by derivation, its main activities?

As in the case of question 3, question 5 is not considered in the present essay.

Question 6. What are the anomalies and unresolved problems that are facing the discipline?

Again, we plead forgiveness and permission to only cite a few of the numerous questions that Churchman has raised in his publications.

Churchman (1984) raised the question of a problem's ontology where he noted an important paradox in the theory of planning: on the one hand, the theory urges us that, before a problem can be solved, it should be defined and formulated. However, he noted that "supposedly problems exist before they are formulated". "How does a nonformulated problem exist?"

Churchman claims that for the class of extraordinary problems (see above) "the preformulation existence is different from that of ordinary problems". "The preformulation existence of extraordinary problems is a moral outrage." In this case, "the planner might pass directly from moral outrage to implementation skipping problem formulation". Churchman raises doubts that the usual problem formulations "capture the meaning of its preformulation existence", in which case he adds "the formulation is wrong and the formulated problem is an illusion and does not really exist". Of course, Churchman is not the first one to raise the issue of realism in science. What is significant is that he is the premier philosopher of the science of management and of management science, a discipline that is better known for its pragmatism than for its incursions in the philosophy of science.

Churchman (1959) spoke about the need to consider measurement in a new light. He characterized measurement as a "decision making activity ... designed to accomplish an objective". In particular, he reminded us that a measurement also depends on how the domain over which the measurement is taken is defined. As an example, a definition of poverty will depend on whether or not we want to extend the measurement to deprivations other

2. THE PARADIGM OF THE SCIENCE OF MANAGEMENT AND OF MANAGEMENT SCIENCE

than economic, such as poverty of education, opportunity, expression or representation.

Churchman (1961) showed that contemporary accounting conventions concerning the method of allocating overhead costs and the misuse of opportunity costs leads to erroneous management decisions when "knowledge of the correct organization for the firm" is unknown, where by "correct organization" he means that the global optimum can never be ascertained.

An important factor in capital budgeting relates to the use of the appropriate discount factor. As we all know, taking a positive discount factor unduly reduces the value of the future to the point where present American managers are obsessed with the short term, and overemphasize short payback periods. Churchman suggests the use of a different social discount factor in management decision making to give the future its due, particularly in cases when the value of assets for future generations is at stake. This different discount factor should be applied when valuing the old-growth forests, clean air, water and land, or any assets related to the environment and to our common future. The value for future generations is of primordial interest when considering the conservation of natural resources and of animal species that are considered "endangered" or in the process of extinction. It is also of importance to provide some methods of valuing cultural assets such as that of the historical and artistic heritage of a nation which are represented by its priceless treasures. To assess the importance of future generations readers should refer to Partridge (1980).

Throughout his work, Churchman (1968b) pleads for the need to take a systems approach to management problems. All his professional life, he fought against the reductionist paradigm, which fragments problems into local and partial jurisdictions and prevents the formulation of global solutions for complex problems. Churchman (1979) warned us against committing the "environmental fallacy" which not only results when a system's environment is overlooked, but also takes place when we fail to consider the shortterm and long-term responses of outside systems upon our own. A paradox arises because, as we try to avoid the "environmental fallacy" and start taking more systems into account, we soon learn another lesson. Churchman reminds us of Anaxagoras' theorem:

> "In everything there is everything" or "In every problem are to be found all problems" which means that, if a problem is unraveled far enough, we will encounter all other problems."

Thus, we are left with the dilemma that if we push the system's boundaries too far, we face the problem of having to consider too many systems and the situation becomes too complex and unsolvable. On the other hand, if we do not push the boundary far enough, we face the "environmental fallacy" when insufficient relevant systems are taken into account (Churchman, 1979; van Gigch, 1984).

In The *Enemies of the Systems Approach*, Churchman (1979; van Gigch, 1980) showed that the systems approach has enemies such as politics, morality, religion and aesthetics. We will close this section by "translating" for the uninitiated what we think Churchman meant by aesthetics, as an "enemy" of the systems approach.

The ideal inquiring system is one that combines the *love of truth* with the *love of good* and *love of the aesthetic* (or *beautiful*). If in designing a system we forget *beauty*, then the design is unaesthetic. *Aesthetics* is hard to define but for Churchman it is synonymous with "intuition" (a way to reveal truth unclouded by doubt). In his mind, aesthetics can also be understood as a form of "awareness" of the three pillars of wisdom, i.e. *faith* represents the need for a guarantor in the systems planning process, *hope* the "belief in the desirable without perceived evidence" and, finally, *love* represents joy, peace and happiness.

How can our "enemies" become our friends? To answer this question, Churchman returns to an old theme he treated in *Challenge to Reason* (Churchman, 1968a), where he showed the dualism between realism and idealism, between reality and vision: to know what reality is like, one needs to place oneself in our enemy's position and to understand the world from his/her vantage point:

> "Rational humans need to leave the body of rationality and place the self in another body, the "enemy", so that reality of the social system can unfold in a radically different manner. From the vantage point he/she can observe the rational spirit and begin to realize not only what has been left out of it, but also what the spirit is like, especially its quality of being human." (Churchman, 1979)

In the process of understanding reality by becoming the enemy, "the hero" of the systems approach "[should not] give up his/her vision even though in reality it fails over and over again". "The vision is at the same time the sign, the dream, in the direction of which our hero counters the enemy". "The visionary must at one and the same time live his visions and the reality of the collective conscious" (Churchman, 1979).

See also last two essays in this volume.

2.6 FINAL REMARKS

In conclusion, today, we honor C. W. Churchman for his lifelong accomplishments on the occasion of his 90th birthday. We thank him for being that "visionary", for rekindling our faith in the "rational spirit", for giving us the renewed strength to "improve the human condition" and to show us the how we can redirect our discipline to accomplish "the dream". We will continue to honor his accomplishments for many years to come.

REFERENCES

Ackoff, R.L., 1979a, "The future of operational research is past", *Operational Research Society Journal*, **30**(2):93–104.
Ackoff, R.L., 1979b, "Resurrecting the future of operational research", *Operational Research Society Journal*, **30**(3):189–199.
Churchman, C.W., 1959, "Why measure?" in: *Measurement: Definition and Theories*, C.W. Churchman and P. Ratoosh, eds., Wiley, New York, pp. 83–94.
Churchman, C.W., 1961, *Prediction and Optimal Decision*, Prentice-Hall, Englewood Cliffs, NJ.
Churchman, C.W., 1968a, *Challenge to Reason*, McGraw-Hill, New York.
Churchman, C.W., 1968b, *The Systems Approach*, Delacorte, New York.
Churchman, C.W., 1971, *The Design of Inquiring Systems*, Basic Books, New York.
Churchman, C.W., 1979, *The Systems Approach and its Enemies*, Basic Books, New York.
Churchman, C.W., 1984, "Churchman's conversations", *Systems Research*, **1**(1):3; (2):89–90.
Churchman, C.W., Ackoff, R.L., Arnoff, L.E., 1957, *Introduction to Operations Research*, Wiley, New York.
Fuller, S., 1999, *Thomas Kuhn: A Philosophical History for Our Times*, University of Chicago Press, Chicago.
Herrscher, E.G., 1989, "Design of the modern inquiring system. Part VI: Hermann Hesse (1877–1962)", *Systems Research*, **6**:267–270.
Kuhn, T.S., 1970, *The Structure of Scientific Revolutions* (rev. edn), Chicago University Press, Chicago.
Kuhn, T.S., 2000, *The Road Since Structure: Philosophical Essays, 1970–1993, with an Autobiographical Interview*, J. Conant & J. Haugeland, eds., University of Chicago Press, Chicago.
Mitroff, I.I., Churchman, C.W., 1992, "A manifesto for the systems sciences: outrage over the state of science", *General Systems Bulletin*, **22**(1):7–10.
Morgan, G., 1980, "Paradigms, metaphors, and puzzle solving in organization theory", *Administrative Science Quarterly*, **25**(4):605–622.
Partridge, E., ed., 1980, Responsibilities to Future Generations: Environmental Ethics, Prometheus Books, Buffalo, NY.
Pavesi, P.E. & Pavesi, P.F.J., 1989, "Design of the modern inquiring system. Part V: Etienne Bonnot de Condillac (1715–1780)", *Systems Research*, **6**:176–178.
Pavesi, P.E. & Pavesi, P.F.J., 1991, "Design of the modern inquiring system. Part VII: Ramon Lull (1232–1315) and his Ars Magna", *Systems Research*, **9**(1):85–92.

Snell, J.L., 1988, "Design of the modern inquiring system. Part IV: Warren S. McCulloch (1898–1969)", *Systems Research*, **5**:359–361.

van Gigch, J.P., 1980, "Book review: C.W. Churchman's *The Systems Approach and its Enemies*, *Human Systems Management*, **1**:275–277.

van Gigch, J.P., 1984, "Epistemological questions raised by the metasystem paradigm", *International Journal of Man-Machine Studies*, **20**:501–509.

van Gigch, J.P., 1988a, "Design of the modern inquiring system. Part I: R. Descartes (1596–1650)", *Systems Research*, **5**:267–271.

van Gigch, J.P., 1988b, "Design of the modern inquiring system. Part II: The contemporary computer", *Systems Research*, **5**:267–271.

van Gigch, J.P., 1988c, "Design of the modern inquiring system. Part III: A.L.C. Destutt de Tracy (1714–1836) and the Idealists", *Systems Research*, **5**:357–359.

van Gigch, J.P., 1989, "The potential demise of OR/MS: consequences of neglecting epistemology", *European Journal of Operational Research*, **42**(3):268–278.

van Gigch, J.P., 1991a, "Design of the modern inquiring system. Part VIII: Symbolism and Ernst Cassirer (1874–1945)", *Systems Research*, **7**(4):298–301.

van Gigch, J.P., 1991b, *System Design Modeling and Metamodeling*, Plenum, New York.

van Gigch, J.P., 1992, Economic versus organizational efficiency: a question of tradeoffs in strategic and environmental management. Presented to the EUROCARE December Board Meeting Conference, Goteborg, Sweden.

van Gigch, J.P., 1997, "The design of an epistemology for the management discipline which resolves dilemmas among ethical and other imperatives", *Systemic Practice and Action Research*, **10**(4):381–394.

van Gigch, J.P., 2003, *Metadecisions: Rehabilitating Epistemology*, Plenum/Kluwer, New York.

van Gigch, J.P., Le Moigne, J.L., 1989, "A paradigmatic approach to the discipline of information systems", *Behavioral Science*, **34**(2):128–150.

van Gigch, J.P., Pipino, L.L., 1986, "In search of a paradigm for the discipline of information systems", *Future Computing Systems*, **1**(1):71–97.

Chapter 3

CHURCHMAN'S CONTRIBUTIONS TO THE ADVANCEMENT OF MANAGEMENT SCIENCE

AGRELL,P.S. AND LEONARZ,B.

Abstract: The present paper documents the impression made by C. West Churchman when he visited that country in the 1990's. It also shows the influence that CWC had on the Swedish systems movement. This summary written fifteen years ago emphasizes the thinking revolution which CWC's ideas brought about. These ideas are still relevant today. For a summary of CWC's epistemological views refer to van Gigch (2003) who presents these ideas in the context of a rehabilitation of epistemology as an indispensable tool to trace the sources of knowledge. Paper presented at the l2th Trienal Conference on Operations Research in Athens, 1990, IFORS'90

3.1 HOW IT ALL STARTED

In the late sixties Churchman's book *The Systems Approach* was being used in the final course for first-year students of business. It formed the methodological basis for innumerable course-papers on a wide variety of topics. Over the last two decades thousands of students of business administration at Stockholm University have been introduced to methodological questions through the writings of C. West Churchman[1].

---

[1] Most of them seemed to like it. Many were even enthusiastic about it, much to their own surprise. This interest in questions of epistemology has also proved to be valuable to those who later enter a profession as an applied systems analyst.Due to the great number of students (usually several hundred beginners each term) most of the instruction was given as lectures. However, this soon proved to be a bad way of conveying the spirit of this book. The heart of the matter was missing. There is no room for argument and dialogue in a lecture hall with two hundred and one people at a time, two hundred of which are being

3.2 ENTER MITROFF, TUROFF AND SWEDISH DEFENSE RESEARCH ESTABLISHMENT

Independently of one another several of the teachers at the department were suddenly captivated by an article in a philosophical journal by Ian Mitroff who was unknown to us at the time. Like us, he wondered where the fun had gone, why dissertations and academic papers are often so one-sided and boring. Why is the humour of science eradicated from its written documents? Unlike us he had the courage to write a paper about this (Mitroff, 1972). Eventually we found out about this mutual interest and decided to do something about it.

At the same time there was a growing interest in Futures Studies and the Delphi Method among others at the department. This interest was shared by The Swedish Defense Research Establishment. They had the money and we had the academic status. Together we planned to invite Murray Turoff to come and discuss Delphi with us. By coincidence it appeared that Turoff had a friend named Mitroff who might also like to come along. They were both invited and both of them accepted the invitation. This proved to be a turning point for everything that had to do with methodology at the department.

We had a wonderful week filled with lectures, seminars and long discussions all through the long light Nordic summer nights. One of the topics that Mitroff and Turoff brought up was that of technology assessment. They did so on the basis of an article of theirs (Mitroff and Turoff, 1973) that is at the same time an application and a summary of Professor Churchman's *The Design of Inquiring Systems*.

After that week we coined the phrase "Churchman is Churchman and Mitroff is his prophet".

---

exposed to these ideas seriously for the first time. These lectures were much appreciated by students and lecturers alike. They left the impression that questions of methodology are debatable and that the debate can be exciting. Long live the dialogue! The book was subsequently translated into Swedish. The translation was a very good substitute for the original. A summary of the basic ideas was made available to beginning students to prepare them for what was to come.The only affordable solution we could find was to have two lecturers. One talks for a while and the other observers the audience to see what seemed to be understood and what seemed to be causing problems. Then the observer puts those questions to the lecturer that he thinks the students would have put, had they had enough time for reflection (which they do not have in the lecture hall). He starts a dialogue with the lecturer to have these points clarified. Then the two lecturers switch roles. Sometimes they have arguments between themselves in front of the audience. A lecture like this would go on for about three hours with a couple of breaks.

3.3 THE DESIGN OF INQUIRING SYSTEMS

The introduction given to us by Mitroff and Turoff made us want to study their sources, in particular the above-mentioned book by C. West Churchman. We formed a seminar group consisting of teachers and doctoral students from the department (remember, this is not a department of philosophy but one of business) and some of our new friends from The Swedish Defense Research Establishment. We met regularly for a whole year working through this book, trying to see its implications for our own research problems and those of our students.

3.4 GETTING THE STUDENTS INTERESTED

We found the description of the various types of inquiry (Leibnizian, Lockean etc) of *The Design of Inquiring Systems* to be an excellent tool for explaining what is meant by "research style", both to ourselves and more importantly to our students.

We knew we had been confined to a fairly narrow methodological workspace, like a small office room with no doors or windows. Now we were given the tools for adding windows and doors to this room, even for making a plan of all the rooms on the same floor in the vast methodological palace we somehow knew was there all the time. This was a great relief.

We brought this material to the courses or discussions on methodology that we had with doctoral students and advanced undergraduate students. After having digested it, many reacted to it by being angry with us: "Why didn't you tell us this from the start? Why have your been withholding this from us?" Then they learnt another thing: teachers are also learners, possibly even mostly learners.

3.5 PROFESSIONAL LIFE AFTER UNIVERSITY

With or without a doctoral thesis, many of the students found employment as planners and systems analysts in private companies or in the public sector. Let us be precise about what happened in the Swedish Defense Research Establishment (FOA) where one of the authors has his chair. This development is typical of Swedish applied think tanks, at least we hope so, although the methodological awareness is relatively high at FOA.

3.6 SATISFYING CERTAIN NEEDS

A research institute at the service of clients is in a squeeze. The work is interdisciplinary in all important projects, so the academic standards to be adhered to come from more than one discipline. This usually means that the rites of all these disciplines cannot be followed. Some have to give way to others even for no other reasons than those dictated by practical considerations such as resource restrictions and time limits. Moreover, many methodological issues in applied systems analysis have no backing in any academic discipline. So, even for a qualified researcher the academic tribes are more of a menace than a support.

The clients also have their demands. They want academically acceptable quality but at the same time they do not welcome all sorts of results. Surprising research results are normally even less welcome in an administrative setting than in the academic world.

FOA, like many others, suffered an implementation crisis in the seventies. A change of paradigm was necessary. Steps in that direction were taken and there is still a dynamic development going on. Recruitment and course-planning together with the University of Stockholm helped create a new generation of analysts and some new approaches to planning and systems analysis. In particular, satisfying the following needs was greatly helped by the new ideas:

3.6.1 The need to have many approaches to choose from

Consultancy is different from making normative methodology and scanning is different from evaluation, evaluation ex post is different from evaluation ex ante etc. The researcher is required to choose an approach that suits the situation without loosing his integrity or independence as a researcher. The five philosophically inspired paradigms of *The Design of Inquiring Systems* and the "four enemies" of *The Systems Approach and its Enemies* demonstrate a number of justifiable approaches, none of which, not even the Singerian, being once and for all the best. Choice and variation is permissible.

3.6.2 The need to understand different ways of reasoning

Elsewhere (Agreli, 1983) one of the authors has described a conflict between a Kantian model-builder and a Lockean empiricist where one does not understand the ways of reasoning of the other. You have to understand all the actors not only to make them understand your results, but for all kinds of cooperation, for example in order to present concepts for learning,

3. EXPANDING CHURCHMAN'S PHILOSOPHICAL DISCOURSE

communication and even for thinking. As a researcher you have to speak the other's language without premature consent. Churchman's books made us realize that adaptation and integrity can be there at the same time.

3.6.3 The need to resolve technical issues

A typical situation is the conflict between the innovators and those defending the old system. By a Churchman-inspired Mitroff technique (Mason and Mitroff, 1981) the conflict is made explicit all the way down to basic paradigms. This is no easy task and the Mason/Mitroff text-book could not always be followed. However, with a Churchman-based insight about coexisting paradigms the analyst's capacity for solving conflicts is markedly better than it would otherwise have been.

3.6.4 The need to change "bad" research styles

At one time exaggerated unstructured empiricism and conceited advisory roles were experienced at FOA. The solution to this problem was not to advise a general change of style but to attack the generality and rigidity of their application. There were both a general development towards more Singerian or Linstonian (Linstone, 1984) styles and an increased interest in a varied application of different approaches.

3.6.5 The need to strengthen the professional pride of applied researchers

Especially newly recruited analysts need to know who they are and what choices are open to them. They are forced to leave their habitual roles of specialists within an academic discipline. They have to face new undefined obligations and freedoms. They have a more multi-faceted work than before and a less firmly defined support. They experience a risk for being alienated. But they also have the possibility of reaching great professional pride They may learn that, unlike what was the case at the university, they need to master not only one of the paradigms defined in *The Design of Inquiring Systems* but all five of them. Investigation is not half-way research. It is more than research. The next step will be to explain even outside FOA that investigation is harder than research.

3.6.6 The need to counter academic criticism

Now and then critics from academic disciplines, knowing little about the world of client cooperation, claim that FOA sells subjectively adapted answers to please the clients and to maintain a budget. This kind of criticism should be taken seriously and this is also done. It cannot be excluded that there might occasionally be an unfair adaptation to the clients' opinions and strivings. However, in ventures of common learning there is an element of adaptation of styles, languages and starting-points which cannot be criticized on these grounds. As stated above an adaptation of approach is possible and even necessary and this can be accomplished without compromising the integrity, the results or the truth-values of the researchers. The new (Churchman) philosophy gives not only new options of choice but also a language to be explicit about the chosen perspectives and epistemologies. This should be an academically respectable way to counter superficial criticism and to invite a more profound and constructive one.

3.7 CONCLUSION

Many applied researchers in Sweden are indebted to C. West Churchman for being better prepared to clarifying their professional identity, to defending and promoting their profession, to understanding and being tolerant towards different ways of reasoning, to entering new kinds of ventures with a sense of freedom and fun.

REFERENCES

Agrell, Per S., 1983, "Facts, methods, programmes and paradigms", *European Journal of Operational Research*, **14**:335-340.
Churchman, C. West, 1968, *The Systems Approach,* Delta, New York
Churchman, C. West, 1971, *The Design of Inquiring Systems,* Basic Books, New York.
Churchman, C. West, 1979, *The Systems Approach and its Enemies,* Basic Books, New York.
Linstone, Harold A., 1984, *Multiple Perspectives for Decision Making,* North Holland/Elsevier, Amsterdam/New York.
Mason, Dick O. and Mitroff, Ian I., 1981, *Challenging Strategic Planning Assumptions,* Wiley, New York.
Mitroff, Ian I., 1972, "The mythology of methodology", *Theory and Decision*, **2**:274-290.
Mitroff, Ian I. and Turoff, Murray, 1973, "Technological forecasting and assessment: science and/or mythology?", *Technological Forecasting and Social Change*, **5**:113-134.
van Gigch, J.P., 2003, *Metadecisions: Rehabilitating Epistemology,* Kluwer/Plenum, New York and London.

Chapter 4

EXPANDING CHURCHMAN'S PHILOSOPHICAL DISCOURSE
New Perspectives

AGRELL,P.S.

Abstract:	Real world problems always present the duality of object and subject and, as a consequence, their management exhibits a particular complexity where analytical tools must take into account purposeful approaches as well as the cultural context in which actions take place. We envisage "beauty" in this sequential contextualization.
	These systems require a double control:
	A *mega-structure*, visible in its organizations and authorities, which provide purposes and restrictions.
	A *meta-structure* which provides inquiring systems and their rules of logic. The mega- and meta-structures compete although the meta-structure is in the service of the mega-structure. That's the order of power.
	The mega- and meta-structures are not isomorphic and the *meta-structure* is not a part of the mega-structure. A *meta-structure* has to be a subtle and heterogeneous meta-system, with sayings of its own, which has to act on all
Key words:	Systems approach, systems theory, meta-system, mega-system, epistemology, guarantor, combinations of method.

4.1 INTRODUCTION

West Churchman opposed positivistic modelling as a worldview for systems science. He emphasized the need to reconcile qualitative and qualitative methods in systems science. He introduced a large spectrum of different methodologies; he related those to ethics and to the history of philosophy. He told us about how to reconcile differing perspectives and he gave us the freedom to go ahead exploring new paradigms.[1]

It is impossible to give an account for the CWC's (C. West Churchman's) overall influence and development growing from his ideas. He created a new systems culture and I will first mention a few of the important systems thinkers from this culture that have been influential to my work. Harold Linstone[2] and Ian Mitroff[3] gave examples of differing perspectives as well as advice about how to proceed with new approaches. John P. van Gigch developed and made Churchman's guarantor concept useful and practical.[4] Mike Jackson and Amanda Gregory taught me ways to reconcile conflicting perspectives.[5] See also Chapter 3 herein.

Yehezkel Dror's distinction between *mega*[6] and *meta* policy is not only valid for high level governance. It is also transferable to lower level design and analysis issues. His *mega* dimension offers a useful simplicity not to forget in any systems analysis. Dror's specific metaconcept appears to be well in accordance with the more general one provided by John van Gigch.

Aside from the present author, other Swedish systems thinkers are also interested in the application of explicit guarantors to multi-perspective views. I mention, Darek Eriksson,[7] Kristo Ivanov[8] and Abdul Khakee[9].

---

[1] Most important for me has been Churchman, West C., 1971, *The Design of Inquiring Systems*, Basic Books; and 1968, *The Systems Approach*, Delta.

[2] Linstone, Harold A., 1994, *Multiple Perspectives for Decision Making*, North Holland.

[3] Mitroff, Ian I. and Linstone, Harold A., 1993, *The Unbounded Mind*.

[4] van Gigch, John P., 2003, *Metadecisions: Rehabilitating Epistemology*, Kluwer/Plenum, London and New York; and 1991, *Systems design Modelling and Metamodelling*, Plenum, also earlier books.

[5] See for example Gregory, Amanda and Jackson, Mike, 1992, "Evaluation methodologies: A system for use", *J Opl Res Soc.*, **3**(1) and Gregory, A., 1996, "The Road to Integration", *Omega*, **24**(3).

[6] The expression *grand policy* is also used. Dror, Yehezkel, 1986, *Policymaking under Adversity*, Transaction Books.

[7] Eriksson, Darek, 2003, "Identification of normative sources for systems thinking: An inquiry into religious ground-motives for systems thinking paradigms", *Systems Research and Behavioural Science*, **20**.

[8] Ivanov, Kristo, 2004, "In memory of C. West Churchman (1913-2004)", *Journal of Organisational Transformation & Social Change*, **1**(2-3).

[9] Khakee, Abdul, 2003, "The emerging gap between evaluation research and practice", *Evaluation*, **9**:3.

4. EXPANDING CHURCHMAN'S PHILOSOPHICAL DISCOURSE 35

I regret that the contacts between French and English-speaking epistemologies are relatively weak. I would have liked to see a co-operation between Churchman and French systems philosophers like Edgar Morin, Jean-Louis le Moigne and Michel Foucault. I try to imagine how Churchman would have resolved the Foucault paradox of *The Discourse*, that it is so powerful yet arbitrary. He would surely have had a view on how to find a guarantor system for it, though of course Foucault would not have agreed.[10] Now this short paper will be my effort to say something more precise about the possibilities for relevance and qualities in Churchman's kind of discourse, the *systems approach*.

4.2   A TAXONOMY OF PERSPECTIVES

As many others I feel like stating my own systems paradigm though a priori such a thing is not welcome. It is to impose your own thinking onto someone else. So I had better start this venture with a very good excuse. Well, first West Churchman opens the option. "The systems approach begins when first you see the world through the eyes of another", he writes as a first conclusion by the end of his famous *"The Systems Approach"*. So, I now offer my eyes to the courageous reader, and I promise you, for the sake of symmetry, that I regularly borrow the eyes of others, even from very remote friends. Another excuse for my initiative is that I shall try to be relatively concrete and specific. I shall try to push the systems science frontier one step further in the direction given by West Churchman and his able group of disciples.

There are more motives for drawing general systems views. Aesthetics is one, both in design and in applications. The specific qualities of systemic overviews, in contrast to pragmatic consensus, are another good reason for those. They provide explicitness so that supplementary exploration may be done in an orderly and describable way. Such overviews help to kinds of complete knowledge since they help to transcend into the unknown. A systems view makes a list of reminders without claiming any absolute truth. A systems view makes it possible to set a focus and to set priorities with an explicit reference to a context. What is excluded becomes visible. This is honest policymaking and it is different to the kinds of superficial consensus we see all too often in our political bodies all over the world, not always by conscious corruption but often by ignorance.[11]

---

[10] Foucault, Michel, 1970, *L'ordre du discourse*, Collège de France, denies both the existence of an external guarantor and a true essence of a discourse.

[11] I prefer just to give a general reference to the occidental evening press.

This general plead for systems views holds with a generality which covers both subject views and object views[12,] that is both for the strategy, e g the discourse or analytical process, and for the vision. Now, the structure I am to draw shall be about and for the analytical process. That is what makes it Churchman. I describe a way to express perspectives in his sense, not just an ontology for possible objects.

The start of my approach was a taxonomy of acts, first drawn and presented at local seminars around 1980, and in 1981 an embryo was presented at the EURO V. There are available publications about it in the EJOR 1982[13] and in a PhD theses from Stockholm University 1991. The start was to see the difference between the *approach* and the *tool* in analysis. This was not generally done at that time. Instead there was an obfuscation which led to a general decline and disrepute of operational research in many countries. This was before the rise of the so called soft approaches[14] and long before the Total Systems Intervention methodology.[15] Also, at that time, there was an adulation for computerized megamodels in my own defence planning environment, whereas I pleaded for pluralism and specific smaller models. All this led me to elaborate in theoretical terms the difference between the *tool* and the *approach* of a systems analysis. Out came not only the definition of two concepts but a symmetric 2x4 matrix. The elements of the matrix are acts of learning and understanding and together they form an extremely flexible pattern to describe analytic approaches. The elements are not forces and they are not tendencies. My experience is that in practice the acts may be distinguished conceptually though normally they are performed in overlap. By this rough definition we see that the levelling of the acts is simply a matter of physical and temporal inclusion. We may compare with Dror's mega-systems, and with Foucault's sets of discourses and events, but so far we have no metaphysics and no meta-system. This will be added in what follows.

Let me, without further references and without any more historic, present my taxonomy as it stands now.

PHENOMENA: object & subject
TOOLS: creativity & analysis
APPROACHES: logics & interaction

---

[12] Miller, James, 1978, *Living Systems*, Mc Graw-Hill, is a good example of an object view. It is about surviving systems and their dynamics without being explicit about how to use this model in policy analysis or in research.

[13] Agrell, Per Sigurd and Vallée, Robert, 1985, "Different concepts of systems analysis", *Kybernetes*, **12**.

[14] The most celebrated summary of those is surely the Rosenhead, Jonathan, 1989, *Rational Analysis for a Problematic World*, Wiley.

[15] In 1991 appeared both the *Critical Systems Thinking* and the *Creative Problem Solving*, both by Bob Flood and Mike Jackson, both with Wiley.

4. EXPANDING CHURCHMAN'S PHILOSOPHICAL DISCOURSE 37

CULTURE: myths & rites

Phenomena[16] stand for the acts of finding information, observation we can also say, but without forgetting the phenomenological complexity. Observations are naive often enough; often we can not afford doing better. In other cases they are combined, indirect or more firmly oriented by the help of models and other tools. We can even say that the observations are included in the applications of the tools. The purpose of information may be specified by requirements of quality; e g by validity, reliability and different kinds of statistical risk-levels for example. Whether the focus is to be on the actors or on an issue expressed (object or subject focus) is another major purpose to decide. These purposes guide the actors in the phenomenological acts, though they basically belong to a set of stakeholders. So far there is still nothing metaphysical about it. However, when these purposes are to be fullfilled well, we shall need a Guarantor or a meta-system in John van Gigch's sense. We need sciences of observation e g statistics, psychology, sociology, ethnology, law and maybe others. J v Gigch would add guarantors of the guarantors, and of course those are needed, but I shall not repeat his teachings here. I need now his first metalevel in order to eliminate a set of common pitfalls and to show the need to consider the set of different meta-systems within the one and same systems analysis project.

Tools stand for the choice of foci and for the fusion of information. A *tool,* here and now, is more complex than the observation in the sense that it embraces the observation. This is true even if the observation also may be very complex. A tool in this our information management-sense deals with something more limited and more specific though, than what we shall call an approach in the following paragraph. In Drors terminology I say that the *tool is* the mega level of the *observations.* Purposes may then be to discover, to combine, to model, to arbitrate, to optimize. We may include to evaluate, e g "to find a numeric expression for", but not assess, which would be something more complex and more executive including a mix of methods and probably also a good deal of tacit knowledge and intuition. The *guarantors* we need have to deal with the right level of complexity, e g with algorithmic issues or with the creativity of the human mind. This is not the observation already focussed and not the approach of the whole project with its social and political implications. The guarantors we need fall into two classes, the one for rule-bound activities and the other for the intuitively

---

[16] The reader is invited to add Edmund Husserl's phenomenological view to this short writing about phenomena especially his thoughts about the intentionality of knowledge which is combined with care for a truth concept. He respects the integrity of our different sciences. His distinction between perceptions and facts is also essential for us here. Husserl, Edmund (1900), Prolegomena zur reinen Logik. Jean-louis le Moigne would be a more modern proponent of intentional phenomenology. See for example his (1995), La modélisation des systèmes complexes, Dunod, Paris.

based creativity. Mathematics, operations research and computer sciences are examples of the first cathegory. Psychology, sociology and Edward de Bono's lateral thinking methodology[17] belong to the other side, which is much too neglected in science as well as in management and policy.

Approaches stand for the complex act of a whole project normally performed by a team and a set of other stakeholders including one or more clients. It uses tools, mostly tools in combination, but all its constituent acts can not be defined as an application of a tool. This complexity makes both goal settings and epistemology so different from those on the tool level. *Approaches* becomes a mega level of the *tools* and of the *observations*. They may have purposes as design, to initiate a discussion, to give warnings, to make aware, to give a syntax or a semantic, to explain, to make intellectually tangible, to identify problems, to lay out aspects, to give arguments, to test coherence, to criticise, to invite criticism, to explore, to give frameworks, to specify conditions, to express purposes and objectives, to create consensus, to allocate responsibilities, to make agendas, to clean up a discussion, to claim excellence, to decrease an anguish. All this is approach not tool.

A nice classification of epistemologically different *purposes* comes from Steen Hildebrandt where three types of problem (decision, behaviour, system) are matched with three methodologies (thinking, communication, search-learn).[18] Nine roles are defined:
- Systemizer,
- Communicator
- Co-manager
- Diagnostician
- Process consultant
- Learner guide
- Expert, Sparring partner
- Experimentalist

What *guarantor/meta-system* we need depends not only on these purposes. It depends also on the cultural context and on the situation. As the history of philosophy shows, there is amazingly little of lasting truths about guarantors and epistemology, but we have the truths of today in academia. That is their *raison d'être*. We have departments of management science (though differing in opinion) who do their best. We have likewise departments of business administration, of sociology of modern organization, of modern history and of modern anthropology. This list would be more a definition of a need than a real guarantor. It is a description of a meta-system though.

---

[17] See for example de Bono, Edward, 1973, *The Use of Lateral Thinking*, Harper.
[18] Hildebrandt, Steen, 1980, "The changing role of analysts in effective implementation of operational research and management science", *EJOR*, **5**.

4. EXPANDING CHURCHMAN'S PHILOSOPHICAL DISCOURSE 39

Culture stands here for the human way of behaviour and as a context to the methods exercised. *Culture* becomes the mega-level of the other categories. The visible side of culture, even in occidental administrations, I have heard being called the rites and they have a backing in myths. The myths contain the ruling perspectives and those are not controllable as the lower complexity levels. Control goes the other way. Violations of the norm are punished, by exclusion often enough. Foucault writes about this[19] and I guess that Churchman and van Gigch could call this a kind of a cynical guarantor or meta-system for culture. The power of the culture may extend to all the other levels and there is no general rule for how this appears in different cultures.

Culture, even your own culture, is observable; it is not mere abstraction, and if you observe or challenge it, you set it on to the level of observations. We may in this way see the taxonomy as a closed a circle or a helix.

All these activities on the four complexity levels have a purpose of producing information. At the process level we may also speak about a production of knowledge, and all happens by different perspectives. Within those we have objectives, which have a relation to wider purposes. All these objectives of acts are something essentially different from the object-systems objectives which we may study as well.

We have now seen a need to give to each level its meta-system; backing or guarantor would say West Churchman. We need them both for our design and for our quality assessments. But, will the managing meta-system of a total approach be synergetic and compatible with its respective meta-systems for tools and for observations or will they create inbuilt contradictions and biases? Will it create the confusion which philosophers like Immanuel Kant[20] and Edmund Husserl[21] fear so much? Churchman is not worried about this. He refers to the eclecticism of his teacher Singer, a straight discovering pluralism. Mitroff and Linstone also advises a free and flexible effort trying what may seem to fit[22] and they might feel uneasy about the planned and structured approaches of many others. Rolfe Tomlinson[23] believes that the problem can be overcome in defendable ways as long as there is a coherent goal-setting. Steve Cropper is more pessimistic. He writes about a necessarily specific context which goes with all tools and all methods[24]. Michel Foucault would solve this high level meta-system problem by calling

---
[19] Foucault, Michel, 1970, *L'ordre du discours*, Collège de France.
[20] Kant, Immanuel, 1787, Prolegomena to the *Kritik der reinen Fernunft*, 2nd edition..
[21] Husserl, Edmund, 1900, Prolegomena to the *Logische Untersuchungen*.
[22] In their *The Unbounded Mind*.
[23] Tomlinson, Rolfe, 1900, "Of tools, methods and methodology", In Eden & Radford eds., *Tackling Strategic Problems*, Sage.
[24] Crooper, Steve, 1990, "Variety, formality and style", In Eden & Radford eds., *Tackling Strategic Problems*, Sage.

it a *discourse* and by stating that it is a matter of social conventions and that no true solution should be sought. Gerald Midgley[25] and Mike Jackson[26] seem to lean on him while making explicit how to match the sometimes conflicting underpinnings, e g the external conditions, of a discourse. I think that all of them are true, but remembering Kant I would like to add my view that the organisation of a purposeful project must first of all have a clear organization of its meta-systems and their respective domains.

The relations between levels and between acts are vital parts.[27] There is the important relational issue of how our acts are appreciated. How seriously are they taken? Are they given a response at all? Are the results takes as truths, conjectures, illustrations or provocations? This makes an enormous difference and it is delicate. For example, the appreciation is what may save modelling and quantification. With the right appreciation, tools of this kind may be put into a defendable methods context. Even if we rarely can take models and figures as truths they may be taken as conditional propositions of some sense. The kind of sense here may take many forms. Jean-Claude Moisdon et al gives some very worthwhile examples in their *Du mode d'existence des outils de gestion.*[28] This view on relations will also affect the format of allocation of human responsibilities in our projects. You will define fair and efficient responsibilities of course, but also demand from each partner that she has a tacit responsibility to understand and use his colleagues with good will and keen senses.

4.3　A TEST

Most of my own projects have followed the pattern described, but that proves nothing of use to the reader even if they have been rewarding, as for example the ones about military command systems.[29] What would be more interesting is if the offered pattern could be more generally applicable. Intersubjectivity may be the word. So, rather than to describe a project of my own, I choose to see how someone else's successful method would fit to my world view. Especially it will be interesting to see how the mixes of different guarantors and meta-systems may be managed. I look for the possible, for what is possible in different situations.

---

[25] Midgley, Gerald, 2000, *Systemic Intervention*, Kluwer.
[26] Jackson, Mike C., 2004, *Systems Thinking*, Wiley.
[27] There is a book in honour of Niklas Luhman which is extremely clear about this: Bakken, T. and Hernes, T., eds., 2003, *Autopoietic Organization Theory*, Abstrakt forlag, Kopenhagen.
[28] Moisdon, J-C., 1997, *Du mode d'existence des outils de gestion*, Seli Arslan.
[29] Agrell, Per Sigurd, 1979, *Vett och vilja i värdering av ledningssystem*, FOI, Stockholm.

4. EXPANDING CHURCHMAN'S PHILOSOPHICAL DISCOURSE

A first nice and relatively simple example can be the Karl Popper's *Conjectures and Refutations* methodology.[30] Here we have a totality of a process, an approach with two parts, the creative conjecture and the analytic testing. Trial and error we could also say. Here the purposes of the totality and its parts are easy enough to see: Scientific *discovery* supported by *design* and *trial*. We have also three different meta-systems with their respective epistemologies and quality criteria. The ones for trial are the most elaborate and clear mainly thanks to the nicely structured domain of probability and statistics. The other meta-systems are more fuzzy but nevertheless firmly dependent on local *discourses*. Together however they work as long as the second phase *appreciates* the first one for what it is worth. Together this Karl Popper's design is widely accepted for many kinds of situation.

Let us next illustrate by a more elaborate example: John Friends Strategic Choice Methodology.[31] His approach contains the tools of *shaping, design, comparison* and *choice*. This is both a general management method and an optionary software (STRAD) for it. The method, with or without its software, produces an overview of a development project. It helps in timing and in setting priorities.

The purpose of *shaping* is to discover the relevant perspectives and uncertainties and to enable different systems delimitations flexibly. It is to create an overview and to discuss relevant foci on that basis. It is to imagine. The guarantor, or metasystem, would come from the area of creativity mentioned above. It could also come from cybernetics and from the biologically inspired systems theory.[32]

The purpose of *design* is to produce coherent strategies starting with more elementary information. Options from a series of defined dimensions are combined. There is a combinatorial display problem to solve which is an algorithmic procedure. Engineering and design sciences are the guarantors.

The purpose of the *comparison* phase is to provide a ranking of suggested strategies. Quality criteria are that it shall be possible to enter assessment dimensions and that a traceable assessment calculus can be made. The extent to which values are exchangeable are to be controlled as in the French multicriteria school.[33] So, this French multicriteria school should be the guarantor of this phase.

The purpose of the last phase, the Choice, is to produce a visible overview as a basis for an executive control of the projects advances. Both the foci and the assessments are to be reconsidered here and method as well

---

[30] Popper, Karl, 1934, *Logik der Forschung*.
[31] In Friend, J. and Hickling, A., 2005, *Planning under Pressure*, Elsevier.
[32] See for example Wiener, Norbert, 1950, *The Human Use of Human Beings*, Miflin and Miller, James. G., 1978, *Living Systems*, McGraw-Hill.
[33] Roy, Bernard, 1985, *Methodologie multicritère d'aide à la decision*, Economica, Paris.

as its software are flexible. The quality and guarantor would have to treat both the dialectical logics of iterations and the human matter of easy perception.

This is not all. This is only the tool level. Its acts work together, but still they have particular epistemologies and they are different. As objects of description they may be distinguished but not separated. We have also a coordinating methods level to consider, the *approach*, the one above the *observation* and *tool* levels. Its purpose, we stated it in the introduction, is to help with overviews and priorities in a development project.

The purpose of the total method is, I quote John Friend, *to offer practical support whenever you face a tangle of tough decisions which are of a developmental nature.* This is an extrovert purpose. Flexible abilities to deal with multiple uncertainties and fast moving events are also a requirement (fulfilled) as well as an ability to make a coherent and efficient synthesis of the four subroutines. What metasystem would guarantee all this? Surely not any of the preceeding tool oriented ones, and already in this observation we have something interesting. We see that not all insights translate between the methods levels. I am not ready to express the synthesizing guarantor better than to refer to John Friends book with Allen Hickling.[34]

A guarantor of the Strategic Choice *approach*, its overriding synthesis that is, would have three sources: Social Science empirics knowing real decision-making, a broad knowledge of modern operational research, even of those not included explicitly in the Strategic Choice method, and finally some real testing and experience with managers and planners.[35]

To conclude this paragraph of test I wish to point out that among the *tools* we have those specifically for *creativity* versus *analysis* and the approach has got facets both of *logics* and *interaction*. So we may recognize the general taxonomy of this article in John Friends methodology. This does not mean that he has been influenced by it. On the contrary! It is since there have been independent origins of the two methodologies that they support and acknowledge each other.

4.4 A DISCUSSION

I wish the meta-system's structure of a project to have power. It is to be considered as a group of clients in dignity comparable with the paying client. This is necessary for the honourable non-prostitute survival of the analyst professions. This is also in the interest of a serious client. He would want

---

[34] Friend, John & Hickling, Allen, 2005, *Planning Under Pressure*, Elsevier.
[35] We find all this in the *Planning Under Preassure* and in a special issue of the *Planning Theory*, **3**(3), November 2004.

advice, knowledge and information with relevance, as well as defined and defendable other qualities. He would want it in all time perspectives, also for future co-operation with analyst professions.

Looking at the control aspect of a project we may feel troubled by the double heading. We have control by clients and their purposes which translate into a structure of purposes and objectives in the project. This control is always taken care of, but we have also a control of quality by a more or less explicit meta-system which comes from the methods sciences. This duality is a known and inevitable conflict even if all agree in principle about defendable quality standards. I do not have the methodology for the arbitration of such conflicts, but I am convinced that even from the perspective of this dilemma the clear organization of the respective meta-systems will be an advantage. It will help the analyst for example in the rather frequent situation where he argues for a relatively costly quality standard with a client.

For many managers and even analysts this multi-perspective taxonomy will be too complicated, not digestible. This is serious objection, but on the other hand, the world is not easy, and with over-simplification we run the risk that common sense, superficial consensus or one of West's "enemies"[36] will take over the management process.

In a scientific perspective we may ask what is original with the taxonomy. First then, it came at the right moment, in the early 1980's when soft and hard methods still had a kind of primitive fight. The Churchman/Singer eclectic message had not had much impact before Russ Ackoff 1979,[37] Harold Linstone 1984,[38] Ian Mitroff in several publications shortly after that and Steve Crooper in 1990. Now, in the year 2005 it is obvious that an eclectic play with guarantors is necessary. Still the eight fields matrix may be considered as new as well as the *appreciation* concept and the distinction between meta and mega in our systems context. This is a specification out of John van Gigch's meta system theory which in its turn is a deduction out of West Churchman's guarantor concept.

---

[36] Churchman, W., 1978, *The Systems Approach and its Enemies*, Basic Books, N Y.
[37] Ackoff, Russel, 1979, "The future of OR is passed", *J Opl Res Soc.*, **2**.
[38] Linstone, Harold A., 1984, *Multiple Perspectives for Decisionmaking*, North Holland, NY, Amsterdam, Oxford.

4.5 CONCLUSION

I think now that we have seen that it may make sense to apply John P. van Gigch's concept of meta-thinking[39] to an epistemologically heterogeneous methodology, and that this is true not only for his own and for my research. We see that the meta-systems needed are really systems with some complexity and not guarantors with an easy to define essence. The reader is supposed to see that the meta dimension and the mega dimension of a project are not the same, even if some of the control comes from an interdependence between the levels and from the interdependence of actions within these. To believe in such a confusion would be to despise the vast and deep accumulated knowledge from all the *tool* level professions. All control does not come by the organized hierarchy. All acts and *discourse events* have also their own epistemologies. To balance those is quality management and in this we have freedom to choose our objectives, our quality criteria and our guarantors and we should do this explicitly for reasons of democracy and scientific scrutiny.

---

[39] van Gigch, John P., 2003, *Metadecisions: Rehabilitating Epistemology*, Kluwer/Plenum, London and New York.

SECTION B

SKEPTICISM, TRUTH, CERTITUDE AND WICKED PROBLEMS

Chapter 5

CERTITUDE IN A POST-MODERN WORLD?
A Coherent Evolutionary Story

BAUSCH ,K.

Abstract: In our post-modern world, we recognize that we construct our knowledge. From this, we too easily conclude that anything goes - that we cannot be sure of anything. The real conclusion is that dualities of true-false and body-mind are not absolutes, but are just useful mental constructions.

If we start from the position of body-mind continuity, we can construct a coherent, evolutionary story that situates our knowledge and certainty. The story starts with the knowing of a simple cell. It ends with grounds for practical certainty. Using the related theories of self-reference and self-organization as guides, we trace the story through an evolutionary history of representations, second-order reflections, and natural epistemology.

Key words: postmodern, iterative questions and communication

5.1 THE LEGACY OF POST MODERNISM

Post modernism attacked dogmatic absolutisms by deconstructing them, taking them apart piece by piece and showing the emptiness of their dogmatic claims. This is the principal accomplishment of post modernism. As a result, there are no longer any big stories that claim both popular and rational belief.

Herein lie a blessing and a curse. The blessing is that bloated repressive ideologies can no longer claim rational justification. In other words, all ideologies are suspect. The curse is that any reform movement is denied rational footing for its programs. This is the situation at an academic level of analysis.

At the level of popular perception, only novel ideas that challenge tradition meet general suspicion. Traditional ways of acting and believing are enshrined as common sense and dissent is ridiculed. While such prejudice against basic innovation has been a staple of human history, two facts of modern life aggravate our present situation. One is the accelerating and steamrolling effect of technological change. The other is the evisceration of the authority of thought especially that of socially conscious scientists and systemic thinkers.

At a time when we most need to think outside the box of business-as-usual, intellectuals are artificially pitted against each other as so-called "think tanks" funded by special interests construct spurious arguments to dispute the consensus of most unbiased scientists. Mainstream media are harnessed to portray the viewpoints of the interests that own them. Politicians are beholden to the pocketbooks that finance their campaigns. Even universities teach the ideas that bring in funding. In vaunted "democracies" such as the United States, consensus is manufactured with the whole panoply of institutional persuasion.

Persons trying to make social and ethical cases in this situation have to create them out of whole cloth. They cannot build upon a commonly accepted story or upon accepted criteria for establishing validity. They can make points around the edges of a problem using some version of the scientific method, but have to make the heart of the cases by appeals to religious, ideological, traditional, ethnic, and/or propaganda-driven stories. Social scientists/activists lack a convincing story of their own; do not even have one that can win acceptance among scientific peers. They are at a distinct disadvantage in trying to effect changes in heart and behavior when their appeals to minds and hearts are drowned in a cacophony of cynicism and institutional propaganda.

We need a shared understanding of where our knowledge comes from and how we come to trust it if we are to stand as a people with a confidence that is not the result of mass persuasion. At the present time, having a coherent, constructivist, epistemological story is important not just for the advancement of science and academic knowledge. It is vital for the survival of free political thought and decision-making.

5.2 ORIGINAL KNOWLEDGE

Humberto Maturana and Francisco Varela offer a sound, probably irrefutable, foundation for embodied cognition: "knowing is doing" (Maturana and Varela, 1987, pp. 29-30). At the most basic level, organisms cope. This coping involves staying alive, reproducing, and interacting with their environments. In doing this they know themselves and their environment. This knowing is learned survival. It involves no extraneous meaning, information, representation, or communication.

This "knowing is doing" is the essence of an organism's autopoiesis (Greek: self + doing), its continuing survival. All of us organisms do our survival. The process is self-referential in the extreme. If we attempt to capture the unconscious experience of an organism from the inside, life is a moment-to-moment adapting of interior states that keeps it together.

In the *Tree of Knowledge*, Maturana and Varela portray this inside organic experience, depicting life and knowledge as having no conscious awareness of representations. They develop an entire scenario for the evolution of thought on this basis that extends from "knowing is doing" to the emergence of an Observer possessed of human language. This Observer in turn, from an assumed external viewpoint, constructs explanations that propose to explain the processes of life and knowing.

5.3 THE PARADOX OF THE INCONGRUENT PERSPECTIVE

In proceeding in this manner, Maturana and Varela strive to avoid the paradox of the incongruent perspective, in which we impose our external perspective upon the inner perspective of organisms. It is impossible for organisms to understand the procedures of reproduction, bifurcation, and increasing complexity that we ascribe to them. This paradox entangles all explanations that try to explain organisms in terms of categories that are imputed to be at work within them.

In human situations, the paradox is observed in the differing attributions given to behavior by the actors and observers of that behavior. When a man is driving in traffic with his mate, for example, the man is intent on negotiating traffic while his mate often thinks that he is driving recklessly for some motive, such as, to prove his masculinity, to impress her, or to punish her.

5.4 ENACTION

We share this basic type of cognition with all other organisms as we enact our lives. This kind of knowledge is captured in the expression, "We make our path walking" (Varela, Thompson, and Rosch, 1993, p. 136). In other words, we move ahead into uncharted territory and what path there might be lies entirely behind us.

We really do exist by enacting our lives. In the moment, we may surround ourselves with expectations, but we are really progressively re-creating ourselves - making our path walking. In the enaction modality, we can immerse ourselves in a reality where self and environment are mysteriously one, as witnessed in the Zen koan, "not two".

In the enaction view of life and cognition, there is a fundamental skepticism about the accuracy of representations and the truth of propositions. This skepticism finds expression in Nietzsche. He says that propositional truth is the weakest form of knowledge. "From generation to generation, it passes on erroneous articles of faith that include: that there are enduring things; that there are equal things; that there are things, substances, bodies; that a thing is what it appears to be" (Bickerton, 1990, p.82).

Autopoiesis is arguably the correct viewpoint for understanding the self-reference of life and cognition. The difficulties encountered in trying to express autopoietic existence, however, are practically insurmountable if we stay within unconscious, non-articulated communication. We may be able to experience such unarticulated self-reference in enaction, which is sometimes called "going with the flow," but scientific thinking about autopoiesis requires that we assume, however fictionally, the viewpoint of an external observer.

5.5 SELF-ORGANIZATION

Autopoiesis stands alone among theories of self-organization in its insistence upon the perspective of the organism. Luhmann works from a premise of self-reference, but explains how those systems work with differences to generate and validate their cognitive maps. Other theories look upon autopoietic systems from the outside and construct theories to describe what they see. These theories go by many names: dissipative structures (Prigogine}, component-systems (Csanyi), self-modifying systems (Kampis), cognitive equation (Goertzel), complexity (Kauffman), nonlinear dynamics (various), synergetics (Haken), etc. Others like Derek Bickerton reconstruct knowledge's evolution.

5.6 THE EVOLUTION OF REPRESENTATIONS

The beginnings of representation arise in the very life process of an organism, as it continually reproduces itself sloughing off elements and incorporating new ones. This is not representation in the sense of "a mental

5. CERTITUDE IN A POST-MODERN WORLD?

sign that re-presents something else in the manner of accurate one-to-one correspondence". It is rather the establishment of differences.

The process of differentiation expands in parallel with the complexity of an organism as it grows by structurally coupling with its environment. Differences gradually morph into representations as nervous systems develop.

5.7 REPRODUCTION AND DIFFERENCES

An organism does not persist as a "thing", if we mean by that term "something that maintains an identical materiality." Living things reproduce themselves moment-to-moment as they ingest new molecules, structurally couple with their environments, and slough off old molecules. They are enduring "structures" because they maintain a flexible fidelity in their moment-to-moment reproduction. Their reproductive activity is their structure.

This ongoing reproduction is also ongoing knowledge because it involves a constant distinguishing between "self" and environment and between "good" things in the environment with which it structurally couples and "bad" things that it shuns.

Cells distinguish between ions that they incorporate, such as, sodium and calcium, and poisonous ions that they avoid such as cesium and lithium. How do they do this? In the originating instances, they use trial and error (unassisted natural selection) in which successful cells join with environmental influences that do not interfere with their essential processes. Using the relative simplicity of their reproductive and metabolic processes, they identify not only their "selves" but also those parts of the overall environment that become their "reality." In this way, species generate their "species-specific realities," which can be defined as "what it is useful for the species to know about what is out there" (Bickerton, 1990, p. 82).

5.8 SORTING DIFFERENCES

Bickerton opposes the conception that language is "primarily a means of communication". He defends the theme that language is primarily "a system of representation, a means for sorting and manipulating the plethora of information that deluges us during our waking life" (Bickerton, p.5). He traces the evolution of categories in species, from simple neuronal patterns, to Primary Representational System (PRSs), to Secondary Representational Systems (SRSs), to protolanguage, to language.

5.9 TRANSMISSION AND CATEGORIES

As organisms grow more complicated, they have to share information and coordinate activities among their parts. Simple sensing mechanisms coalesce into sophisticated ones. Highly structured cognitive states and behaviors derive "through evolutionary processes, from far simpler ones" (Bickerton, p. 77).

When living things are disturbed, they react to stay alive. But they do not react to everything in their environments. To stay alive, organisms selectively react to differences that they perceive as differences. Some differences, those it does not sense, make no perceptual difference for an organism; they are not information for it.

What is it, then, that makes a difference make a difference for an organism? The quality of an organism that leads it to sense a difference is its ability to react to that difference. Living things react only to the things they sense (information). Conversely, if an organism cannot react to something in its environment, then that something is not information for it.

Sensing and reacting are the poles of an organism's activity. To connect those poles, an organism must have some memory reference, some way of representing its activity. Reacting requires three things: sensing information, representing it in some way, and behaving on the basis of those representations.

5.9 PLANT PROTO-REPRESENTATION

Plants have inferior representational powers, compared to animals, because they lack nervous systems. Plants, however, are not without representations. They remember their responses to night and day, to seasons, to humidity, etc. In their water and nutrient circulatory systems, they distribute and deposit molecular messengers (cf. Rossi, 1993), things such as the enzymes cAMP, ATP, and ADP.

The "endocrinal" communication system of a plant is rather slow. A plant's representational system is in large part global. An enzyme released into its circulatory system activates the DNA of each cell individually, of course; but it effectuates changes in the sprigs of a bush and its leaves in about the same way at about the same time.

5. CERTITUDE IN A POST-MODERN WORLD? 53

Some plant behavior, however, is not global. The "sensitive plants", that Bickerton focuses on, react to touch in a localized manner and with some dispatch. These plants lack neural activity. Their sensoria are very narrow, being limited to those elements of the environment that affect them and that they can handle. Within such limits, sensitive plants, such as sundews, have cells and/or systems of cells that have ability to both discriminate touch and react to it.

The sundew is a sensitive plant that eats insects. It has leaves covered with sticky hairs. When a fly is stuck on those hairs, the leaf gradually closes and the plant consumes the fly at its leisure. Bickerton interprets this as an information-behavior transaction because the cells that gather the information about the fly are distinct from the cells that cause the leaf to close (Bickerton, pp. 78-79).

Plant discrimination does not reach the level of perception because neural activity is absent; but it is "proto-representation" because it is the evolutionary basis of the representation found in nervous systems.

The ability of cell A to distinguish two states creates a distinction (prey present vs. prey absent) and a faculty of sensitive distinction that presage further evolutionary distinctions and more complex distinguishing faculties. The ability of cell B to react to the distinction made by cell A creates an adaptation that increases an organism's survival options. These increased options open more possibilities for more sensitive distinctions by cell A and its likely cohorts.

5.10 ANIMAL REPRESENTATION

Nervous systems enable animals to process environmental information with dispatch and specificity. They augment endocrinal processes with electrical ones. They send messages faster and with more specificity.

Messages can be sent along nerves for fairly long distances from one specific location to another in an animal's body. Those messages are recorded and encoded in patterns of neural and neuropeptide activity. They are "meaning" for animal sensoria. They are relayed back and forth, to and from the relevant cells in an animal's body to enable its sensing, recording, and reacting to information.

The connections between sensitivity and behavior are patterns of accumulated wisdom contained in an animal's neural/neuropeptide system. Webs of neural and messenger molecule interaction are interposed between an animal's sensing and its behavior. The patterns in these interacting webs represent both stimuli and responses. These patterns match up and form an overall representation that is the animal's response to the stimulus. They are not necessarily representations of an external world; they are patterns of behavior that keep the animal alive.

5.11　PRIMARY REPRESENTATION SYSTEM (PRS)

Anemones shrink away from two predatory species of starfish and ignore five other non-predatory species. Such activities require a web of interneurons connecting sensory and motor cells because information from different sources (external, interoceptive, and proprioceptive) has to be coordinated. These neural webs encode internal and external states into neural codes. In Bickerton's words: "Nervous systems translate both internal and external states into quantitative terms, represented by variations in the firing rates of the relevant neurons" (p. 83).

Within a PRS, these representations correlate. They enable an animal to behave in ways that are appropriate to its life situation. The actual correlations involve multiple factors and a bewildering (as yet undeciphered) mechanism that matches frequencies, intensities, and sequences of neural and neuropeptide discharges. The process of correlation, however, is not conceptually difficult. It can be represented as a simple 2x2 matrix in which hunger and prey, presence and absence are the only variables that influence a predator's activity.

	Presence	Absence
Hunger	1	2
Prey	3	4

In case (1 & 3) the predator would spring into action. In all other cases, (1 & 4), (2 & 3), and (2 & 4), it would not attack. The predator would attack only if hunger and prey were both present. If either hunger or prey were not present, it would not attack.

More complex animals have the ability to make varied responses to slightly different stimuli. They can effectively vary their responses because they incorporate articulated senses and an ability to collate information from numerous sources. A lizard, for example, alters its behavior as it stalks a fly on the basis of spatial measurement and directional sensors and a sense of the fly's alertness. Lizards have a degree of autonomy. Their behavior is no longer totally predictable; it is co-determined by external factors and by activities within their brains.

5.12 CATEGORIES

When an animal perceives something, it selects what seems to be the appropriate category from the set of all categories in its reference system. The categories and the category system are adaptable because species seek to extend their representations in ways that will give them an advantage. With humans, the search for regularity and order is compulsive and can be a matter of self-awareness.

Some categories are innate as is revealed in a baby monkey's instinctual fear and avoidance of snakes. "Snake" is a fuzzy category in its innate form: waving a iece of rope or hose can excite a naïve baby monkey. "Snake" can, however be sharpened by experience, then it becomes a learned category.

Categories can be extended to other realities by verbal or preverbal metaphor. The categories that animals and species create are intimately related to the kinds of differences that they can react to (knowing requires doing). If they cannot react to a difference, that difference is not information for them. They build their categories according to their evolutionary requirements: what they eat, what other things they need, what they fear, what their preferred strategies of survival are, and how those strategies interact with the strategies of other animals and species.

On what basis does a species or an animal discriminate its categories and how does it relate those categories to its life choices? Consider a wildebeest that does not flee from every lion that it sees; it just gets frightened and flees when a lion is a threat. How does it do that?

Certain things in a wildebeest's environment are vital to its survival. Recognizing a charging lion is essential to its life; feeling fear (or panic) in that situation is also necessary. The wildebeest's flight from a charging lion involves several categories and subcategories especially "charging lion", "fear/panic", and "run". These categories are integral to the wildebeest's "getting away from the lion".

In the service of survival outside objects are grouped together if they require similar behavioral responses. "Bird", for example, would be "something that lays round things you can eat if you can find them"; "tree" would be "something you can climb if pursued", etc. In general,

> "the set of categories that constitutes the PRS of any species is determined by the evolutionary requirements of that species: what it eats, what other things it needs, what it fears, what its preferred strategies for survival are, and how those strategies interact with the strategies of other species." (Bickerton, 1990, p. 91)

Categories have an evolutionary history. They accumulate as patterns of activity in ganglia and ever-larger nervous systems. At each level of this process, responses to situations, with their corresponding categories, are integrated into larger patterns of neural activity. Representations and categories that satisfice for survival at one level are co-opted and adapted for survival at subsequent levels.

5.13 NEGATIONS

Negations are crucial to categorization especially for humans.. If we need a tree to climb to get away from a bear, we do not look for some object with the essence of "treeness". We simultaneously eliminate any possible alternatives to the first thing that we can climb. We are able to do this because our minds work "with large numbers of weak parallel processors rather than a single strong processor" (Bickerton, 1990, p. 93).

In addition, negations provide the basis for survival categories like "own species/not own species", "predator/non-predator", "prey/non-prey", etc. Such categories would naturally create object trees of the following general constitution:

species
                       / predator
non-species   / prey
                      \ non-predator
                                                    \non-prey, etc.

Using dichotomies something like these and fuzzy parallel processing, pigeons can identify categories like humans and tree, subcategories of different kinds of people and trees, and even trees with just a few branches in a painted cityscape. Vital categories like "trees" and "humans" have been necessary for the survival of pigeons for millions of years. They are probably hardwired into pigeon neural circuitry. Other recognitions would be learned and held in ontogenic memory.

5.14 REPRESENTATIONS OF SPACE, ACTION, SOCIETY

Animals do not develop categories of objects in a vacuum. They sense and categorize objects in relationships. They construct and memorize cognitive maps of their surroundings. They negotiate their survival activities in relation to those cognitive maps. They inherit some aspects of such relationships like the three-dimensionality of experience and distance measurement; they memorize, in their ontogeny, important stable relationships like invariant physical constraints; and they concentrate on

5. CERTITUDE IN A POST-MODERN WORLD? 57

their survival activities moment to moment. All these elements of an animal's spatial cognitive map are only one component of its larger cognitive map, its PRS.

Within an animal's PRS, there are also categories for different kinds of movement. There are cells that represent particular aspects of movement: some indicate movement to the right, others indicate approaching movements, etc. Larger cell groupings indicate things like quickness of movement, danger of movement, accessibility of prey, etc. Among primates, complex patterns of neural activity represent things like "hand grasping".

Representations of objects in neural patterns can be considered not only as proto-concepts; they can also be considered as proto-nouns. Representations of embedded activities like "cow grazing" or "hand grasping" can be considered as forerunners of proto-verbs. All these proto-concepts become more numerous and distinct in social relationships.

Social relationships are both catalysts and objects of category formation among primates. The continuous interactions among parents and siblings are also embedded in neural representations based upon categories (social, behavioral, spatial, and material). As an animal becomes more social it articulates categories in its PRS. In a circular process, articulated categories enable complex social relations, which create better articulated categories, etc.

Social interaction leads to coordination of effort, competition, infighting, and a recognition of individual differences in terms of wiliness, persistence, strength, aggressiveness, productiveness, etc. Patterns of interaction between individuals also develop. All these activities, individual differences, and patterns of relating are represented in the neural patterns of individual animals, social groups, and species.

5.15   SIGNALS

Animal signal calls step out of the PRS and into the realm of Secondary Representational System (SRS). An example of signal calls is found in the signal communication of vervet monkeys, which has been extensively studied. Vervet monkeys use three distinct calls to alert each other of the presence of pythons, martial eagles, and leopards, which are their natural predators. These signal are holistic in nature, they refer only to specific things, such as a martial eagle. They are not modifiable to indicate a crow or a robin. (Bickerton, p.15).

5.16 PROTOLANGUAGES

Protolanguages have a much greater lexicon than signal calls, while lacking the syntactic features of true language. We revert to protolanguage sometimes. When we are really stressed, we might speak without syntax, saying things like: "Sick", "Tired", "Not hungry", "Bed". Somewhere in our evolutionary history, this kind of communication came into being.

Protolanguages can be identified today in the lexical communication of some great apes and of human children under the age of two.

The linguistic attainments of the great apes are remarkable. Roger and Deborah Fouts, the long-time researchers with Washoe, summarize their findings of chimpanzee language as follows:

"Chimpanzees can pass this [sign] language on to the next generation ... They can use it spontaneously to converse with each other as well as with humans ... They can use their signs to think with, as evidenced by their private signing ... They can have an imagination ... They have good memories ... They may even be able to perceive seasonal time." (Fouts and Fouts, 1993, p. 39)

The same authors recorded over 5200 instances of chimpanzee-to-chimpanzee signing. They analyzed the content of that signing as follows:

"The majority of signing by the chimpanzees occurred in the three categories of "play", "social interaction", and "reassurance"; these accounted for over 88 per cent of the chimpanzee-to-chimpanzee conversations. The remaining 12 per cent was spread across the categories of "feeding", "grooming", "signing to self", "cleaning", and "discipline". (Fouts and Fouts, 1993, p. 33).

These findings document not only the achievements of great ape communication, but also its limitations.

The differences between animal communication and human language are qualitative, and discontinuous. Even the most sophisticated systems of animal communication (such as those demonstrated by Washoe and her companions) are limited to a finite number of topics on which information is exchanged. Human communication, on the other hand, is practically infinite. Animal systems have very limited ways of combining message components, while human language has strict and flexible rules of unlimited combinations. Where other animals communicate only about things that have evolutionary significance for them, human beings communicate about anything.

Children under the age of two display a similar lack of syntax. They are limited to words and phrases, such as: "Mama", "Eat", "Chair", "Doggie",

5. CERTITUDE IN A POST-MODERN WORLD? 59

"House", "Apple", "Thank You", "TV". Sometime about the age of two, children develop syntax and begin saying sentences like: "Mama go house", "Doggie watch TV".

5.17 LANGUAGE

The transition from protolanguage to syntactic language is a quantum leap. Bickerton proposes that the syntactic tree mirrors the "A or not A" way that the brain organizes its categories. As our ancestral brains reached a sufficient degree of sophistication in our evolutionary past, under the pressing urgency for more flexible communication, we inserted the bifurcating structure of our representational system into our protolanguage. This process was facilitated by a concurrent evolution of our respiratory system that permitted us to exhale through our mouths.

Five million years ago, the PRSs of primates had the necessary infrastructure for the beginnings of secondary representation networks (SRSs), that is, protolanguages and language itself. Advanced primates at that time had protoconcepts that could serve as nouns and verbs. Language of some sort was an evolutionary certainty given selective pressures favoring more plastic responses and a viable way to embody an SRS. In Bickerton's words:

> "All that was then required for the emergence of at least some primitive form of language was ... some set of factors that would make the development of secondary representations advantageous, and some means through which such representations could be made both concrete and communally available." (Bickerton, 1990, p. 101)

5.18 METAPHOR

We have seen that human cognition has its evolutionary basis in the learning processes and cognitive maps of animals prior in the evolutionary tree and the history of the genus homo. As a result of this, we have a basic-level preconceptual structure based upon inherited "capacities for gestalt perception, mental imagery, and motor movement. When we developed language we began to expand our meanings into domains where we have no clearly discernible preconceptual structure. We imported those meanings via metaphor" (Lakoff, 1980).

We need metaphors. Without them, we could not make sense of our worlds. We do not even grasp many realities without the metaphors that give them context. Consider the use of metaphor in simple questions and answers like the following:

>Are you *in* the race on Sunday (race as CONTAINER OBJECT)
>Are you *going* to the race? (race as OBJECT)
>Did you *see* the race? (race as OBJECT)
>The *finish* of the race was really exciting. (finish as EVENT OBJECT within CONTAINER OBJECT)
>There was a *lot of good running* in the race. (running as SUBSTANCE in a CONTAINER)
>I couldn't do *much sprinting* until the end. (sprinting as SUBSTANCE)
>*Halfway into* the race, I ran out of energy. (race as CONTAINER)
>He's out of the race now. (race as CONTAINER OBJECT)
>(1980, p. 31).

In our metaphors, we find the cultural assumptions of our social perceptions and activities

> "Metaphors have entailments through which they highlight and make coherent certain aspects of our experience. A given metaphor may be the only way to highlight and coherently organize exactly those aspects of experience.
>
> Metaphors may create realities for us, especially social realities. A metaphor may thus be a guide for future action. Such actions will, of course, fit the metaphor. This will, in turn, reinforce the power of the metaphor to make experience coherent. In this sense, metaphors can be self-fulfilling prophecies." (1980, p. 156)

Julian Jaynes (1976) argues the importance of metaphor for describing and understanding even ourselves. He examines the language of Homer to ascertain the roots of ideas such as soul, heart, spirit, mind, and emotion. In his extrapolation of these meanings, he draws upon and extends the work of Bruno Snell. Snell observed concerning "psyche," that Homer used it to designate a breath or a fluid that escaped a body at death. "It forsakes man at the moment of death, and ... flutters about in Hades; but it is impossible to find out ... the function of the psyche during man's lifetime (Snell, 1982, p. 8).

The words used by Homer have developed as metaphors to express our semantic pride in being intellectual creatures. Psyche, which we take to mean soul (or mind) meant "breath". Phrenos, which developed into our ideas of heart and courage, referred to sensations in the chest. Thumos is the word that lies behind our ideas of an emotional soul; it was originally just a word to describe agitation or willful motion. And noos, the word behind our exalted ideas of gnosis, referred to what one could see, what we metaphorically term a "field of vision". In brief, all the words we use to describe psychological reality are metaphors built upon bodily experience. When we reason, we draw analogically on the felt interrelations of our body and the relation of our body with various external objects.

5.19 WORLD AND UNIVERSE

Using this armamentarium of categories and syntax, we build up our "world", which is everything that we conceptualize about the inside and outside of our lives. We can distinguish this "world" from the "universe" that we conceive as everything there is and that we hypothesize to have self-organized to its present state. "World" is the more autopoietic concept that supposes we will never have objective knowledge. "World" is a subset within the "universe" that contains all the concepts we might have about the "universe".

It is useful to keep this distinction, the distinction between self-reference and self-organization, and the basic distinction between the autopoietic and the objectivist viewpoint in mind as we develop the pivotal ideas in the next section.

5.19 SECOND-ORDER CONSIDERATIONS

With our attainment of language, we set out on a long quest to know the world in ever-finer detail. We have divided up the world for millennia in an effort to quiet our unease at its immensity, and to control it for our advantage. This kind of knowing can be categorized as first-order cognition.

As we have sharpened our metaphors and syntax, we have also turned to language to understand our language. This is second-order cognition. We use linguistic constructions to analyze linguistic constructions. This second-order cognition deals with ideas like: intentionality, transcendence, possibility, selection, contingency, negation, self-reference, information, meaning, communication, material implication, non-referential information, cognitive equation, pre-existing structure, and bits.

5.20 INFORMATION AND MEANING

As we live our lives, we are constantly faced with different situations, some of which are routine and some novel. Routine situations do not present us with information, that is, with differences that make a difference

(Bateson, 1972, p. 315). Novel situations do confront us with information. In these novel situations we are overburdened with possibilities. What will happen if I do this? ... if I do that? What will I become in the process? If we decide one way, we have to deny ourselves the adventure of going another way.

Transcendence is built into this experience because we are always attending to things that are beyond our boundaries. Not only that, we attend to outside things with an awareness that they are outside us. This outside/inside awareness is accompanied by an actual/ potential awareness. We experience what we are at the moment (our actuality) and virtually what we might become (our potentialities) should we choose one or another of our present options.

To continue living, we have to reduce the complexity that confronts us by selectively attending to one possibility and negating the others. By that selection, we selectively integrate our actuality with the transcendence of some of our other possibilities. We have no guarantees in our selection; we never know if we are actualizing the right possibility.

As we continue to make these kinds of decisions, we build up our meaning, our meaning of life. When we make a successful selection (linking or separating from environmental elements), we add that selection to our meaning (inner circularity). In this way we build up the complexity of our autopoietic reproduction.

5.21   INFORMATION AND COMMUNICATION

Information as something we might send in a message is a selection from the complex of our thoughts and our ways of expressing ourselves. Communication is a complex interaction involving selecting (1) the information we want to express, (2) some way to utter that information, and (3) some way to understand the utterance that might be received.

The common metaphor for communication in our society is that of the pipeline wherein communication is the transmission of information from one person to another. Some of the implications of this metaphor are:
Information is a thing that can be sent.
Utterance is a simple act of sending.
Reception of information is a simple process.
There is a simple identity of what is sent and what is received.
Communication is a two-part process of sending and receiving.

All of these implications of the transmission metaphor are misleading (Husserl; Luhmann, 1990, pp. 25 ff.). First of all, using "thing" as a metaphor for information obscures the fact that information is a selection that reduces complexity. Information as it is experienced, therefore, is a contingent and partial event. Even if we call an information-event a "thing", it is neither substantial nor permanent nor well defined.

Second, utterance is not an act of mere sending. Utterance involves selection of some meaning to share and some code to share it; it involves, therefore a selection of a selection. Utterance is the proposal of a selection.

Third, the completion of communication requires more than a simple reception; it requires understanding. The nature of this understanding needs to be probed in the nature of the communication itself and not merely semantically.

Fourth, the identity of what is sent and what is received is not a simple matter. Any achieved identity is guaranteed not by the content of the information, but by the whole context and the communication process itself.

Fifth, communication is a three-part process and not a two-part process. It involves three selections: information, utterance, and understanding. As has already been described, information is a selection that reduces complexity in our lived experience. When we communicate, we choose from among the information selections that we have already amassed and do something with them (act them out; utter them in some way). Utterance, therefore, is the second selection. Understanding (the third selection) underlies the whole process as the pre-existing recognition of social situation and is the outcome of every successful communication.

Each of these selections is made on the basis of difference and unity. The overall unity is the system-environment in which we define ourselves self-referentially as different from our environment. Information is the selection from difference that enhances a system's unity. Utterance is the selection from a system's store of information and coding. Understanding is a selective action that distinguishes between information and utterance.

5.22 MATERIAL VS. FORMAL IMPLICATIONS

George Kampis provides two important, interrelated distinctions to this discussion: material versus formal implications and referential versus non-referential information (Kampis, 1993, pp. 135–150).

The distinction between material and formal implications is seldom recognized. Material implications (Aristotle's material causes) refer to the actual interactions that make something happen. Formal implications are the causes that we attribute to the happening. Formal explanations average out material and chemical interactions. They simplify reality for us and increase our control over it. They bring us so much comfort that we mistakenly equate them with the workings of natural things. They do not, however, match precisely how things happen.

With this distinction, we can make new sense of the phenomenon of bifurcations and dissipative structures. We experience bifurcations at a macroscopic level in terms of the formal implications that we attribute to a system. From our perspective, a system begins to behave according to different (formal) rules. From a microscopic level, the level of material causality, there is no discontinuity because the implications at that level, which are largely unknown to us, continue doing business as usual. This distinction also helps us understand the paradox of the incongruous perspective in which we attribute choices to organisms of which they are not conscious.

5.23 REFERENTIAL VS. NON-REFERENTIAL INFORMATION

Non-referential (symbolic) information is expressed in signals, symbols, protolanguage, and language. It is the kind of information that expresses formal explanations for material processes. Referential (non-symbolic) information generates the material processes. This is the kind of information that is shared in the Primary Representational System (PRS) described by Bickerton. It is passed through webs of neural and messenger molecule interactions that coordinate animal sensing and behavior.

Referential information has no process-independent scaffolding and no symbols. The non-symbolic signs of this information are neural patterns within organisms that interact with each other in ways that maintain, alter, or replace existing event-patterns to produce new event-patterns. The non-symbolic signs of this information are simply interacting components that constitute their ever-evolving meaning in their relationships. Such signs have no permanence or independence; they have the nature of molecular signs. They do embody semiosis, but they exhibit no distinction between object, sign, and interpretant. Only observers can overload them with such distinctions.

In the perspective of referential information, biological and mental processes are manifestations of a more general semiosis that starts with the emergence of meaning in replication. The semantics of this semiosis issues from material implications that cannot be reduced to description and syntax. The mind processes referential (non-symbolic) information in a replication process where the original information content disappears and new information is created. In the transition from sensation to perception, for example, the continuity between sensation and perception is maintained in patterns, which "are 'fixed point' solutions of a self-modifying process" (Kampis, p. 145). This self-modification is a material process that cannot be cast a priori in a formal manner. Even in this first instance, however, a pattern of behavior is enacted that endures as a kind of memory.

5.24 THE COGNITIVE EQUATION

Ben Goertzel develops a mathematics that depicts the growth process of self-organizing systems. His cognitive equation models a process, in which a self-organizing system expands its processes while maintaining its essential circularity. His brief, informal expression of the cognitive equation is:
1. let all processes that are "connected" to one another act on one another.
2. Take all patterns that were recognized in other processes during Step (1), let these patterns be the new set of processes, and return to Step (1) (Goertzel, 1994, p. 152).

This is the continual process in which the system endures as a learning system that remembers its patterns of behavior. It can also be expressed as follows: At time **t-1** a system has set of essential patterns and components. At time **t**, it mixes these essential patterns and components with elements of its environment. At time **t+1**, it reconstitutes itself with essential and compatible patterns that it recognizes in the time **t** mix.

This equation maps creative processes on all level of reality, including the social. The creative progress of science, for example, begins with recognized patterns (theories), which are brought into question by novel information. In the uncertainty generated by non-conforming information (e.g., blackbody radiation), scientists allow "all the processes that are "connected" to one another act on one another" (Goertzel), and recognize a new pattern of information (Planck's constant, and eventually quantum physics). Then the process repeats itself; e.g. quantum physics confronts thermodynamics and generates the science of nonequilibrium thermodynamics.

In Goertzel's conception, a single self-reproducing pattern cycle constitutes a simple cognitive map. This self-reproducing pattern cycle begins the process toward permanence and reliability of other more-complex patterns. These patterns are representations in a sense that is less strong than a re-presentation of a pre-given world. They are patterns associated with a stimulus. They sever some aspects of reality from an amorphous background and thereby create meanings, cognitive maps, belief systems, and virtual worlds. These meanings (for animals) are maintained in nervous systems that acquire accustomed patterns of functioning (those patterns which have proved successful).

5.25 PRE-EXISTING STRUCTURE

James Gibson presents an ecological perspective on the nature and processes of perception, information, and representation. He stands in opposition to those thinkers who hold that our minds create our experiences. He proposes convincing evidence that we perceive pre-existing structure in the world.

In the context of the ideas presented so far, that structure would be the result of self-organizing processes in our environments. The objects in our environment have organized themselves in the course of evolution. Our relationships to their affordances has been shaped by innumerable self-correcting expectancies of the course of organic, animate, human, and personal evolution. In this perspective, our concepts do represent pre-existing structure, but in a manner that is far from simplistic.

5.26 INFORMATION AS BITS

In contrast to the preceding second-order theories, mathematical information theory is almost totally first-order and pragmatic. It assumes in opposition to Nietzsche that "there are equal things" and that information itself consists of totally generic things called "bits." A bit is a binary (yes/no) choice. The quantity of information in a system is the number of bits "which have to be made to achieve a unique selection from the possibilities" (Checkland, 1981, p.315).

The philosophical underpinnings of this kind of information theory are Cartesian, atomistic, and Newtonian. They are Cartesian because they assume a mind/body dualism in which objective things are re-presented to minds that exist independently from material reality. They are atomistic because they assume that the object, sign, and interpretant of communication have reality independent of each other. They are Newtonian because they ignore nuances in individual communications and do not account for self-organization.

Information theorists have built an imposing mathematical edifice upon this elementary conception of the bit. The burgeoning information technologies attest to the power of this mathematics. They have revolutionized the ways that we communicate and organize our lives. They have enabled us to model reality in ways that were inconceivable fifty years ago.

5.28 REFLECTION

The theories of enaction and bits stand at opposite poles of contemporary cognition theory. They offer contrasting satisfactions for our desire to know. Enaction theory describes an existential situation that resonates with our experiences of walking into the unknown. Bit theory offers pragmatic tools that help us to cope with the explosions of information that surround us. Enaction points to our inward experience. Bit theory points outward and objectifies reality so that we can attempt to control it. Both theories simplify our cognitive life. Both are valuable for us. But they are polar opposites in many ways.

Radically embodied cognition, as represented by the enaction theory of Varela, minimizes the role of representation in our knowing and living. The "bits" of information theory manipulate representations as things. The theory of embodied cognition denies the Cartesian split between mind and matter. Information theory assumes that split.

These opposed theories are alternative selections made by us in these later stages of our self-organization. Both of them simplify aspects of our lived knowledge. They are complementary.

There are also the connections between them that we have just discussed. Between the poles of enaction and bits, the evolution of categories and of second-order reflection can be understood through the cognitive equation and through the interrelated theories of self-organization that are called various names: autopoiesis, component systems, complexity, chaos, nonequilibrium thermodynamics, dissipative structures, nonlinear dynamics, and synergetics.

5.29 A SELF-REFERENTIAL EPISTEMOLOGY

Natural epistemology is a necessary consequence in any theory that denies a priori assumptions. An assumption utilized in such an epistemology is recognized as being (a) the at-one-time hypothesis of someone (b) that has been tested and (c) has been found to be reliable. In a systemic, evolutionary account of the social world, the norms for judging a theory's validity have to self-consciously reflect the evolutionary processes that generate them.

The previous sections portrayed the natural evolution of cognition through stages from "knowing is doing" to knowledge as bits. They addressed the question "how do we know?"

In this section we address the question, "how can we trust our knowledge?" We start from the positions that all our knowledge is self-reflective and that we have no direct access to the outside world. We consider the inherent circularity of knowledge and how its circle is broken. We explain the autopoiesis of knowledge and develop a self-referential epistemology. We relate self-referential epistemology to Quine's naturalized epistemology. We generate norms of validity on the basis of redundancy, test, and other criteria. Finally, we propose a mimetic resemblance between self-referential logic and the scientific method.

5.30   BACKGROUND

We live in an age of intellectual turbulence that invites new formulations of old conundrums. Within this turbulence, the emerging sciences of self-organization open vistas that were invisible thirty years ago. In bio-socio-cultural disciplines, mechanical and phenomenological approaches vie for dominance. Neither approach shows much interest in self-organizing systems. It would seem that the time is ripe for synthesis. In the arena of information science, systemic thinking shows us the way.

5.31   THE CIRCULARITY OF KNOWLEDGE

Conceptual knowledge is circular. We know everything in reference to something else. The science of semantics tends to make this a vicious circle. It contends that signs can be interpreted only in reference to each other. They do not directly signify anything in the external world. (cf. Lacan, p. 150).

The inescapable circularity of knowledge creates problems in knowledge theory. In the grand perspective of accumulated human wit, our knowledge produces all the units that it uses. In evolutionary history, our knowledge begins with the original formation of living cells. How do we advance given this circularity? How do we break out of the circle?

5.32   ASYMMETRY

We break the circle by creating asymmetries. We assert our will to power and simply declare that something is so. (When faced with a logical impasse, we just do something.) We tie our knowledge down to something that we project outside our self-enclosure. When we test our assertion repeatedly and find that it works for us, we add it to our reality bank.

It is necessary to project differences, facts, and themes on the basis of basically arbitrary selections (Luhmann, p. 466). These projected differences create the fictions that are needed for coping with the complexity of the world. Nietzsche and Goertzel find that logic is created by projecting such asymmetries. Nietzsche says that our will to power creates logic by declaring that unique things are equal. (Nietzsche, 1968, p.277) Goertzel says, "Logic is a lie, but a necessary one" (Goertzel, p. 58).

5.33   THE HERMENEUTIC CIRCLE

Hermeneutic science attends to how we interpret texts. It addresses the question: How can we understand a text if all the criteria for understanding are interior to the text? It answers this question by employing the hermeneutic circle.

The hermeneutic circle is a virtuous circle. It is a process of assuming a general idea of the whole text and then continually adapting its meaning as one confronts detailed parts of the text" (cf. Polkinghorne, p. 227). Ricouer explains the circle in terms that Luhmann and Goertzel would appreciate: To break the circle, we guess first and then validate our guesses, changing them if necessary (cf. Ricouer, p. 91).

Projected asymmetries break the circularity of knowledge. In the fabric of self-reference, they generate the realities and themes that enable interaction and evolution. The disciplined fictions of logic help us to a measure of referential transparency. Taken together, external and internal asymmetries create our worlds. They also place us on the edge of chaos where major breakthroughs in conceptualization and intimate nuances of understanding are possible.

5.34   CIRCULAR EPISTEMOLOGY

It is often objected that any natural epistemology involves a vicious circle in reasoning and creates self-sealing doctrines. Those objections as applied to self-organizing knowledge can be met as follows:

All knowledge derives from circular, autopoietic processes.

Epistemology cannot escape circularity because it is also an autopoietic creation.

Autopoietic knowledge is not self-sealing because it must pass the justifiable criteria that are proposed by non-self-referential epistemologies.

5.35 REDUNDANCY

The circularity of knowledge precludes grounding its validity in final elements because all available elements have been produced autopoietically. Our theories and our very selves result from autopoietic processes. We cannot validate the authority of our theories from a position outside the autopoietic circle. Our knowledge is not grounded in direct observation of any object or process either. How then do we come to trust our ideas about our worlds and ourselves?

Luhmann explains that trust on the basis of redundancy. We trust ideas that have worked for us in the past, that are tried and true over and over, and that are shared by our social group. Those ideas that have worked in human history and the evolution of our race have the deepest redundancy for us, and they are most deserving of our trust. This redundancy is further reinforced in ideas that were evolved in our pre-human evolution. Some of these ideas would be: "things fall", "food is good", and "love children".

Goertzel specifies how the elements of knowledge become redundant within the framework of his mathematical psychology. For him, a fact is an "interpretation which someone has used so often that they have come to depend upon it emotionally and cannot bear to conceive that it might not reflect "true" reality" (Goertzel, p. 59). Facts are patterns that are produced by consciousness whose job is to produce reality. They are generated in a dynamic feedback process that cycles patterns through higher-level (unconscious) cognitive processes and middle-level perceptual (linguistic and phenomenological) processes. "A pattern only acquires the presence, the solidity that we call "reality", if it has repeatedly passed through this feedback loop" (Goertzel, p. 109).

5.36 THE AUTOPOIESIS OF KNOWLEDGE

With the autopoietic processes of knowledge, we attain, retain, test, and extend our cognitive maps. We continually reproduce our knowledge by incorporating new patterns while retaining unity in every reproduction. As autopoietic systems, we continually dissolve and reconstitute ourselves and our knowledge while maintaining accompanying self-reference (Luhmann). In Goertzel's framework, a self-generating system reproduces at time t patterns that existed among its components at time t-1. In the redundancy of this autopoietic reproduction, a system gains a sense of self.

Autopoietic reproduction is not merely redundant. Self-reference is not "a complete duplication of whatever functions as the self at any time" (Luhmann, p. 460). Self-organizing systems metabolize with outside energy

5. CERTITUDE IN A POST-MODERN WORLD?

and information. They expand their circle by coupling with environmental information. A difference in its environment becomes information for the system when it selects an action that accommodates that difference.

5.37 EPISTEMOLOGY OF SELF-REFERENCE

Self-reference itself is a thematic asymmetry that we impose upon our reality. This imposition has major epistemological ramifications. If self-organization produces us and the world as we know it, then our knowledge continues that process of self-organization and our criteria for trusting that knowledge have developed in the same way.

Our epistemology, therefore, in the broad context of evolution, is also self-organized. It consists of pragmatic rules that have been developed and repeatedly tested for utility and logical coherence. These rules carry great weight because of the redundancy provided by their continual testing and their prominence in our cultural belief systems. In the final analysis, however, self-referentiality and circularity lie behind any rules we derive. Our rules originate as guesses and gain compliance because of their utility.

Any epistemology, that aims for universality must apply to itself the standards that it applies to its field of study. If scientists of social theory find contingency and self-reference in the evolution of social processes, they are bound to admit that their theories (which are particularized manifestations of evolution and social processes) are contingent and self-referential.

Even pragmatically universal principles, such as those proposed for communicative action by Habermas, are based upon the contingent evolution of our organisms and our language systems. Habermas, of course, recognizes (as does any post-modernist) that his standards or truth, rightness, and truthfulness are contingent creations of human history. Classical science fails to recognize this reality. It is caught in an insoluble paradox. It claims to be complete but it is "incapable of describing the scientists who produce it or of postulating existence without scientists to observe it (Artigiani, p. 51).

5.38 NATURALIZED EPISTEMOLOGY

Self-referential epistemology bears many similarities to Quine's "naturalized epistemology" (1969, pp. 69-90). The principal similarities are denial of general or absolute norms, recognition of circularity, and emphasis on holism.

Quine holds that there is no role in epistemology for a philosophical theory that is independent of the sciences (cf. Hookway, p, 465). Solomon supplies a definition of this naturalized epistemology:

"Epistemology is a science that identifies generalities (i.e., general laws) about the effectiveness of particular reasoning processes and methods that are applicable across all contexts - such as subject domain, stage of development of discipline, goals of inquiry, persons and groups conducting the inquiry." (Solomon, 1995, p. 353)

Recent thinkers

"have broadened the scope of their naturalistic epistemological projects to include social (encompassing anthropological, historical, institutional, political, sociological) factors influencing reasoning. This broadening of scope is no longer viewed as controversial, but instead as a new and fruitful set of directions for research, dubbed "social epistemology." (Solomon, 1995, p. 354)

Self-referential epistemology extends naturalized epistemology further. It integrates systemic and evolutionary processes into social epistemology. In this, it follows Quine, who suggested an "evolutionary epistemology" built upon the blocks of inductive reasoning shaped in evolution (cf. Quine, 1969, p. 90).

Luhmann adds an important new consideration. "Quine ... clearly emphasizes the connection between the "naturalization" of epistemology and the acceptance of circularity, but he fails to see that reality is also structured circularly, independently of knowledge" (1996, p. 615). Because our reality, including our social and psychic realities, is the product of circular processes of self-reproduction, there is a conformance between our reasoning processes and the processes of nature, including our own evolution.

5.39 ACKNOWLEDGING ROOTS

Luhmann believes that the "conventional theory of science in the natural sciences does not reckon with the epistemological problems of global theories" (1995, p. 615). A global theory must justify not only some particular facts but also its own theoretical framework. The importation of unexamined standards into such systems is unwarranted and debilitating. In particular, "concepts like emergence, self-reference, and entropy/negentropy ... [have] a position of prime importance, which theories of science must honor, because they concern the genesis both of systems and of the possibility of observation" (Luhmann, 1995, pp. 480-481).

5. CERTITUDE IN A POST-MODERN WORLD? 73

The circularity of epistemology that is present in global theories is forced upon us. "One cannot avoid them [circles]. One can sharpen them as a paradox and leave it at that. But one can also build them into the theory of science, for they contain precise instructions for self-control" (Luhmann, 1995, p. 482). The logic of this science goes as follows:

The minimal requirement of theories is that they must "always be formulated so that their object is subject to comparison.

"If they themselves appear among their objects, they must subject themselves to comparison.

"As their own objects, they must continue to function under the pressure of comparison.

"Whatever is attained in the system ... [in the way of knowledge] must also prove its worth in the theory, however unpleasant (e.g., relativizing) the result of the self-comparison may turn out to be" (Luhmann, 1995, p. 482).

Self-referential circularity requires every science to acknowledge its roots. It requires physicists, for example, to acknowledge that they themselves and the way they conduct physics are physical processes. Also, "if one knows that all judgments are based on previously established categorizations, that is, rest on pre-judgments, then research into prejudice must recognize itself a research about itself" (p. 482).

5.40 VALIDATING KNOWLEDGE

Luhmann offers several ways that we use to assure ourselves that our knowledge is true. First of all, one supplies reasons for knowledge. Then since these reasons "merely transform the circle into an infinite regress", one places one's hopes in approximating reality evermore closely. "If one in turn justifies the reasons and keeps every step of this process open to critique and ready for revision, it becomes more improbable that such an edifice could have been constructed without reference to reality. The circularity is not eliminated. It is used, unfolded, de-tautologized" (1995, p. 479).

Luhmann also answers a similar question: "How can one guarantee that observation maintains contact with reality when it claims to be knowledge, even scientific knowledge?"

"First, one can move the site of knowledge claims from psychic systems to social systems. Social systems can be made independent of individual motives and reputations. The knowledge of social systems "can be subjected to its own conditionings, perhaps in the form of "theories" and "methods". Knowledge can also be evaluated on its productivity and its ability to generate new knowledge. Second ... One could force global theories "to test on themselves everything that they determine about their object" (1995, pp. 484-485).

5.41 FOUNDATIONS OF HUMAN SCIENCE

John Warfield works within the tradition of Charles Peirce. Peircean science incorporates the knower into the definition of knowledge. It deals with the triad: object, concept, and knower (not with just the dyad of object and concept). It deals overtly with the blind spot of conventional Cartesian science: the embodiment of the knower. It does not assume a disembodied or cosmic mind that can objectively declare about truth and falsity.

In this tradition Warfield finds four universal priors at the foundation of all science: the human being, language, reasoning through relationships, and means of archival representation. A solid and productive science must attend to all of these priors. The natures of human beings, their abilities, and their limitations determine what science is possible and what its limitations are. Language, both specialized and everyday, needs to be understood in its relationships to other languages and to the things it represents. Science must have referential transparency; that is, it needs to make clear its relationships of meaning and logic. Finally, science must attend to the understandability and reliability of how it is recorded.

The "object language" of science is mathematical logic. It functions with utmost generality. It makes possible structures of deep and long inference and assists the human mind to overcome the limitations of short-term memory. It needs to be supplemented with a natural language that serves as its metalanguage.

In ordinary language, a word creates a single "fuzzy set" expression for the patterns that are related to its occurrence. "In other words, everything which relates to the word 'Ice' is part of the meaning of the word" (Goertzel, p. 69). Warfield proposes that definitions by relationship can supply semantic precision for sociological concepts. In these definitions, a concept's relationships to other factors are stated in terms of comparison, influence, time, space, and mathematics. He contends that the validity of any science depends upon the capacity of the scientific community to construct Definitions by Relationship for the full complex of relevant concepts involved in the science. Warfield actually constructs definitions by relationship for the key concepts of generic design. In so doing, he furnishes a model for other areas of social science.

The physical sciences have several methodological advantages over the social sciences.

5. CERTITUDE IN A POST-MODERN WORLD? 75

1. They deal with discrete (numerable) entities or with ones that can be easily quantified. Therefore,
2. They can easily use numbers to order entities in a universally understood way.
3. Physical sciences do not have to attend to complicated definitions and histories of their concepts.
4. Physical science is standardized.

The human sciences do not have these advantages. They deal with qualities that are not clearly distinguishable. The cool clarity associated with numbers cannot be transferred to those qualities merely by attaching quantifications to them. The social sciences lack standardization. They must attend closely to their core definitions and their referential transparency. In addition, they must attend to the abilities and limitations of the human beings doing social science and they must be clear about their use of object language and ordinary language.

Fortunately, science and mathematics do not require numbers. They merely require referential transparency. To attain referential transparency social scientists must resort to strings of long and deep logic to compensate for their lack of access to the "cool clarity" of numbers. They must specify the foundation, theory, methodology, and applications of their science. They must explicate the circular relations between

1. the postulates that intervene between foundation and theory,
2. the selection criteria that are used to create methodology from theory,
3. the roles and environment in which the methodology is applied, and
4. the evaluation feedback that applications supply toward improving the design.

5.42 CRITERIA OF VALIDITY

Epistemology establishes criteria for judging the validity, reliability, and probability of statements and systems of belief. Some criteria are almost universally accepted: coherence, compatibility with empirical results, openness to test, and ability to predict results. Some additional criteria are the following:

1. Habermas says that the ideals of truth, truthfulness, and rightness are essential to both language and the social framework. In particular, these qualities are the essential, although counterfactual, bases for deciding matters of value and justice.
2. Luhmann too proposes quasi-absolutes that must be served by theories. He says that problem of double contingency is solved when individuals grant freedom, intelligence, and personhood to each other. According to

him, human culture rests upon the fragile foundation of that mutual respect. Under the presumption that the preservation of human culture is an essential good, these qualities of respect for individuality, freedom, and intelligence become essential goods that must be served by theories.
3. Complexity theory sees a goal that is approached by all self-organizing systems - a position on the edge of chaos. Goertzel expresses this goal as maintaining a balance between circularity and dialogue.
4. Evolutionary theorists see physical and mental reality as part of a whole. They emphasize stewardship to the environment and future generations.
5. Theorists of systems design emphasize the need of comprehensiveness in obtaining input from all the stakeholders of a situation.

From this list, it is evident that the criteria for judging theoretical validity go beyond the criteria of the scientific method.

5.43   SUMMARY

In self-referential epistemology, organisms use intuition (the recognition of pattern) in their every selection. They judge the "truth" of their selections on their productivity: whether they die, survive, or thrive. Organisms that live on the edge of chaos, balancing circularity and dialogue attain consistency in this kind of truth.

This epistemology illuminates blind spots that occur in mind-body (Descartes) and text-reality (Saussure) dualisms. It shows why we need our bodies to know. It explicitly states the kinds of language and reasoning that we need in social science. It employs criteria that have proven effective in the survival and progress of our race.

Epistemology, as implicit in systems theory, derives theoretical, practical, and ethical criteria of validity from the self-organization processes that create us. It displays a transparent evolutionary logic that connects us with our roots. It traces a self-organizing trail in which "Knowing as doing" progresses to become our communication, cognition, and epistemology.

The validity claims of human cognition are extensions of elementary pattern recognition, selection, survival, and productivity. Truth as internal coherence is an extension of autopoietic pattern recognition. Truth as authenticity (and truthfulness) is an extension of selections that conform to inner patterns. Truth as beauty is an extension of the coherence of newly emergent patterns with existing patterns. Truth as rightness is a selection that honors the solution to the problem of double contingency. Truth in the pragmatic sense is an extension of the tests of survivability and productivity. Other standards for true/ethical judgment and action, such as enlightened self-interest, concern for future generations, respect for the environment, and even "is it fun?" can easily be traced to evolutionary forebears.

5.44   NEW HORIZONS

Artigiani recounts a brief history of this century's scientific epistemology. Early in the century, scientists pursued Mach's ideal goal of science: to produce a "mimetic reproduction" of nature. Because of the anomalies of relativity and quantum physics, scientists gave up that ideal. Heisenberg, for instance, said "The atomic physicist has to resign himself to the fact that his science is but a link in the infinite chain of man's argument with nature, and that it cannot simply speak of nature itself" (1955, p. 15). Bohr said, "We are suspended in language"; and "physics concerns what we can say about it in language" (quoted in Artigiani, p. 41). Toward the end of the century, Prigogine's theory of dissipative structures re-conceptualized physics in terms of evolutionary processes.

In this post-modern physics, nature observes itself in a manner analogous to that used by a scientist in a laboratory. "As science observes nature, nature is involved in continuous observation of itself" (p. 51). The correspondences are:

> instrumental readings ~ perturbations;
> laboratory conditions ~ environment;
> alternative organization possibilities ~ bifurcations;
> measurements ~ emergent structures.

In natural observation, evolving systems experience perturbations, react to them, and record their successful reactions as emergent structures.

At the end of the 20th century, the new science of emergent reality fulfills Mach's injunction (to produce a "mimetic reproduction") in a totally unexpected way. The method of science is the method of nature. "By proposing new models and metaphors, post-modern science mimics the processes by which nature self-organizes" (p. 55).

For Treumann, evolution is explained by "deterministic chaos, which is nothing else but ... information dynamics" (p. 90). He finds a delightful surprise for science in the theory of information dynamics. He sees a common pattern at work. Reductionistic science finds itself dealing with deterministic chaos because it cannot avoid errors in measuring its subject matter. Deterministic chaos in nature arises because "nature is itself supplying these errors, because nature is itself measuring all the time" (Treumann, p. 94).

In other words, the measuring of science mimics the measuring of nature. As a result, science's description of the world is not only an idealization made by human minds. It is also a correct idealization because "[Nature] corrects itself and generates the chaos we are confronted with daily" (p. 94). The method of science is the method of evolution and, not surprisingly, it produces ever-more-precise representations of life.

5.45 PRACTICAL RAMIFICATIONS

At a time when the world needs strong theoretical guidance, scientists and philosophers propose fractious conceptions of truth and validity. Many post-modernist thinkers go beyond demonstrating that we have no grand cultural metanarrative. They actually revel in the idea that we should not have one

In this paper, I take an opposing stand. I believe that we can generate consensual grounds for scientific and systemic beliefs. This consensus needs a unifying structure. I believe the coherent story told above provides a first iteration for that structure. The time has come for us to cease dwelling in silos of pet theories. There is a time for diversifying and a time for uniting. Now is the time for synthesis.

REFERENCES

Artigiani, R., 1993, "From epistemology to cosmology: Post-modern science and the search for new cultural maps", in: Laszlo and Massulli, 1993, pp. 29-60.
Bausch, K., 2001, *The Emerging Consensus in Social Systems Theory*, Kluwer Academic/Plenum, New York.
Bickerton, D., 1990, *Language & Species*, The University of Chicago Press, Chicago.
Checkland, P., 1981, *Systems Thinking, Systems Practice*, Wiley & Sons, New York.
Csanyi, V., 1989, *Evolutionary Systems and Society*, Duke University Press, Durham, N.C.
Eigen, M., 1992, *Steps Toward Life*, Oxford University Press, New York.
Fouts, R.S. & Fouts, K.H., 1993, "Chimpanzees' use of sign language", in: *The Great Ape Project*, P. Cavalieri and P. Singer, St. Martin's Press, New York.
Gibson, J.J., 1986, *The Ecological Approach to Visual Perception*, Lawrence Erlbaum, Hillsdale, N.J.
Goertzel, B., 1994, *Chaotic Logic: Language, Thought, and Reality from the Perspective of Complex Systems Science*, Plenum Press, New York.
Habermas, J., 1984, *The Theory of Communicative Action*, Volume One: Reason and the Rationality of Society, Beacon Press, Boston. (Original work published 1981).
Haken, H., 1981, "Synergetics: Is self-organization governed by universal principles?", in: *The Evolutionary Vision*, E. Jantsch, Westview Press, Boulder.
Hookway, C., 1994, "Naturalized epistemology and epistemic evaluation", *Inquiry*, **37**:465-485.
Jaynes, J., 1976, The *Origin of Consciousness in the Breakdown of the Bicameral Mind*, Houghton Mifflin, Boston.

Kampis, G., 1993, "On understanding how the mind is organized: Cognitive maps and the 'physics' of mental information processing", in: Laszlo and Masulli, pp. 135-150.

Kauffman, S.A., 1993, *The Origins of Order: Self-Organization and Selection in Evolution*, Oxford University Press, New York.

Lakoff, G. & Johnson, M., 1980, *Metaphors We Live By*, University of Chicago, Chicago.

Laszlo, E. & Masulli, I., 1993, *The Evolution of Cognitive Maps*, Gordon and Breach, Philadelphia.

Luhmann, N., 1995, *Social Systems*, Stanford University Press, Stanford.

Maturana, H.R. & Varela, F.J., 1987, *The Tree of Knowledge: The Biological Roots of Human Understanding*, New Science Library, Boston.

Nietzsche, F., 1974, *The Gay Science*, (W. Kauffman, Trans.), Random House, New York (Original work published 1887).

Polkinghorne, P., 1983, *Methodology for the Human Sciences*, SUNY, Albany.

Prigogine, I. and Stengers, I., 1984, *Order Out of Chaos*, Bantam Books, New York.

Quine, W.V., 1969, *Ontological Relativity and Other Essays*, Columbia University, New York.

Ricouer, P., 1979, "The model of a text: Meaningful action considered as a text", in: *Interpretive Social Science: A Reader*, P. Rabinow & W. Sullivan (eds.), University of California, Berkeley.

Snell, B., 1982, *The Discovery of Mind in Greek Philosophy and Literature*, Dover, New York.

Solomon, M., 1995, "Naturalism and Generality", *Philosophical Psychology*, 8(4):353-363.

Treumann, R.D., 1993, "The cognitive map and the dynamics of information", in: Laszlo & Masulli, pp. 77-98.

Varela, F.J., Thompson, E., & Rosch E., 1991, *The Embodied Mind: Cognitive Science and Human Experience*, The MIT Press, Cambridge.

Chapter 6

ADDRESSING COMPLEX DECISION PROBLEMS IN DISTRIBUTED ENVIRONMENTS
An Integrated Inquiring Systems and Sensemaking Perspective

PAUL,D.L.

Abstract:	This paper examines four teleconsultation projects from a Distributed Singerian-Churchmanian Inquiry System (DSCIS) and sensemaking perspective. This application of DSCIS to address wicked decision problems in remote health care delivery is the focus. Health care providers regularly face wicked decision problems in the diagnosis, treatment, and monitoring of their patients' health. Wicked decision problems in health care delivery are a result of the complexity of human anatomy and psychology, where each individual presents health care providers with unique problems due to comorbidity of conditions, differences in age, and variations in genetic and environmental factors. Drawing on Mason and Mitroff's framework, DSCIS are examined in terms of the characteristics of the individuals involved, the inquiry process, modes of presentation, and the social context in which such systems exist. This research provides insights into and contributes to a better understanding of how DSCIS can effectively and efficiently address wicked decision problems.
Key words:	Decision Theory: Wicked Problems; Health-Care delivery Systems; Systems Science; C. West Churchman (1913-2004): Design of Inquiring systems.

6.1 INTRODUCTION

Mason and Mitroff (1973) identified five constructs by which an information system can be analyzed: the psychological type of the individuals involved, the class of problems, the type of inquiring system (method of evidence generation and guarantor of evidence), the modes of presentation, and the organizational context. They specified one type of inquiring system—Singerian-Churchmanian—as being able to effectively and efficiently address both wicked decision problems and uncertainty. Singerian-Churchmanian inquiry systems (SCIS) are systems of inquiry that generate exoteric knowledge by sweeping-in and synthesizing additional data, information, and perspectives until they have the proper means to address the problem type at hand (Mason and Mitroff, 1973; Courtney, Croasdell, and Paradice, 1998; Courtney, 2001).

In the over thirty years since that seminal paper, the movement towards virtual organizing has resulted in Distributed SCIS (DSCIS), where information technology is substituted for proximity. The result is that SCIS participants no longer have to be collocated, and the number of potential participants and the variety of the expertise and experiences available have increased. How SCIS—whether distributed or collocated—address uncertainty is fairly straightforward. Uncertainty involves structured problems where the probability of the possible states of the world are known (Mason and Mitroff, 1973), and SCIS address uncertainty by sweeping in additional data, information, and information processing capabilities (Galbraith, 1974).

How SCIS address wicked decision problems is less clear. Wicked decision problems are unstructured problems where one or a combination of the possible sets of nature, actions, outcomes, or utility of such outcomes is unknown, and where the relationship between actions and outcomes are nonlinear (Mason and Mitroff, 1973). Additional information processing capacities alone are not enough to address wicked decision problems because more information may not improve the understanding of the problem as the problem already has too many possible and plausible meanings, not too few (Weick, 1995). Instead, wicked decision problems require additional, different capabilities: those of *sensemaking*.

Sensemaking is a complex, multifaceted, subjective social process of equivocality removal, where social realities and intersubjective meanings are generated by the merging or synthesizing of individual thoughts, feelings, and intentions (Weick, 1979; 1995). It involves the enactment of sensible environments and is focused on and by extracting salient cues. It is an ongoing social process grounded in the construction and maintenance of the sensemakers' individual identities. It is retrospective and driven by plausibility rather than accuracy.

6.1.1 Research Objective

The purpose of this paper is to increase the understanding of how DSCIS address wicked decision problems, and what impacts the effectiveness of such systems. This is accomplished by applying Mason and Mitroff's definition of a management information system as a framework to understand DSCIS in general, and then analyzing how DSCIS address wicked decision problems from a sensemaking perspective. In this paper, how one type of inquiring system—Distributed Singerian-Churchmanian—generates evidence and guarantees such evidence in order to address a particular class of problems—wicked—in an organizational context of self-directed, virtual teams, is researched.

6.1.2 Research Setting

This paper focuses on the application of Distributed SCIS to address wicked decision problems in remote health care delivery. Health care providers need increased information processing and sensemaking capabilities to address the uncertainty and wicked decision problems they regularly face in the diagnosis, treatment, and monitoring of their patients' health. Wicked decision problems in health care delivery are a result of the complexity of human anatomy and psychology, where each individual presents health care providers with unique problems due to comorbidity of conditions, differences in age, and variations in genetic and environmental factors (Eddy, 1984, 1990; Eddy and Billing, 1988). The geographic isolation of remotely located health care providers has historically limited the resources they have available to address uncertainty and wicked decision problems. Technology barriers have limited remotely located health care providers' information processing and sensemaking capabilities to those available in the community, or to additional capabilities available via telephone or facsimile.

Advances in information technology have changed this. Telemedicine, "the use of electronic information and communications technologies to provide and support health care when distance separates the participants" (Institute of Medicine (IOM), 1996, p. 1) has the potential to improve the ability of remote health care providers to address uncertainty and wicked decision problems by increasing remotely located primary care providers' information processing and sensemaking capabilities, as well as their access to information. Telemedicine increases remote care providers' access to specialists, sub-specialists, medical literature, and databases. One particular application of telemedicine, *teleconsultations*, appears to have significant potential to address wicked decision problems faced in remote health care delivery. Teleconsultations are real-time videoconferences between geographically separated health care providers—usually remotely located or rural primary care providers with the patients and their families often present and specialists or sub-specialists located at major medical centers (IOM,

1996). Telemedicine, utilized for teleconsultations are, among other things, *Distributed Singerian-Churchmanian Inquiry Systems.*

Drawing on the findings resulting from the intensive study of four teleconsultation projects, and integrating these findings with concepts from the sensemaking, inquiry systems, and information systems literature, this research sheds light on how DSCIS address wicked decision problems. Thus paper utilizes a method grounded in intensive case studies because the characteristics of sensemaking and DSCIS require an intensive research method capable of capturing the context of the situation at hand (Starbuck, 1992; Weick, 1995). This paper first presents a brief overview of DSCIS and sensemaking, and identifies the similarities between the two. The methodology and results are then presented and discussed. The paper closes with a discussion of the implications for both research and practice.

6.2 DISTRIBUTED SINGERIAN-CHURCHMANIAN INQUIRY SYSTEMS AND SENSEMAKING

6.2.1 SCIS: A Brief Overview

The purpose of a SCIS is to produce exoteric knowledge to solve the problem at hand. An advantage of SCIS over other types of inquiry systems is that they are capable of addressing both wicked decision problems and decisions involving uncertainty. Knowledge is a collection of information, an activity, or potential. It includes ethics, morals, and aesthetics in addition to scientific and specialized knowledge (Churchman, 1971). Knowledge is generated in SCIS by continual learning and adaptation through feedback. This continual, multilevel learning is generated by sweeping in technical, organizational, social, and personal perspectives, and results in a synthesis of multiple viewpoints (Courtney, Croasdell, and Paradice, 1998). This sweeping-in requires including individuals of different psychological types, and both personalistic and nonpersonalistic modes of presentation.

The system of measurements selected for the SCIS are critical because they are used to create insight and build knowledge. Knowledge is connected to measurable improvements, yet the guarantor of SCIS results is problematic because the measure of performance is:

> (T)he "level" of scientific and educational excellence of all society, a measure yet to be developed (where) the subject in the inquiring system's finding may not be the real subject which a specific question about nature has raised" (Churchman, 1971, pp. 200-201).

Such measures are constantly evolving; as a result, the output of a Singerian-Churchmanian inquiry system "is to be taken as is" (Churchman, 1971, p. 202) in the short-term, and guaranteed by agreement among the different participants, where new variables and laws are swept in to provide guidance and overcome inconsistencies.

6.2.2 Sensemaking: A Brief Overview

Sensemaking is a complex, multifaceted, subjective social process of equivocality removal, where social realities and intersubjective meanings are generated by the merging or synthesizing of individual thoughts, feelings, and intentions (Weick, 1979; 1995). Sensemaking has seven key properties. First, sensemaking is focused on and by extracting salient cues. Cues are the building blocks of sensemaking activities and include narrative information carried in stories, art, and graphics, as well as facts and hard data. Both modes of presentation and environmental conditions can affect what cues are or are not noticed, which may in turn significantly impact the effectiveness of sensemaking activities (Weick, 1979; 1995; 1999).

Second, sensemaking is retrospective and requires reflection and deliberation, where qualitative and quantitative evidence about the environments enacted are generated and refined in an effort to make sense of the situation at hand. Sensemakers do this by drawing on their experiences, expertise, cognitive models, scripts, and emotions from the past and develop or enact sensible environments (Weick, 1993; 1995; 1999). Third, sensemaking is a social process sensitive to the social context in which it exists because social support, consensual validation, and shared relevance all tend to have important influences on sensemaking activity effectiveness, and social context is mediated by talking, conversation and discourse (Weick, 1979; 1985; 1999). Fourth, sensemaking involves the enactment of sensible environments under conditions where there are multiple plausible but equivocal meanings and environments. Enactment—the creation of one's own sensible environment—is an active process where sensemakers' activities define and iteratively refine their environment (Weick, 1995; 1999).

Fifth, sensemaking is grounded in the construction and maintenance of the sensemakers' individual identities, and is derived out of the need for self-improvement and self-validation. Personal identity is the sense of who one is in a particular setting and is constituted out of the process of interaction (Weick, 1995; 1999). Sixth, sensemaking is driven by plausibility rather than accuracy. Sensemaking involves the telling of a good story, and accuracy may inhibit developing a good story and enacting sensible environments.

However, plausibility is itself constrained by one's own experience and stake in events, the social context, the recent past, the familiarity of the scenarios, the relevance of actions taken, and visible cues (Weick, 1995; 1999). Finally, sensemaking is an ongoing process where sensemakers are continuously bombarded with more information and additional experiences. Sensemakers are able to identify the need for sensemaking only when they make an effort to put boundaries around some portion of the continuous flow of events and convert them into discrete events (Weick, 1995; 1999).

6.2.3 SCIS and Sensemaking

Both SCIS and sensemaking have a number of similarities in terms of the problem types addressed, their processes, human factors, and their outputs. Both can address wicked decision problems and decision-making under uncertainty. Both produce similar outputs — exoteric knowledge that is used to solve the problem at hand. Both have similarities in their processes in terms of generating and guaranteeing evidence, and both are social processes that are sensitive to the experiences and education of the participants involved. Subjectivity is inherent in both the processes and the outputs, and both take a holistic perspective of the problem at hand, where complexity is not decomposable, and where the crucial design task is to identify the problem faced and recognize its uniqueness. In terms of Mason and Mitroff's framework, both sensemaking and SCIS require the input of individuals with different psychological types, which in turn requires different modes of presentation depending on what the different personality types consider data and information. In this research, both operate in a context of virtual, self-directed teams. Such teams are being required to act as distributed SCIS and engage in sensemaking to address even more wicked decision problem types with fewer resources and greater time constraints than ever before. The sensemaking literature provides insights into how DSCIS can more efficiently and effectively address the increasingly wicked decision problems faced.

6.3 METHODOLOGY

This research was grounded in the intensive study of multiple cases, relying on issue-focused semi-structured interviews of key informants as the primary means of data collection. Multiple case studies enhanced validity and generalizability and provided the thick description needed to develop a deeper understanding of DSCIS in order to better understand and explain them (Benbasat, Goldstein, and Mead, 1987; Eisenhardt, 1989; Orlikowski,

6. ADDRESSING COMPLEX DECISION PROBLEMS IN DISTRIBUTED ENVIRONMENTS

1993; Yin, 1994). Grounded theory methodology enabled theory to be generated and tested by collecting and inductively analyzing the data through constant comparisons within and across cases (Glaser and Strauss, 1967; Martin and Turner, 1986; Strauss and Corbin, 1998).

6.3.1 Case Description

Four teleconsultation projects at three different sites were studied. Each involved a relationship between a health sciences center (HSC) and a rural health care facility. Projects involving HSCs were selected because they represent the majority of non-military telemedicine projects (IOM, 1996; Office of Rural Health Policy, 1997), and they did not have many of the legal, cultural, financial, and other non-technological barriers associated with some teleconsultations (Paul, Pearlson, and McDaniel, 1999). Each teleconsultation project involved virtual, autonomous teams where participation and membership was voluntary.

To allow technological and procedural bugs to be addressed and the novelty of telemedicine to pass, each of the telemedicine projects chosen had been operational for a minimum of four months. Site I was the only site where two different teleconsultation projects were located. One project involved pediatric oncology while the other involved infectious diseases. Site I used VTEL equipment. Site II used PictureTel equipment to initially screen potential bone marrow transplant oncology patients and later to monitor such patients. Site III used a system it designed and built to provide access to multiple medical specialties. Table 6.1 provides a summary of each site.

6.3.2 Data Collection

Fifty-two health care professionals were interviewed, and the interviews were audiotaped and transcribed. Semi-structured interviews of key informants provided thick and richly textured data (Eisenhardt, 1989; Martin and Turner, 1986; Orlikowski, 1993), and focused on the actual usage of the telemedicine equipment. Key informants were selected based on their current or past direct involvement in their organization's teleconsultation projects and their availability, and were members of one of three groups—health care providers (physicians, physician assistants, or nurse practitioners), administrators, and information technology professionals.

Table 6.1: Site Information

	SITE I		SITE II	SITE III
Clinical Application	Pediatric Oncology	Infectious Diseases	Bone Marrow Transplant—Oncology	Multiple Specialties
Date Started	1995	1995	1996	1989
Clinical Activity	Monitoring of Pediatric Leukemia Patients	Diagnosis, Treatment, and Monitoring of Multiple Drug Resistant Infectious Diseases Patients	Initial Consultation To Determine Patient Physical and Psychological Suitability For Bone Marrow Transplant Procedure and Patient Monitoring As Needed	Diagnosis, Treatment, and Monitoring of Multiple Patient Conditions Including Surgery, Pediatrics, Neonatology, Dermatology, and Nephrology
Project Status	*Discontinued After Four Months*	Ongoing	Ongoing	Ongoing
System Utilization	One Time/Week	Multiple Times/Wk	One Time/Week	One to Two Times/Week
HSC Participants	Pediatric Oncologists	Infectious Diseases Specialist, Others	Oncologists	Primary Care Physicians (One With Surgical Training)
Remote Site Participants	Registered Nurses	Primary Care Physicians	Oncology Sub-Specialists, Nurses, Case Coordinator	Multiple Specialists and Sub-Specialists
Site Location	Southwestern United States		Western United States	Southwestern United States
HSC Affiliation	HSC I		HSC II	HSC III
Distance To HSC	200 Miles		300 Miles	400 Miles (200 To Nearest HSC)
HSC Equipment Location	Telemedicine Studio	Studio Later Desktop Unit	Administrative Area	Telemedicine Studio
Telecom Link	Full T1		Full T1	Full T1
Remote Facility	Hospital		Physicians' Clinic	Hospital
Remote Equip. Location	Administrative Conference Room		Administrative Area (Was In Reception Area)	Administrative Conference Room
Remote Facility Telemedicine Equipment	VTEL Videoconference Equipment. Includes Xenon Light Source and Attachments, Electronic Microscope and Stethoscope		PictureTel Videoconference Equipment	Own Non-Proprietary Videoconferencing System. Includes Xenon Light Source and Attachments.

6.3.3 Data Analysis

Data from four teleconsultation projects were analyzed and reanalyzed and common patterns across the research sites were found. The inquiry system, sensemaking, and information systems literature were utilized to help in understanding the findings. This process involved the constant revisiting of the data and the research literature. What emerged out of this iterative process was a grounded theory of how DSCIS can effectively and efficiently address wicked decision problems. The next section presents the findings of this study.

6.4 RESULTS

6.4.1 Addressing Wicked Decision Problems before Telemedicine

Wicked decision problems were not a new phenomenon to remotely located health care providers; however, prior to the advent of telemedicine, economic or technology barriers limited remote care providers' capabilities to their own abilities or the abilities of other primary care providers in the community. Patients' economic status often prevented their traveling to an HSC to seek specialty care, and there were limitations on the effectiveness of communicating with specialists by telephone or facsimile. Despite the remotely located providers' best intentions, these limitations negatively impacted the quality of patient care. A remotely located primary care physician involved in the multiple specialties teleconsultation project described the situation.

Question: How were some of these cases handled prior to telemedicine?

We just took a wild-ass guess. We did, because a lot of these folks can't get up there (to the HSC). You know, you can try calling, but it's not the same thing as looking at it, and so a lot of it was just... You just used your best judgment and went on.

6.4.2 Addressing Wicked Decision Problems with Telemedicine

Telemedicine was utilized as a distributed Singerian-Churchmanian inquiry system in a diverse range of clinical applications involving the diagnosis, treatment, and monitoring of patient health. For example, the first teleconsultation in multiple specialties project occurred shortly after the telemedicine equipment had been installed. The teleconsultation's main purpose was a technical system test involving a patient who could not afford to travel back to HSC III for a surgical follow-up. Just prior to the teleconsultation, a baby was born at the rural hospital. The baby was having problems, and the rural physician was having difficulty interpreting (or making sense of) the baby's x-rays. A neonatologist at the HSC was consulted. She made a diagnosis after examining the x-rays, and the baby's condition was stabilized. This teleconsultation was representative in that it assisted the decision maker's effort to interpret the situation at hand, resulting in the reduction or elimination of unnecessary treatments or emergency evacuations.

6.4.3 Telemedicine as a Distributed Singerian-Churchmanian Inquiry System

This section examines the four teleconsultation projects from a DSCIS and sensemaking perspective, where the sensemaking literature provided insights to the effectiveness and efficiency of DSCIS. Drawing on Mason and Mitroff's framework, DSCIS were examined in terms of the characteristics of the individuals involved, the inquiry process, modes of presentation, and the social context in which such systems existed.

6.4.3 Characteristics of the Individuals Involved

SCIS rely on a wide breadth of inquiry to generate evidence, and involves sweeping-in additional concepts and variables to generate new hypotheses and evidence when the initial hypotheses are deemed too inconsistent with the initial evidence generated. This sweeping in of additional variables and perspectives can be accomplished by including individuals of different psychological types, making the usefulness of telemedicine as a DSCIS sensitive to both the quality and variety of others' expertise. The creation of intersubjective sensemaking is grounded in the construction and maintenance of the sensemaking individuals' identities, which means that the training, experience, and cognitive complexity of the group members are important. The lack of diversity of individuals' identities limits intersubjective sensemaking capabilities and replaces them with generic subjective sensemaking capabilities relying on established roles, scripts, and organizational routines instead of feelings, thoughts, and intentions (Wiley, 1988; Weick, 1995). Generic subjective sensemaking may be sufficient when situations faced are similar to those faced before, but in novel situations intersubjective sensemaking is needed. Therefore, the greater the variety and quality of the applicable expertise and experiences of those involved in sensemaking, the greater the intersubjective sensemaking capabilities of the group as a whole.

The infectious diseases teleconsultation project enabled a rural physician to utilize varied expertise in two manners. First, she put together a group of doctors who had different types of expertise relative to her problem. Second, she utilized a different group of specialists who had different methods of treating the infectious diseases. The result was that the participants in this teleconsultation project believed they had developed an expertise not available elsewhere.

6.4.3.1 Levels of Expertise

However, each party needed to have at least a minimal level of expertise and experiences. Without it, the usefulness of telemedicine as a DSCIS was limited. For example, in both the pediatric oncology and the bone marrow transplant teleconsultation projects, specialists at the HSCs relied on the expertise of the remote care providers to act as their sense of touch in performing physical examinations. In the case of the bone marrow transplant teleconsultation project, this was not a problem because the remotely located oncologists could substitute for the sense of touch and provide the information the bone marrow transplant specialists needed. However, the lack of requisite expertise on the part of the nurses at the remote site was a major reason why the pediatric oncology teleconsultation project was discontinued. The pediatric oncologists relied on the nurses to possess a certain level of expertise to generate the needed cues that could not be effectively transmitted via the telemedicine equipment. The lack of such expertise significantly inhibited the pediatric oncologists' sensemaking capabilities. In this case, the nurses lacked the training (requisite level of expertise) to perform such an exam adequately.

6.4.3.2 Technology/Expertise Trade-offs

Under certain conditions, the technological capabilities of the telemedicine equipment could compensate for a site's lack of a requisite level of expertise. The potential for such a circumstance occurred in the pediatric oncology teleconsultation session. One of the earliest signs of a relapse in pediatric leukemia is the appearance of abnormal clusters of cells in the eye, and the pediatric oncologists needed to be able to examine the whole area of the retina. One means of doing so was by utilizing a *direct opthamaloscope*—with which only a small portion of the retina could be seen at any time—and maneuvering it around so that the retina was eventually completely viewed. In the teleconsultation sessions, the oncologist had to rely on the remote site's nurses to manipulate the instrument so that the retina was completely examined. However, the pediatric oncologists were not confident the nurses were able to adequately control the direct opthamaloscope because the nurses had not been trained to use it. The pediatric oncologists were concerned they were missing important visual cues critical for sensemaking, and they stopped performing retinal exams as part of their teleconsultation sessions. A technological solution to overcoming the remote site's nurses lack of expertise in operating a direct opthamaloscope existed in the form of an *indirect opthamaloscope*. However, indirect opthamaloscopes were very expensive, and HSC II did not purchase one.

6.4.4 The Distributed Singerian-Churchmanian Inquiry System Process

This section examines DSCIS processes in order to better understand how evidence was generated and how the generated evidence was guaranteed.

6.4.4.1 Generators of Evidence

The generator of evidence of a SCIS is the sweeping in of additional variables and perspectives. SCIS rely on constant refinement—both qualitatively and quantitatively—to generate evidence (Singer, 1959; Churchman, 1971). Refinement requires deliberation and reflection. Sensemaking is the enactment of sensible environments, and is retrospective in that people's experiences and expertise are utilized to make sense of the situation at hand. The individual must draw on their cognitive models, scripts, and emotions from the past and develop or enact sensible environments. Therefore, the generator of evidence when telemedicine was utilized as a DSCIS was deliberation and reflection.

This was exhibited in the manner in which the telemedicine sessions were organized and the technology designed and utilized. The teleconsultation sessions were organized and the equipment deployed in a manner that minimized distractions and did not interfere with the DSCIS processes. This was accomplished by designing the teleconsultation sessions in a manner that made them as similar as possible to the face-to-face consultations with which the physicians were familiar, and designing the equipment to enable the health care providers to focus on the medical activities in which they were engaging. Sensemaking is sensitive to the context in which it occurs. Engaging in the consultations virtually was a significant change from how physicians had traditionally consulted with each other. At least initially, changing other aspects of the consultation process in telemedicine was not desirable because the physicians may not have been receptive to that much change at one time. The Director of Telemedicine at HSC III explained his philosophy about their teleconsultation sessions.

> The secret is organization—not equipment—and you can't get too tied down on equipment. The equipment has to have the criteria that it has no detractors. Detractors from what? Detractors from what we're trying to do. We're trying to give consultative service.

Constant Refinement of Evidence Generated

The evidence initially generated by SCIS are constantly refined and adapted by continual learning through feedback. In the case of telemedicine, this meant the parties involved—in particular the remote care providers—had to be able to expand their expertise and experiences. While the usefulness of telemedicine as a DSCIS was positively associated with the variety and quality of applicable expertise and experiences of the individuals involved, the ability of telemedicine to sustain itself was dependent on its ability to enable those involved in the teleconsultation sessions to grow by gaining additional expertise and experience. Health care providers most likely would face increasingly wicked decision problems as their patients aged and medical knowledge continued to expand. They needed to continue to upgrade their capabilities not only by increasing their access to outside expertise and experiences, but by upgrading their own expertise and experiences.

This was accomplished through the educational component of the teleconsultation sessions. This learning aspect enabled the remote care providers to address more of the local health care problems without having to utilize the telemedicine equipment to consult with specialists. A remote physician participating in the multiple specialties teleconsultation project felt that the learning effect taking place during teleconsultation sessions enabled remote care providers to address a greater variety of their patients' health care problems on their own. He estimated the learning effect resulted in a 30% decrease in teleconsultation sessions over time. This enabled remote care providers to address health care problems locally they previously would have had to transport to the HSCs. In these cases, the remote care providers utilized the telemedicine equipment to seek assistance for situations with which they were not familiar that previously would have been handled by transporting the patients to the HSCs. This resulted in an increase in the number of sessions but a drop in the number of patients transferred to the HSCs.

The educational component of the teleconsultation sessions was perceived as a major benefit of the teleconsultation sessions in that it included gaining tacit knowledge not available in the literature. A remote site oncologist believed that one of the unanticipated benefits in the bone marrow transplant teleconsultation project was that he developed a better understanding of the high dose chemotherapy and stem cell transplant procedures participating.

Changing the Problem Space

Another means of generating evidence in a SCIS is to change the nature of the basic problem by making structured problems more wicked, or by providing more structure to wicked problems (Mason and Mitroff, 1973). For example, at one point cancer was a generic term for a number of cellular pathologies considered terminal. As medical knowledge progressed, more specific types of cancer with different treatment regimens and survival rates were identified. In other words, a Singerian-Churchmanian inquiry system restructures the problem space.

The use of telemedicine in emergencies was an example of transforming a wicked decision problem into a structured decision problem. Emergencies tend to be wicked in that the history of the individual involved most likely is not known, nor may all relevant medical conditions be known. The exact injury may not be known, and the more serious the injury, the more likely the patient is unable to communicate essential information. Further, the nature of emergencies is such that time is of the essence. Telemedicine was not utilized much as a DSCIS in emergencies because the decision was made to turn the wicked decision problems of an emergency into a structured decision problem. The decision the remote care providers had to make was whether to treat the patient locally or transport them to the nearest HSC. To make that decision, the remote health care provider had to answer two questions: First, are the facilities available locally to provide the type of treatment necessary? Second, is the expertise and experience available locally to provide the necessary treatment? If the answer to either of these two questions was no, then the decision was the patient needed to be transported. This structuring of a wicked problem was highly effective in that the HSC involved with the multiple specialties teleconsultation project experienced approximately two emergencies per year which were in the gray area about whether or not they should be handled over the telemedicine system.

6.4.4.2 Guarantor of Evidence Generated

The guarantor of evidence of SCIS output is to be taken as is—i.e. plausible but not for certain. Plausibility was the means by which the generated evidence was guaranteed. The providers involved in telemedicine determined the sufficiency of the output of their deliberations by assessing the plausibility of their interpretations. The concept of plausibility as the guarantor of evidence generated was derived from the sensemaking literature. A property of sensemaking is that it is driven by plausibility rather than accuracy. Accuracy may be beneficial but it is not necessary at the time of sensemaking activities. The building block of sensemaking is action—the enactment of sensible environments—and the search for accuracy may actually be an undesirable activity because such a search inhibits action.

Plausibility as the guarantor of evidence generated was illustrated in the way the teleconsultation session utilized x-rays. All the telemedicine sessions studied were utilizing radiographic images in their teleconsultations, even though the images transmitted did not meet American College of Radiology standards. The physicians involved still found the quality of the transmitted images more than sufficient for their sensemaking needs. The Director of the Bone Marrow Transplant Unit at HSC II used the equipment to evaluate CT-scans, while a pediatric oncologist at HSC I utilized the transmitted images to confirm the presence of a tumor in one of their patients. Image quality was sufficient for the health care providers' needs despite the resolution not meeting professional body standards because the providers were not utilizing the radiographic images to make definitive diagnosis; rather they were utilizing them to judge the plausibility of their suspicions or beliefs. Therefore, the quality of the radiographic images had to be sufficient to support or call into question their interpretation of the situation at hand.

6.4.5 Modes of Presentation

DSCIS face the challenge of effectively communicating the different types of information and evidence necessary. The greater the quality and variety of the experiences and education of the individuals involved in SCIS, the greater potential to address even more wicked decision problems. Therefore, involving individuals with different personality types is important, but what the different personality types consider data and information varies widely. Such information includes narrative information as well as quantitative data. The ability to communicate these different types of data and information requires different modes of presentation.

6.4.5.1 Presentation of Cues

Sensemaking emphasized the importance of cues as the building blocks of sensemaking activities. These cues include narrative information carried in stories, art, and graphics in addition to facts and quantitative data. Slight changes in the cues noticed and their interpretation can have a significant impact on sensemaking activities. To support the extraction of cues, the telemedicine projects studied all designed the sessions to avoid distractions and be similar to traditional face-to-face consultations. This was important because context can have a significant impact on sensemaking activities, and the teleconsultation sessions studied tried to make the context as similar to face-to-face consultations as possible.

The effectiveness of telemedicine as a DSCIS was thus dependent on the type of cues necessary for sensemaking activities. Not surprisingly, telemedicine was an effective DSCIS when visual cues were important. However, telemedicine was not a useful DSCIS for all types of visual cues. A remote physician involved with the multiple specialties teleconsultation project believed that the key was where a picture was worth a thousand words. He explained:

> If you're just giving data and the guy can comment on the data about a patient, then it's really not that necessary, but when ... you try to describe a skin lesion—you can say maculopapular until you're blue in the face, but people don't really know what you're trying to say. But when you show them pictures, that says it all.

Indeed, those applications of telemedicine which utilized the teleconsultation sessions primarily for information processing which could have been performed over the telephone or via facsimile were generally discontinued. A pediatrics oncologist noted that they were getting good—but not additional—information from the discontinued pediatric oncology teleconsultations.

6.4.5.2 Personalistic and Nonpersonalistic Communication

Sensemaking requires the communication of both personalistic and nonpersonalistic information, and a concern with distributed SCIS is that much of the richness or intimacy of interpersonal interaction is lost in the virtual world, resulting in a more generic subjective sense of the world. The teleconsultation sessions studied were able to support very personal and emotional interaction between patients at remote sites and specialists at HSCs. The Director of the Bone Marrow Transplant unit at HSC II described the most important part of the initial consultation with the prospective transplant patient, which could be effectively performed via telemedicine, as follows.

> The critical part was the connection, the discussion with the patient, looking them in the eye, getting the pieces of information that you could not quite get off the paper.

Distributed SCIS were able to communicate soft data such as the information contained in a naturalistic or narrative mode such as stories and metaphors. This was important because a patient trying to communicate an illness in effect is trying to communicate a story, and telemedicine was able to facilitate such storytelling. A remote physician involved with the multiple specialties teleconsultation project gave an example where he had a patient with a fatal hepatorenal syndrome and a teleconsultation session was used to reassure a patient's family they were making the correct but emotionally difficult decision to let a family member die.

6.4.6 The Social Context of Distributed Singerian-Churchmanian Inquiry Systems

The sensemaking literature highlighted the importance of the social context in which inquiry occurs. Sensemaking is a social process, even when performed by an individual not part of a group because even individual reflection and deliberation is affected by interactions with others. It was therefore critical for telemedicine to be effectively utilized as a DSCIS that the remote health care providers were treated with respect, especially in cases where there were significant professional status differentials between the high status HSD sub-specialists and specialists and the low status rural primary care providers.

Further, it was critical that the specialists and sub-specialists trusted the expertise of the remote care providers involved in the teleconsultation projects. While there may have been physicians at the HSCs who had concerns about the abilities of rural physicians, they were generally not involved in the teleconsultations. A consistent finding was the specialists and sub-specialists involved in the teleconsultation projects respected the abilities of care providers at the remote site. One oncologist even described himself as being a "partner" with HSC II's sub-specialists in the treatment of the remote site's patients. The pediatric oncology teleconsultation project was the only project studied where the HSC specialists expressed doubts about the expertise of the remote site's care providers and this was also the only project studied which was discontinued. While there were a number of reasons for the project's demise, the pediatric oncologists not trusting the remote site's nurses' expertise was a major contributor.

6.5 DISCUSSION

The purpose of this paper was to increase the understanding of how DSCIS can effectively and efficiently address wicked decision problems. This was accomplished by applying Mason and Mitroff's definition of a management information system as a framework to understand DSCIS in general, and then analyzing how DSCIS address wicked decision problems from a sensemaking perspective. Drawing on the findings resulting from the intensive study of four teleconsultation projects, and integrating these findings with concepts from the sensemaking, inquiry systems, and information systems literature, this research sheds light on how DSCIS address wicked decision problems. Table 6.2 presents a summary of the findings.

Table 6-2. Summary of Findings

I. Information System Construct—Distributed Singerian-Churchmanian Inquiry System Description	
Sensemaking Properties	Implications
I. Class of Problem—Wicked Decision Problems and Decision-making under Uncertainty	
II. Psychological Type of Individuals—All Psychological Types Must Be Supported	
1. Grounded in the sensemakers' individual identities	▪ Sensitive to the quality and variety of others' expertise. ▪ Technology/Expertise Trade off
III. DSCIS Process—Sweeping-in As The Generator of Evidence (Constantly Refined by Continual Learning) with Science/Education Best Practices As Guarantor of Evidence	
1. Retrospective requiring reflection and deliberation. *2. The enactment of sensible environments.* *3. An ongoing process grounded in the sensemakers' sense of their identities.* *4. Driven by plausibility rather than accuracy*	▪ Process and technology facilitate retrospect and enactment. ▪ Change problem space by providing structure to wicked problems. ▪ Need for self-improvement and positive image of self. ▪ The continuing education component is important. ▪ Plausibility as guarantor of evidence.
IV. Modes of Presentation—Both Personalistic and Non-Personalistic Modes Must Be Supported	
1. Focused on and by extracting salient cues.	▪ Process and technology must facilitate cue extraction. ▪ Communicate both soft and hard data.
V. Organizational Context—Self-directed Virtual Teams	
1. A social process.	▪ Sensitive to the social context in which it exists. ▪ Respect the other participants and trust their

6. ADDRESSING COMPLEX DECISION PROBLEMS IN DISTRIBUTED ENVIRONMENTS

I. **Information System Construct**—Distributed Singerian-Churchmanian Inquiry System Description
expertise.

[Adapted from Mason and Mitroff (1973), Courtney, Croasdell, and Paradice (1998), and Courtney (2001)]

6.5.1 Characteristics of the Individuals Involved

Insights into how the perspectives of different individuals of varying psychological types can be better considered and integrated were presented. The sensemaking literature emphasizes the importance of sensemaking of being grounded in the construction and maintenance of the sensemaking individuals' identities. The findings indicated the effectiveness of the sweeping in of additional variables and perspectives, and therefore the usefulness of a DSCIS, was sensitive to both the quality and variety of others' expertise. Under certain conditions, the technological capabilities of the equipment could compensate for a site's lack of a requisite level of expertise. More frequently, the opposite condition existed; that is, the possession of a requisite level of expertise was utilized to overcome technological shortcomings of the information system.

6.5.2 The Inquiry System Process

Insights into how evidence is generated and guaranteed in DSCIS were presented. The generator of evidence in a DSCIS is the sweeping in of additional variables and perspectives. Singerian-Churchmanian inquiry systems rely on constant refinement—both qualitatively and quantitatively—to generate evidence, and refinement requires deliberation and reflection. Therefore, in a DSCIS, the generator of evidence was deliberation and reflection, and the sessions were organized in a manner that minimized distractions, while the equipment was deployed in a manner in which it did not interfere with the inquiry system processes. Evidence generated is constantly refined and adapted by continuous learning through feedback, and understanding how such learning and feedback is accomplished and nurtured in DSCIS was also gained as a result of applying a sensemaking perspective. Sensemaking is grounded in the construction and maintenance of the sensemaking individuals' identities, and the parties involved had to be able to expand their expertise and experiences.

A fundamental characteristic of a SCIS is that it changes the nature of the basic problem. Sometimes this means structured problems become more wicked, and at other times it means wicked problems become more structured. The sensemaking literature emphasized the importance of enacting sensible environments, and the use of telemedicine in emergencies were examples of enacting a sensible environment by changing the problem space.

Plausibility is the means by which the generated evidence was guaranteed. The participants who were part of the DSCIS determined the sufficiency of the output of their deliberations by assessing the plausibility of their interpretations. While accuracy may be beneficial, plausibility is more critical because the building block of sensemaking is the enactment of sensible environments and the search for accuracy instead of plausibility may actually inhibit enactment. Thus, the DSCIS sessions were used to test the participants' hunches in order to check their plausibility.

6.5.3 Modes of Presentation

Considerations about the modes of presentation were brought to the forefront by utilizing a sensemaking perspective of DSCIS. Sensemaking emphasized the importance of cues as the building blocks of sensemaking activities. Context, and slight changes in the cues noticed and their interpretation can have a significant impact on sensemaking activities. To support the extraction of cues, the telemedicine projects studied all designed the sessions to avoid distractions and be similar to traditional face-to-face consultations. The effectiveness of telemedicine as a DSCIS was dependent on the type of cues necessary for sensemaking activities; however, telemedicine was not a useful inquiry system for all types of visual cues. Distributed SCIS face the challenge of effectively communicating the different types of information and evidence necessary, and the findings indicate that DSCIS were able to communicate information that was not narrative information and quantitative data.

6.5.4 The Social Context

Finally, insights into the organizational context of self-directed virtual teams acting as DSCIS were gained by using a sensemaking perspective. Sensemaking is a social process, and it was critical for telemedicine to be effectively utilized as a DSCIS that the parties were treated with respect, and that the DSCIS participants trusted the expertise of the other parties involved.

6.5.5 Implications for Research

This research has addressed themes identified by Mason and Mitroff as being central to information systems research. This research examined autonomous workgroups, composed of individuals of multiple psychological types, facing wicked decision problems by acting as Singerian-Churchmanian inquiry systems in order to generate and guarantee evidence. Such evidence was presented in personalistic and nonpersonalistic modes. The findings indicated DSCIS generated evidence through deliberation and reflection, and guaranteed the generated evidence by assessing its plausibility, utilizing stories, graphics, facts, figures, and theoretical models. This research has demonstrated self-directed virtual teams utilized as DSCIS, relying on respect and trust as control mechanisms, could effectively and efficiently address the wicked decision problems existing in remote health care delivery. However, DSCIS were not likely to be useful in times of crisis because such times did not allow for deliberation and reflection. DSCIS are capable of enriching the expertise and experiences of the individuals involved, and such systems increased participants' importance by expanding and making more valuable their expertise and experiences.

Given the emergent nature of DSCIS, further research is needed. First, empirical validation and collaboration of these concepts in other settings is needed. Second, the teleconsultation projects studied all relied on trust and respect as a means of control. Teleconsultation sessions involved teams with very stable membership. Virtual teams whose membership is more dynamic needs additional study. Third, only dyadic teams were studied in this research. Additional research involving multiple party sensemaking teams is needed.

6.5.6 Implications for Practice

This research has implications for practitioners as well. First, this research may be useful to organizations in other hypercompetitive environments deploying or considering deploying virtual teams. Health care delivery is a hypercompetitive environment characterized by an expanding knowledge base, increased equivocality and uncertainty, time constraints, and cost constraints, and telemedicine teleconsultation sessions involved temporary virtual teams engaged in knowledge-based activities—sensemaking. Second, this research has shed light on the technology requirements needed to support DSCIS systems. In addition to providing access to the explicit knowledge available in the literature, DSCIS must support cooperation and collaboration. In this case, this involved videoconferencing with additional document and image processing capabilities. DSCIS must be able to support different modes of presentation, including both the personalistic and nonpersonalistic modes of presentation. Third, this research also benefits those installed telemedicine projects whose utilization may be disappointing by providing a different perspective by which to assess their projects.

6.6 CONCLUSION

This paper examined four teleconsultation projects from a distributed Singerian-Churchmanian inquiry system and sensemaking perspective. Drawing on Mason and Mitroff's framework, DSCIS were examined in terms of the characteristics of the individuals involved, the inquiry process, modes of presentation, and the social context in which such systems existed. This research has provided insights into and contributed to a better understanding of how DSCIS can effectively and efficiently address wicked decision problems.

REFERENCES

Benbasat, I., Goldstein, D.K. & Mead, M., 1987, "The case research strategy in studies of information systems", *MIS Quarterly,* **11**:369-386.
Churchman, C.W., 1971, *The Design of Inquiry Systems,* Basic Books, New York, NY.
Courtney, J.F., 2001, "Decision making and knowledge management in inquiring organizations: Toward a new decision-making paradigm for DSS". *Decision Support Systems,* **31**:17-38.
Courtney, J.F., Croasdell, D.T. & Paradice, D.B., 1998, "Inquiring organizations", *Australian Journal of Information Systems,* **6**(1): 3-15.
Eisenhardt, K., 1989, "Building theories from case study research", *Academy of Management Review,* **14**:532-550.
Galbraith, J., 1974, *Designing Complex Organizations,* Addison Wesley, Reading, MA.
Glaser, B.G. & Strauss, A.L., 1967, *The Discovery of Grounded Theory: Strategies for Qualitative Research,* Aldine Publishing Company, New York, NY.
Institute of Medicine, 1996, *Telemedicine: A Guide To Assessing Telecommunications In Health Care,* National Academy Press, Washington, D.C.
Martin, P.Y. & Turner, B.A., 1986, "Grounded theory and organizational research", *Journal of Applied Behavioral Science,* **22**(2):141-157.
Mason, R.O. & Mitroff, I.I., 1973, "A program for research on management information systems", *Management Science,* **19**(5):475-487.
Office of Rural Health Policy-United States Department of Health and Human Services, 1997, *Exploratory Evaluation of Rural Applications Of Telemedicine,* ORHP, Rockville, MD.
Orlikowski, W.J., 1993, "CASE tools as organizational change: Investigating incremental and radical changes in systems development", *MIS Quarterly,* **17**(3):309-340.

Paul, D.L., Pearlson, K.E. & McDaniel, R.R. 1999, "Technological barriers to telemedicine: Technology management implications", *IEEE Transactions on Engineering Management*, 6(2):279-288.

Singer Jr., E.A., 1959, *Experience and Reflection.*, University of Pennsylvania Press, Philadelphia, PA.

Starbuck, W.H., 1992, "Learning by knowledge-intensive firms", *Journal of Management Studies,* **29**:23-36.

Strauss, A. & Corbin, J., 1998, *Basics of Qualitative Research: Grounded Theory Procedures and Techniques*, 2nd ed., Sage, Newbury Park, CA.

Weick, K.E., 1979, *The Social Psychology of Organizing*, McGraw-Hill, New York, NY.

Weick, K.E., 1985, "Cosmos vs. Chaos: Sense and nonsense in electronic contexts", *Organizational Dynamics,* **14**(2):51-64.

Weick, K.E., 1993, "Sensemaking in organizations: Small structures with large consequences", in: *Social Psychology in Organizations: Advances in Theory and Research*, J.K Murnigham, ed., Prentice-Hall, Englewood Cliffs, NJ.

Weick, K.E., 1995, *Sensemaking in Organizations*, Sage, Thousand Oaks, CA.

Weick, K.E., 1999, "Sensemaking as an organizational dimension of global change", in *The Human Dimensions of Global Change*, J.E. Dutton & D. Cooperrider, eds., Sage, Thousand Oaks, CA.

Wiley, N., 1988, "The micro-macro problem in social theory", *Sociological Theory,* **6**:254-261.

Yin, R.K., 1993, *Applications of Case Study Research*, Sage, Thousand Oaks, CA.

Yin, R.K., 1994, *Case Study Research: Design and Methods*, 2nd ed., Sage, Thousand Oaks, CA.

Chapter 7

EPISTEMIC HUMILITY
A View from the Philosophy of Science

MATTHEWS, D.

Abstract: If our knowledge of the world is always filtered, interpreted and (in important ways) 'constructed' by our *a priori* faculties then we can never know things as they truly are and we are forced to accept a degree of humility with respect to our 'scientific' pronouncements.

Key words: humility, filtering and interpretation

7.1 INTRODUCTION

"The most beautiful thing we can experience is the mysterious. It is the source of all true art and science" - *Albert Einstein*

This book makes the argument that Churchman was a sceptic (of sorts) and that the nature of his scepticism is best described as 'epistemic modesty'.

Churchman's intellectual journey through the management and planning sciences is well known. So is his call for a 'systems approach' to world problems (and his corresponding modesty in regard to the writ of his beloved systems approach). This essay, however, does not intend to traverse this well-trodden path. Rather, it seeks to argue for a similar modesty to that of Churchman's (what the author has referred to elsewhere as epistemic humility). It does so by following a different path to humility – a path that takes us through the philosophy of science and its various pitfalls and puzzles.

The reader may recall that before he made his most lasting contributions to the management and planning sciences, West Churchman was the Editor in Chief of the *Philosophy of Science* journal (1948–1957). However, it seems that following his move to Berkeley, Churchman and the philosophy of science community drifted somewhat apart. Churchman, for his part, began his major work on the systems approach, whilst the philosophy of science community busied itself attempting to respond to various scepticisms towards the pronouncements of science. With the benefit of hindsight, however, it could be argued that the two followed a remarkably similar epistemic journey over the ensuing years – a journey that has led to a kind of humility about our knowledge claims that was, surely, unforeseeable amidst the optimism surrounding post-war reconstruction in the 1950's.

It is the intention of this essay therefore, to trace the journey of philosophy of science over the last half a century. This journey begins with what is labelled intuitive verificationism (the intuitive belief that scientific theories can be verified by observation), yet quickly moves through positivism, critical rationalism and social constructivism before suggesting a way forward towards a position of epistemic humility about the pronouncements of science and critical reflection about the methods of inquiry that scientists routinely use. Finally, the essay concludes by highlighting the similarities between the drift towards epistemic humility within the philosophy of science and a similar, Churchman-inspired, drift within the management and planning sciences.

7.2   INTUITIVE VERIFICATIONISM

"Science is derived from the facts." - *Francis Bacon*

A popular conception of science is captured by the slogan 'science is derived from the facts' (Chalmers, 2000). These facts are assumed to be claims about the world that can be directly established by a careful, unprejudiced use of the senses. Thus, science is based upon what we can see, hear and touch rather than personal opinion or speculative conjecture. Under this conception, the laws and theories that make up scientific knowledge are verified by empirical evidence. Once these laws have been verified they can be drawn on to make predictions and offer explanations. This account (labelled here as the verificationist account of science) has a certain populist appeal and can be depicted as shown in Figure 7.1 below.

The crucial step in this process, as the name suggests, is step 4a: verification. Verification relies on the logic of induction, which states that:

"If a large number of A's have been observed under a wide variety of conditions, and if all those observed A's possess the property B, then <u>by induction</u> all A's have the property B."

7. EPISTEMIC HUMILITY

The use of inductive inferences to 'verify' the results of science is often credited to Francis Bacon (1561-1626) and received widespread popularity during the Enlightenment, with Isaac Newton expounding its virtues in his *Opticks*:

> "Analysis consists of making experiments and observations, and in drawing general conclusions from them by induction, and admitting no objections against the conclusions but such as are taken from experiments or other certain truths" (Newton, 1704).

Figure 7-1. The Verificationist Account of Science

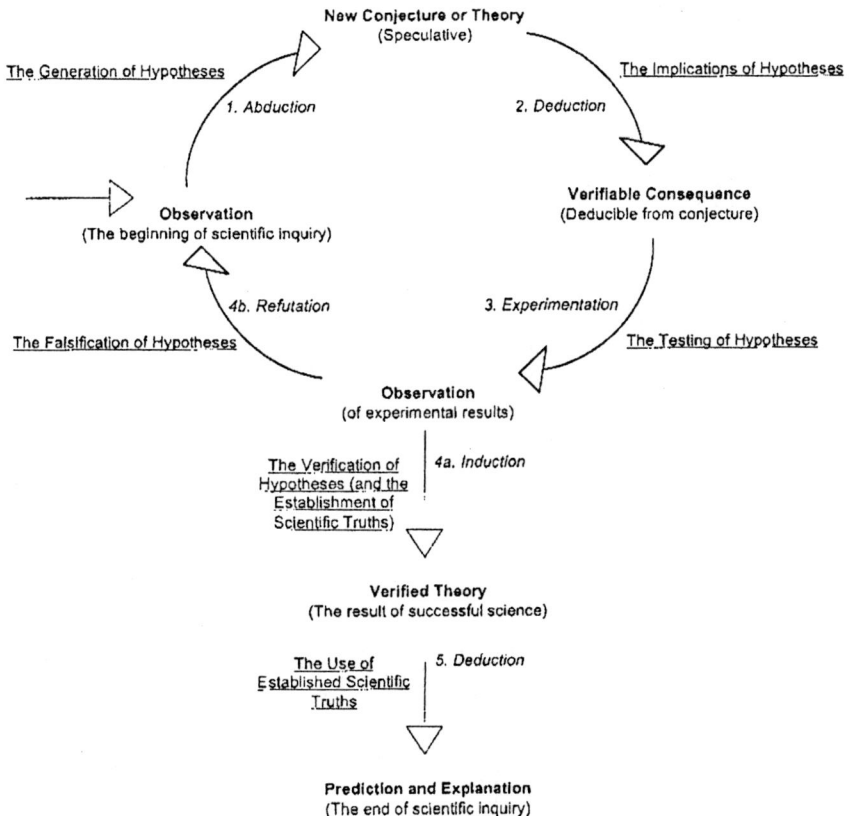

According to the verificationist, scientists formulate theories in order to explain perceived patterns in the world and verify these theories by accumulating supporting evidence. As such, verificationism is an empiricist approach to science. The crucial step in the process, as we have seen, is the one involving the use of inductive logic. According to induction, the principle of repeatability is what puts any knowledge that may be called 'scientific' in a different domain from opinion or taste. In this sense scientific knowledge is often described as 'public knowledge' – we have no choice but to accept that which can be repeatedly demonstrated in experiments. As such, the verificationist uses the label 'science' as an epistemic honorific.

Notwithstanding its intuitive appeal, there are certain aspects of the verificationist account that have attracted critical attention within the philosophy of science. These have to do with the (inductive) logic of verification. After the results of an experiment become known, it is important to realise what has been proven and what has not been proven. What has been conclusively demonstrated is that at time, T, under conditions, C, according to observers, O, experiment, E, yielded results, R. Every opinion about the experiment or theory based on the outcome of the experiment, however, may be disputed. This is due to the fact that induction is not a logically conclusive form of proof.

The study of abductive, deductive and inductive forms of reasoning constitutes the discipline of logic, a detailed account of which is outside the scope of this work. However, what must be stated is that of the three forms of reasoning involved in the verificationist account of science, only deduction can lead to logical proof. An example of a simple logical deduction is:

1. *All papers on philosophy are boring*
2. *This is a paper on philosophy*
3. *Therefore, this paper is boring*

In this argument (1) and (2) are the premises, and (3) is the conclusion. It follows that if the premises are true, then the conclusion must also be true[1]. This is the key feature of a logically valid deduction. Alternatively, a simple inductive argument is:

1. *Many papers on philosophy are boring*
2. *This is a paper on philosophy*
3. *Therefore, this paper is boring*

In this example, the conclusion does not necessarily follow from the premises, even if the premises are true. It is impossible to logically prove (3) from the 'facts' of (1) and (2).

---

[1] What deduction is silent on is the truth or falsity of the premises themselves. Thus, if someone where to dispute either of the premises in the example above, then although they would have to accept the deduction, they could still, nevertheless, dispute the truth of the conclusion.

7. EPISTEMIC HUMILITY

A simple example of an attempt to base a scientific law on inductive forms of reasoning is provided below:
1. Metal x_1 expanded when heated at time t_1.
2. Metal x_2 expanded when heated at time t_2.
...
n. Metal x_n expanded when heated at time t_n.
n+1. Therefore, all metals expand when heated.

However, repeated observations of metals expanding when heated can never 'prove' the conclusion that all metals expand when heated. No matter how large *n* becomes, there can be no guarantee that a sample of metal at some stage in the future will not contract.

The problem with proof by induction was understood as early as 1741 when David Hume published his *Treatise on Human Nature* (Hume, 1741). Briefly summarised: multiplying confirmatory observations will never get us any closer to proof – there is always the possibility that further observation may reveal exceptions that disprove the rule[2]. According to Hume, we discover the 'rules' of induction by observing what we habitually accept. Hume's attack on the inductive method initially fell on deaf ears. Bell (1994) argues that the reason for its *"chilly reception"* was that it threw doubt on two of the most cherished beliefs of the modern world: first, the belief in the certain truth of Newtonian physics, and second, the belief in the ability of the experimental 'scientific' method to yield the true nature of the world. In his treatise, Hume argued that induction may yield probable truth and it was this suggestion that eventually led to the modification of the verificationist account in the 19th century.

The first serious attempt to modify the inductive dogma of Francis Bacon and his contemporaries involved weakening the demand that scientific knowledge be proven to be absolutely true. Under this modified verificationist account of science, scientific knowledge is deemed 'probably true' in light of the evidence. Thus, the vast number of observations of metals expanding when heated warrants the assertion that the claim that 'all metals expand when heated' is probably true. The logic of induction can, therefore, be reformulated as:

---

[2] Bertrand Russell made the point metaphorically in his story of the inductivist turkey (Russell, 1912). This turkey found that, on his first morning at the turkey farm, he was fed at 9am. However, being a good inductivist, he did not jump to conclusions. He waited until he had collected a large number of observations under a variety of conditions of the fact that he was fed at 9am. Finally, he was satisfied and inferred that "I am always fed at 9am". Alas, his conclusion was shown to be false on Christmas morning, when his throat was cut.

"If a large number of A's have been observed under a wide variety of conditions, and if all these observed A's have the property B, then <u>by induction</u> all A's probably have the property B."

A far-reaching (and probably unintended) consequence of the new probabilistic version of induction was the spur it gave to the study of probability theory (Salmon, 1970). Perhaps the most important development for our purposes came from the work of Thomas Bayes. In his famous *Essay Towards Solving a Problem in the Doctrine of Chances,* Bayes (1763) set out to calculate the probability of an 'outcome' given prior knowledge. His subsequent theorem (Bayes Theorem) involves assigning prior probabilities of a given phenomenon occurring before trials are made. From these probabilities, and observational evidence, it is possible to determine the posterior probability that a given event will occur. When used in an iterative fashion, Bayes' formulae 'wash out' any arbitrariness in the assignment of prior probabilities. This recursive use of Bayes theorem is known as the Bayesian method. Because the Bayesian method allows posterior probabilities to 'over-ride' prior probabilities, it provides a link between estimated and actual probabilities of an outcome. It is this link, which gave a new hope for the logic of induction and the verificationist account of science.

Unfortunately for the verificationist, however, this hope was never to be realised (Chalmers, 2000). To calculate the probability of interesting scientific theories requires designing the constituent elements of the outcome. For example, to calculate the probability that Newton's gravitation is true requires assigning the prior probabilities of the gravitational attraction of the Moon to Earth, Sun to Mars *etc*. It is mathematically demonstrable that these prior probabilities will be 'washed out' with repeated use of Bayes formula, however problems arise due to the fact that there can be no reliable way of determining the posterior probabilities necessary to wash them out. One simply cannot devise an empirical test to determine the posterior probability of gravitational attraction between any two objects[3].

In summary, Bayes' formulae cannot fulfill the hope that theories can be inductively secured[4]. This seems to cast doubt on the empiricist's belief that science is derived exclusively from observable facts. According to a group of philosophers and scientists congregating in and around Vienna in the early 20th century this was of such concern that they pursued a major research program aimed at salvaging the empirical basis of science. The size, scope

---

[3] Another problem with the probabilistic form of induction is that attempts to justify it ultimately appeal to exactly the same form of inductive argument as the original (Chalmers, 2000).

[4] They have, nevertheless, been profoundly important in theoretical statistics and usefully applied in modelling many unanticipated phenomena.

7.3 LOGICAL POSITIVISM

> "It is mathematical representation alone that makes possible the notion of objective knowledge." - *Moritz Schlick*

With the fall of induction as a logically valid method on which to base scientific claims to truth, a group of philosophers who became known as the 'Vienna Circle' took up the challenge of finding a replacement. After a significant amount of effort, the school of thought that emerged became known as positivism.

The word positivism comes from the French word *positer*: to posit. It was first employed by Auguste Comte (1893), who, following Hume and the empiricists, argued that science should be based on what is 'posited' to our immediate senses. As such, speculative assertions not reducible to observation must be excluded from science. For this reason, positivism came to mean 'anti-metaphysics[5]'. When formal systems, such as mathematics, are applied to what is 'posited' positivism becomes logical positivism.

The rise of logical positivism in the early 20th century was sparked by the need to understand the impact of contemporary developments in theoretical physics on the philosophy and methodology of science (Ayer, 1959). At the time, physics was in a state of upheaval with Einstein's (1915) General Relativity, Bohr's (1928; 1935) Quantum Mechanics and Heisenberg's (1927) Uncertainty Principle each contributing to what was to become the overthrow of the Newtonian cosmology. In the face of this upheaval, the Vienna Circle aimed to ensure that the empirical foundations of science were not discarded. Indeed, it is within the positivist school that the empiricist tradition reaches its most sophisticated form.

The Vienna Circle was initially known as the Ernst Mach Society after the scientist and philosopher of the same name. Mach believed that 'scientific' knowledge must be considered to be 'true' and in a different domain from that of opinion or taste. Thus, he considered scientific modes of inquiry as being able to produce 'objective' and 'verifiable' 'knowledge',

---

[5] The term 'metaphysics' does not have a precise or agreed upon meaning (for that matter neither does science, as we shall see later). In positivist philosophy of science, metaphysics is used as a pejorative term generally applied to whatever is regarded as non-empirical. Traditionally, however, metaphysics has been regarded as the study of what lies behind the world of appearance. Ironically, a great many people would regard the sciences of physics, chemistry and astronomy as fitting this description.

whereas all other modes of inquiry as only ever producing 'subjective' and 'unverifiable' 'opinion'. Mach was an arch empiricist and argued that the entire content of science should consist of the relationships among the data of our sense-experience (Ray, 2000). Scientific concepts and generalisations, according to Mach, do not exist (ontologically) in their own right. They are merely names of particular empirical objects (a position known as 'nominalism') and are meaningful only inasmuch as they are grounded in observation. Any concept that cannot be reduced to observation, according to Mach, was meaningless. It was this position that Quine (1953) referred to as 'reductionism' in his paper *Two Dogmas of Empiricism*. The members of the Vienna Circle, including A. J. Ayer, Rudolph Carnap, Herbert Fiegl, Hans Hahn, Carl Hempel, Otto Neurath, Moritz Schlick and Friedrich Waismann, were initially bound by this common 'reductionist' belief. Moreover, they believed that all 'scientific' modes of inquiry would one day be unified into a single meta-methodology.

During the 1920's, the logical positivism of the Vienna Circle formed an alliance with the logical empiricism of the 'Berlin School'. The similarities of both schools were far-reaching: both insisted upon empiricism; both emphasised the importance of logic; both looked to the physical sciences as the paradigm of objectivity and both completely rejected metaphysics. The alliance between the two schools was further formalised with the founding of the Journal *Erkenntnis* in 1930, with Rudolph Carnap (Vienna) and Hans Reichenbach (Berlin) the co-editors.

In attempting to maintain the privileged position of the exact sciences, the positivists first made some significant concessions. These concessions were part of an ambitious attempt to completely reformulate the verificationist model and hence circumvent some of the well-known problems associated with induction. Thus, Reichenbach (1920) in the introduction to his book on Einstein's relativity writes:

"Every factual statement, even the simplest one, contains more than an immediate perceptual experience; it is already an interpretation and therefore itself a theory … the most elementary factual statement, therefore contains some measure of theory".

This clear statement of what is now called 'the theory-ladenness of observation' represented a move away from the received (empiricist) dogma of the 19th century that facts always precede theory. Carnap (1928) also endorsed this view in his classic of positivist thought *The Logical Structure of the World*. In response to these concessions, the positivists modified much of the terminology of empiricism. For example, 'facts', which imply some direct, theory-neutral, connection to the underlying 'real world', became known as 'data' (which was considered to only imply observational input).

7. EPISTEMIC HUMILITY

In this sense, the logical positivists began to embrace a kind of phenomenalism, which views propositions asserting the existence of physical objects as analytically equivalent to propositions asserting that subjects would have certain sensations were they to have certain others. As such, the positivists turned their attention to the analysis of these sensations, developing a sense datum theory (Fumerton, 1992)[6].

The positivist adoption of a phenomenalist epistemology raised a whole set of new difficulties associated with the scientific claim of objectivity. According to phenomenalism (and its more recent variant, phenomenology), the only contingent propositions that can be known directly, are those describing the contents of our own minds and if any belief about the 'real world' is to be justified, it must be inferentially justified from what we know about our minds. In order to solve the problem of objectivity Schlick (1918) claimed that by ordering, interpreting, and structuring the data of our sensory perceptions within a rigorous mathematical framework it was possible to 'objectify' perceptions and transform them from 'appearance' into 'experience'. In other words, it is mathematical representation alone that makes possible the notion of objective knowledge. Thus, Schlick (1918) draws the distinction between knowledge of cognition (*erkennen*) and acquaintance with the immediately given sensory perception (*erleben*). The later, since it is momentary, was thought to be incapable of yielding knowledge. According to Schlick, knowledge is possible only when we embed such momentary perceptions within a rigorous mathematical system. The idea being that in the face of the theory-ladenness of observation and hence the subjectivisation of perception (*erleben*), there only remains one objective component of knowledge and that is logic itself (*erkennen*). According to the logicial positivists, the discipline of mathematical physics was the obvious shining example of perceptions being embedded within a rigorous mathematical framework and was therefore paradigmatic of objectivity and rationality and deserved pre-eminence amongst the sciences.

The difficulties associated with using formal systems (such as mathematics) to 'objectify' sense experience began to arise with the fall of the Kantian conception of the transcendental unity of apperception. According to Kant, all rational beings effectively bring to every observation and interpretation of the world the same set of *a priori* concepts. When applied to Kant's spatiotemporal construction, the doctrine of the transcendental unity of apperception effectively states that the space and time underlying the constructive procedures of pure mathematics are the very same space and time within which we perceive and experience nature through the senses (i.e. they are the *a priori* intuitions that form the sensible preconditions of objects of experience). It is this conception of the intuitive

---

[6] Where sense data were viewed as mind-dependent entities.

nature of pure mathematics that enabled Kant (1781) and others to explain how mathematics, in its full precision, is applicable to the chaotic world of sense. However, with the realisation that pure mathematics no longer requires a basis in spatiotemporal construction, but can instead proceed 'formally' via strict logical deduction within an axiomatic system, it was no longer possible to maintain that any mathematical theory has a necessary relation to our sensory perceptions. Hence, there is no longer a single privileged framework. Many such frameworks are possible and some of them were being applied to Einstein's theory of relativity at the time.

The positivist attempt to preserve a basically Kantian conception of knowledge and experience in the face of the collapse of Kant's doctrine of the transcendental unity of apperception created fundamental, and ultimately unresolvable tensions within their entire program. The problem confronting them can be expressed as follows:

1. They wished to maintain the privileged position of the exact sciences (especially mathematical physics) in the face of the rising problems associated with the verificationist account of science and the overthrow of the Newtonian cosmology.
2. In doing so, they accepted Kant's position of the theory-ladenness of observation, and the spatiotemporal construction of mathematics. Thus, it was assumed that mathematics provided the theoretical framework needed to confer objectivity and rationality onto sensory perception.
3. Yet with Hegel and Nietzsche's critique of the transcendental unity of apperception (and with it the collapse of Kant's understanding of spatiotemporal construction) they understood that there is no longer any single privileged framework that can alone perform this 'objectifying' function. On the contrary, every framework appears to exemplify its own particular standards of objectivity and rationality.

In response to this, the Marburg School of Hermann Cohen, Paul Natorp and Ernst Cassirer drew explicitly relativist conclusions (Cassirer, 1923; 1925; 1929; 1942; Schilpp, 1949; Hazelrigg, 1989; Friedman, 2000). Since there is no longer a single privileged framework, each framework may supply its own standards of truth, and hence objectivity. This position led to the 'coherence theory of truth', which views coherence and consistency within a particular framework (or set of assumptions) as sufficient for 'truth'. Since the positivists wished to maintain the privileged position of mathematical physics, however, the relativism of the Marburg School was anathema to them and they went to great lengths to try to avoid the Marburg conclusions. Accordingly, the latter positivist writings revolve around the problem of adjudicating between competing frameworks.

The adjudication problem was exacerbated when Schlick (1915) admitted that both Einstein's relativity and the classical explanations of Lorenz,

Fitzgerald and Poincare could explain the data from the Michaelson-Morley experiment of 1887. The two theories, its seemed, led to all the same empirical predictions. Thus they were deemed to be empirically equivalent[7]. Schlick admitted that positivism had no answer to the question of adjudicating between empirically equivalent frameworks and that Einstein's theory was more likely to be true because it was simpler than the competing aether theory (Schlick, 1915). However, there is no clear reason to believe that simplicity is a reliable guide to truth. As Friedman (1992) has questioned: "why in the world should nature respect our, merely subjective, preference for simplicity?"[8]

Throughout the inter-war years, a vigorous research program pursued by the positivists attempted to solve the problem of adjudicating between empirically equivalent theories of explanation. The outcome of all of this was the doctrine of conventionalism, which states that empirically equivalent theories (such as relativity and the aether theory) are really not two conflicting theories at all. Their disagreement is only apparent and so there is no need to adjudicate between them. Rather than a disagreement over truth, there is only a pragmatic question of convenience. In this sense, the choice is purely a matter of convention.

A logical consequence of conventionalism was that empirical facts were deemed to be the only truth and that the 'cognitive meaning' of a scientific theory consisted solely in its implications for actual and possible observations. Indeed, the positivists acknowledged this with the notorious Verifiability Principle, which was wielded to question the 'cognitive meaningfulness' (or legitimacy) of all discourse regarding unobservables[9]. Happily, the Verifiability Principle could not be sustained. First, because it proved impossible to view advanced theories, such as Einstein's relativity, as mere summaries of actual and possible observations. That is, scientific theories often embody knowledge of unobservable phenomena (in the positivist's language this means that scientists routinely do metaphysics). Second, and perhaps more fundamentally, the notion of 'observable facts' required the possibility of theory-neutral observation, a notion that the positivists had explicitly rejected[10].

---

[7] This is an early statement of the **under-determination thesis**, which states that no body of evidence can support any theory to the exclusion of all rivals.

[8] The retreat to simplicity is a long-standing epistemic cop out. It can be traced to William of Ockham's (1300-1349) principle, known as Ockham's razor, which states that *"when faced with competing explanations, accept the most simple"*, and ultimately to Aristotle who claimed that *"nature operates in the shortest way possible"*.

[9] The Verifiability Principle is remarkably similar to Peirce's (1905) **'pragmatic maxim'**.

[10] The Verifiability Principle also leads to what is now known as the paradox of confirmation. The paradox arises from an attempt to characterise the relation between hypothesis and evidence. The most celebrated example, the raven paradox, begins with the

The story of logical positivism is a story of failure and has been discussed as such by various commentators (Andersson, 1994; Bell, 1994; Friedman, 1992; Laudan, 1996; Ray, 2000; Stroud, 2000). In particular, the positivists failed to develop an account of science based solely on the relationships between the data of sense experience. Accordingly, in the aftermath of the positivist episode, it has been common to afford metaphysical (non-empirical) speculation a place in science. However, this has raised a whole new set of difficulties associated with the status of unobservable entities. Several attempts have been made to clarify the new situation over the last 50 years. Out of these attempts, it may be argued that two broad positions have arisen. On the one hand are those who argue that scientific theories are true by virtue of their correspondence with the 'real', transcendental truth of the matter, whilst on the other, are those who argue that no such correspondence can ever be substantiated and therefore scientific theories are merely 'instruments' for helping us correlate observational data and make predictions. These positions correspond to two separate understandings of truth. Furthermore, they also correspond to two separate understandings of scientific progress: one which claims that current

---

hypothesis "all ravens are black", symbolised as (x)(Rx->Bx). According to Nicod's condition a hypothesis is verified by its positive instances and falsified by its negative ones. Thus the observation *"this is a raven and it is black"*, (Ra^Ba), verifies the hypothesis and the observation *"this is a raven and it is not black"*, (Ra^~Ba), falsifies the hypothesis. The paradox arises from a fundamental principle of confirmation known as "the equivalence condition". According to the equivalence condition, if a hypothesis is confirmed by some evidence statement, *E*, then it is confirmed by any other evidence statement logically equivalent to *E*. In light of the equivalence condition, the logical equivalence of *"all ravens are black"* and *"all non-black things are non-ravens"* implies that these statements are supported by the same body of evidence, *E*. Therefore (and here is the paradox) the observation of a non-black, non-raven (~Ba^~Ra), such as a blue book, is a confirming instance of the hypothesis. Further, the hypothesis (x)(Rx->Bx) is logically equivalent to (x)[(Rx v ~Rx)->(~Rx v Bx)]. Here the antecedent is a tautology, and thus any truth-value assignments that make the consequent true will confirm the hypothesis that all ravens are black. Since the consequent is a disjunct and the extensional deductive model of hypothesis testing requires only one of the disjuncts to be true for the compound to be true, then *"all ravens are black"* is logically verified by any observation that is either black or a non-raven. Surprisingly then, the discovery of a star (or indeed the four-inch refractor used to view the star) confirms the generalisation that all ravens are black. These results should be disturbing to anyone who desires an account of science based upon empirical confirmation. Philosophers and logicians have handled the paradox in a variety of ways. Some have rejected Nicod's condition (which would be disastrous for an empirical/positivist account of science), others have abandoned the equivalence condition. However, many have argued that this paradox shows that scientific observations are systematically defective, since they would not give equal weight to observations, *"ignoring the blue book and attending to the black raven"* (Hempel, 1945; 1965; 1966; Goodman, 1954). For further elaboration of the paradox of confirmation the reader is referred to Trout (2000), from which much of this discussion has been sourced.

scientific theories are 'truer' than their predecessors (in the sense that they more accurately represent the real world); the other which claims rival theories cannot be compared solely on objective, rational grounds and, therefore, current scientific theories are not necessarily 'truer' than the ones they replace. These positions, together, have contributed towards the drift towards epistemic humility within the philosophy of science over the past fifty years and, as such, warrant a separate discussion. They are, respectively, critical rationalism and social constructivism.

7.4 CRITICAL RATIONALISM

"The wrong view of science betrays itself in the craving to be right." - *Karl Popper*

Karl Popper is undoubtedly the giant of the critical rationalist school. Popper was educated in Vienna in the 1920's at a time when the Vienna Circle (and logical positivism) were at the height of their influence. Popper himself tells the story of how he became disenchanted with the positivist school and their belief that science was especially reliable because it was derived from 'empirical data' (Popper, 1979). It was against this setting that Popper developed his critique of the verificationist account of science leading to his, now famous, split with the positivists[11].

Intellectual life in the Vienna of Popper's youth was dominated by science-based ideologies. Much of this was due to the privileged position science found itself in due to the wide-spread intuitive appeal of verificationism. Popper, however, became suspicious of the way in which he saw Freudians, Darwinists and Marxists supporting their theories by an appeal to the same empirical verification. Acceptance of these new 'sciences' had as Popper observed:

"The effect of an intellectual conversion or revelation, opening your eyes to a new truth hidden from those yet initiated. Once your eyes were thus opened you saw confirming instances everywhere: the world was full of verifications of the theory. Whatever happened always confirmed it. Thus, its truth appeared manifest; and unbelievers were clearly people who did not want to see the manifest truth." (Popper, 1969).

It seemed to Popper that these theories could never go wrong because they were sufficiently flexible to accommodate any instance of human

---

[11] Indeed, the ensuing debates between Popper and the positivists (especially Carnap) were to become a feature of philosophy of science for the next four decades.

behaviour or historical change as compatible with their theory[12]. Consequently, although giving the appearance of being powerful theories confirmed by a wide range of facts, they could in fact explain nothing because they could rule out nothing. Popper, on the other hand, thought that a theory with genuine explanatory power would make risky predictions. Predictions that could be tested and, if they did not obtain, would refute the theory. For example, Einstein's theory had the implication that rays of light should bend as they pass close to massive objects (such as the Sun). As a consequence, a star situated beyond the sun should appear displaced from its usual position in the absence of this bending. Eddington looked for this displacement by viewing the star at a time when the light from the sun was blocked out by an eclipse. As it happened, the displacement was observed and therefore Einstein's theory became widely accepted (or verified as the positivist would claim). The point that Popper makes is that the apparent position of the star might not have been displaced. By making a specific, testable prediction that is logically deducible from the general theory, the theory stands to be falsified by observational evidence.

Popper's contribution to the positivist debate was, therefore, to completely discard two of the fundamental tenets of the school: the idea that science did not encompass metaphysics and the idea that theories were scientific if and only if they were verifiable (the verifiability principle). According to Popper, science begins with theories (or 'conjectures' as he put it) construed from imaginative or even mythological speculations about the world. These conjectures provide the starting point for scientific investigation. In this sense, Popper stands in direct opposition to the empiricism of the positivists who denied a place for non-empirical structural conjecture in science. Furthermore, according to Popper, progress is made not by searching for confirming evidence, which can always be found, but by searching for falsifying evidence. It is falsification that progresses science by revealing the need for a new and better explanation. Thus, science progresses by trial and error, or as Popper would have it, by conjectures and refutations. Only the fittest theories survive. Although it can never be said of a current theory that it is true, it can always be said that it is the best available and that it is better than any that have come before. Popper thus argues that the defining characteristic of scientific theories is not their verifiability but their falsifiability (Popper, 1969, 1972).

Popper's falsifiability criteria, thus distinguishes science from other intellectual pursuits, among which he includes pseudo-science and metaphysics. In sharp contrast to the logical positivists, he refused to equate non-science with non-sense and thence nonsense. According to Popper, it is

---

[12] This observation was an early statement of what is now known as the under-determination thesis.

7. EPISTEMIC HUMILITY

sensible to include non-sensible knowledge in science. Indeed, science begins with non-sensible (or rational) conjecture about causes and effects. It progresses through empirical falsification and it never comes to rest in the sense of arriving at truth (as the verificationists believed). Although they cannot be tested scientifically, metaphysical (or pseudo-scientific) doctrines are often meaningful and important. Popper even credited pseudo-scientists like Freud and Adler with valuable insights that might one day play their part in a genuine science of psychology. His criticism was not that pseudo-scientific or metaphysical theories were nonsense, but merely that it is incorrect to believe these theories could be verified by searching out supporting evidence. It is this view of science that led Popper (1969) to describe it as:

"The method of bold conjectures and ingenious and severe attempts to refute them."

The basic function of a scientist, according to Popper, is to:

"test theories, deduce consequences of theories and discover whether these consequences obtain. If they do not, the theory is refuted, if they do, the theory survives. The more tests the theory passes, the more credible it becomes".

A visualisation of Popper's account of science (known variously as the critical rationalist, falsificationist, hypothetico-deductivist or deductive-nomological account of science) is provided in Figure 7.2 below.

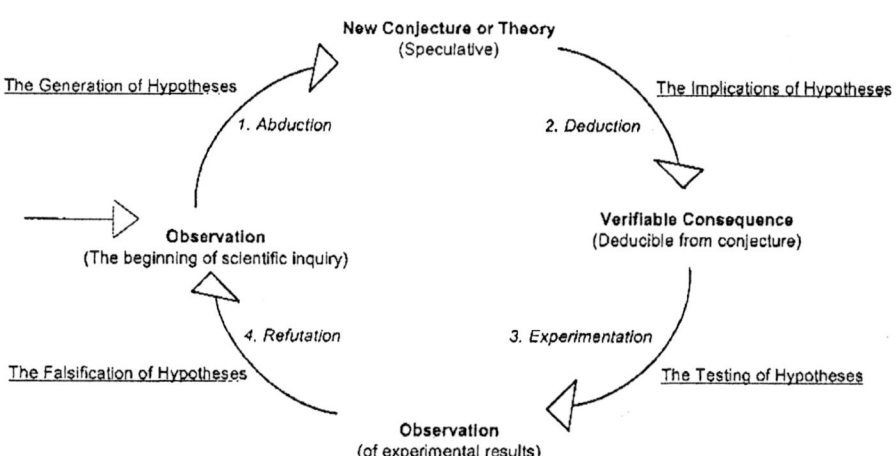

Figure 7-2. The Critical Rationalist Account of Science

120   *Chapter 7*

According to Popper, his account of science has solved the long-standing problem of induction by defining science as an activity that does not involve inductive inferences at all (Popper, 1972). Instead of induction, deduction is used to reveal the consequences of theories so they can be tested and, perhaps, falsified. A feature of Popper's view of science is that no claims are made that the survival of tests shows a theory to be true or even probably true. At best, the results of such tests show a theory to be an improvement on its predecessor. As such, the critical rationalist settles for progress rather than truth. For this reason a hypothesis refuted is more valuable than one that survives the test (Popper, 1972). Popper put the essential point of his critical rationalist account in the following aphorism:

"The wrong view of science betrays itself in the craving to be right" (Popper, 1969).

Thus, it is only after an established theory has been refuted that a new (and better) one is proposed. Scientific progress can thus be visualised as a Burkean spiral, as new theories (with better explanatory and predictive power) replace old ones that have been empirically refuted (see Figure 7.3). It is this conception of progress that made Popper believe that scientific theories approach a 'correspondence' with the truth of the underlying nature of the 'real world'.

Figure 7-3. Scientific Progress According to the Critical Rationalist School

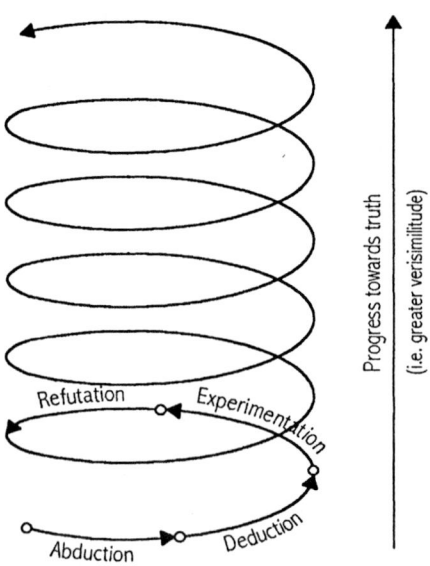

7. EPISTEMIC HUMILITY

According to Popper, a scientific experiment is one in which some significant conjecture is at risk. This implies that every scientific experiment implicitly, or explicity embodies some theory and that this theory stands to be refuted by the results of the experiment (note that it can never be proved). As we have already mentioned, Popper uses the example of Einstein's falsifiable conjecture that (according to his theory of relativity) light passing near the Sun should be bent by it.[13] Popper was impressed by the contrast between Einstein's bold and falsifiable conjecture on the one hand and the pseudo-scientific schools of the Marxists, Darwinists and Freudians that dominated the Vienna of his formative years[14] (Magee, 1985).

Despite the enormous success of the critical rationalist account of science in circumventing the problems of induction (by defining science as an activity that does not involve induction), it suffers from a number of inadequacies. These stem from both the logic of refutation and the socio-historical applicability of the theory in practice.

The first set of objections to the critical rationalist account of science can be raised on logical grounds. Simplistically, the logic of falsification is said to follow the law of Modus Tollens, which has the form:

> Premise 1: If H then O
> Premise 2: But ~O
> Conclusion: Therefore, ~H

Here, the first premise refers to the test statement associated with the falsification and the second to the empirical basis of the falsification. Obviously, given the truth of both premises, the conclusions holds by simple deduction. However, objections have been raised in regard to the truth of both of these premises.

In regard to the truth of test statements, objections have been raised on the grounds that it is always possible to protect a theory from falsification by deflecting the falsification onto some other part of the complex web of assumptions that face the tribunal of observation in every experiment. This position is known as the Quine-Duhem thesis: the thesis that most advanced

---

[13] When the results from the 1919 eclipse experiment became known, Professor Littlewood sent an excited note to Bertrand Russell: *"Dear Russell, Einstein's theory is completely confirmed"* (Checkland, 1981). According to Popper, this sort of rhetoric highlights a profound misunderstanding of the nature of scientific knowledge. Perhaps a better response would have been along the lines of *"Einstein's theory has survived this difficult test".*

[14] Popper's position on such pseudo-sciences is perhaps best presented in his paper *Darwinism as a Metaphysical Research Programme*, in which he claims that evolutionary theory was not a scientific theory capable of passing the test of falsification, but a metaphysical research programme (Popper, 1974).

scientific theories are a complex interlocking set of concepts, observations, definitions, presuppositions, experimental results and connections to other theories and that no single fact is going to be crucial to the survival of the theory. In fact, facts radically under-determine theories, the consequence of which is that it is impossible, on purely logical grounds, to falsify a sophisticated scientific theory experimentally – any observation can be accommodated by making suitable adjustments to the 'web of beliefs'. A corollary of the Quine-Duhem thesis is the underdetermination thesis, which states that when comparing sophisticated scientific theories there does not exist a crucial experiment[15].

Lakatos (1977) made a similar point when he stated:

"Is, then Popper's falsifying criterion the solution to the problem of demarcating science from pseudo-science? No. For Popper's criterion ignores the remarkable tenacity of scientific theories. Scientists have thick skins. They do not abandon a theory merely because facts contradict it. They normally either invent some rescue hypothesis to explain what they then call a mere anomaly, or, if they cannot explain the anomaly, they ignore it, and direct their attention to other problems."

Moreover, he demonstrated how this may work by making use of a hypothetical scenario involving Newtonian gravitation:

"A physicist of the pre-Einsteinian era takes Newton's mechanics and his laws of gravitation, N, the accepted initial conditions, I, and calculates, with their help, the path of a newly discovered small planet, p_1. But the planet deviates from the calculated path. Does our Newtonian physicist consider that the deviation was forbidden by Newton's theory and therefore once established, it refutes the theory N? No. He suggests that there must be a hitherto unknown planet p_2, which perturbs the path of p_1. He calculates the mass, orbit, etc. of this hypothetical planet and then asks an experimental astronomer to test his hypothesis. The planet p_2 is so small that even the biggest available telescope cannot possibly observe it; the experimental astronomer applies for a research grant to build yet a bigger one. In three years time, the new telescope is ready. Were the

---

[15] An experiment E is crucial between T_1 and T_2 if and only if T_1 predicts that E will yield O and T_2 predicts that E will yield ~O. A classic example of this is Foucault's experiment designed to decide between the wave and particle theories of light. According to the wave theory, the velocity of light in water should be less than the velocity of light in air, whilst according to the particle theory, the velocity of light in water should be greater than the velocity of light in air. Foucault's experiment supported the wave theory. However it was noted that in deriving ~O from the particle theory a set of auxiliary assumptions were required, thus, the particle theory could always be saved by altering the auxiliary assumptions. The under-determination thesis, together with the related Quine-Duhem thesis, provides a powerful rebuttal to the critical rationalist account of science.

unknown planet p_2 to be discovered, it would be hailed as a new victory of Newtonian science. But it is not. Does our scientist abandon Newton's theory and his idea of the perturbing planet? No. He suggests that a cloud of cosmic dust hides the planet from us. He calculates the location and properties of this cloud and asks for a research grant to send up a satellite to test his calculations. Were the satellite's instruments to record the existence of the conjectural cloud, the result would be hailed as an outstanding victory for Newtonian science. But the cloud is not found. Does our scientist abandon Newton's theory, together with the idea of the perturbing planet and the cloud, which hides it? No. He suggests that there is some magnetic field in that region of the universe, which disturbed the instruments of the satellite. A new satellite is sent up. Were the magnetic field to be found, Newtonians would celebrate a sensational victory. But it is not. Is this regarded as a refutation of Newtonian science? No. Either yet another ingenious auxiliary hypothesis is proposed or ... the whole story is buried in the dusty volumes of periodicals and the story never mentioned again" (Lakatos, 1970).

Thus, sophisticated theories are not only unverifiable but unfalsifiable as well. Or, in Lakatos' (1970) words: "scientific theories are not only equally unprovable and equally improbable, but they are also equally undisprovable."

The difficulties facing the logic of falsification, however, are not limited to the supposed 'truth' of test statements (Premise 1). The supposed 'truth' of the empirical basis of the falsification (Premise 2) has also come under attack. As we have seen, the 'theory-ladenness of observation' states that judgements direct what a scientist observes and what s/he passes over (Bhaskar, 1986; Hollway, 1989). In other words, what an observer sees is not determined solely by the images on their retinas, but on the experience, knowledge and expectations of the observer (Chalmers, 2000). Consequently, when observation (or experimentation) provides evidence that conflicts with theory, it may be the perception of the evidence that is at fault rather than the scientific theory. Nothing in the logic of the situation requires that it is always the theory that should be rejected on the occasion of a clash with observation.

Thomas Kuhn (1970a) raised a number of examples from the history of science where observations have been influenced by the expectations of the scientist. For example, Kuhn (1970a) tells the story of the 'discovery' of the planet Uranus by Sir William Herschel in 1781. What Kuhn finds interesting about this story is that Uranus was actually 'discovered' (in the sense that it was observed) on at least 17 different occasions between 1690 and 1781 but each time was held to be a star (Andersson, 1994). After Herschel 'discovered' that Uranus was actually a planet, however, astronomers began 'seeing' a planet where previously they saw a star. In order to explain this, Kuhn (1970a) referred to psychological experiments that showed the same drawing being seen in different ways by different observers. A particularly famous example was one discussed by Wittgenstein (1953), which could be seen as either a duck or a rabbit. According to Gestalt psychology, at some point the observer 'sees' the hitherto unseen second image and a 'switch' occurs whereby the second image is the one observed from there on. In a similar way, Kuhn suggests that: "what were ducks in the scientists world before the revolution are rabbits afterwards". Whereas, before Herschel astronomers saw a star, afterwards they began seeing a planet. Thus, observation itself is theory laden.

As we have already stated, whilst the logic of Modus Tollens is beyond dispute, for the conclusion to be true the premises must also be true. However, it seems that test statements (Premise 1) suffer from a lack of crucial experiments and that the empirical basis of the falsification (Premise 2) suffers from the theory ladenness of observation. Accordingly, the entire critical rationalist account begins to run into insurmountable difficulties.

These difficulties are compounded by the embarrassing fact that if critical rationalism had been strictly adhered to in practice, then those theories generally regarded as being among the best scientific theories would never have been developed because they would have been rejected in their infancy (Chalmers, 2000). For example, in the early years of its life, Newton's gravitational theory was falsified by observations of the moon's orbit. It took almost fifty years to deflect this falsification on to causes other than the theory (Chalmers, 2000). Later in its life, the same theory was known to be inconsistent with the details of the orbit of the planet Mercury. Scientists did not abandon the theory for this and it turned out that it was never possible to explain away this falsification - yet the theory remained the dominant scientific worldview for hundreds of years (Chalmers, 2000).

There are numerous other examples of scientists resisting falsifying evidence in support of their own theories, including such revolutionaries as Copernicus, Maxwell and Bohr (Feyerabend, 1975). In fact, Kuhn (1962; 1963; 1970a; 1970b) argues that anthropological studies suggest that scientists rarely, if ever, try to falsify the dominant theories of their disciplines. Rather, these theories become the contextual certainties for the discipline and are only ever rejected when a new theory comes along that 'fits' better with the whole inchoate set of concepts, observations, definitions, presuppositions, experimental results and connections to other theories that make up a dominant paradigm. As such, the psychology of research rarely ever matches the critical rationalist logic of research. Indeed, Kuhn (1970b; 1977; 1998) goes so far as to suggest that the history of science casts serious doubts on Popper's argument that science progresses cumulatively towards truth. According to Kuhn, new theories often solve problems associated with old ones but introduce different problems at the same time. As such, there are usually plusses as well as minuses in every instance of theory change.

Given the logical and historical inadequacies of critical rationalism, it is reasonable to conclude that theory choice is typically made on grounds that include things other than strict logical deduction and empirical testing. This view was held by the pragmatists, William James and John Dewey, at the turn of the 20th century and has received much recent attention within the philosophy of science by the so-called sociologists of science. Accordingly, it is to the sociological view that we now turn, beginning with its foremost exponent – Thomas Kuhn.

7.5   SOCIAL CONSTRUCTIVISM

"Science is fundamentally a social undertaking." - *Thomas Kuhn*

Notwithstanding its initial success, Popper's falsifiability criterion has met an increasing number of difficulties and, as such, has largely been abandoned. To begin with, the ironical observation that if Popper's method had been strictly adhered to, the most celebrated scientific theories would have been falsified (and thus discarded early on in their development) has been used to question the historical validity of falsification (not to mention the sociological workability). Secondly, the Quine-Duhem thesis has highlighted the difficulty in determining which part of the 'web of assumptions' is falsified by a clash with observation, which has, in turn, undermined the applicability of the logic of Modus Tollens to science. No theory, it seems, faces the 'tribunal of observation' in isolation. Finally, even the validity of the tribunal itself has been called into question, with the theory ladenness of observation thesis throwing doubt on the supposed objectivity of observation. It seems, therefore, that Popper's account of science fails to grasp the full complexity of the mode of development of major scientific theories. Since the 1960's it has been common to conclude from this that a more adequate account of science must be *anthropo*logical as well as logical. Such an account would seek to understand the psychological, social and cultural frameworks in which the scientific activity takes place. One of the reasons for this stems from the history of science. Historical study reveals that the evolution and progress of major sciences exhibit a social dimension that is not captured by either the verificationist, positivist or falsificationist accounts (Chalmers, 2000).

The sociological view burst onto the intellectual scene with the publication of Thomas Kuhn's book *The Structure of Scientific Revolutions* (Kuhn, 1962). According to Kuhn, "history, if viewed as a repository for more than anecdote or chronology, could produce a decisive transformation in the image of science by which we are now possessed". The common image of science until Kuhn was one in which science was progressing cumulatively towards greater truth and mastery of the world around us. However, Kuhn derides this view as amounting to: "little more than a tourist brochure". According to Kuhn, the cumulative view of science fails because the selection of what theories constitute part of the cumulative historical narrative is always generated by present science and only those elements of past science that lead to present science are included (Hoyningen-Huene, 1993). As such, the cumulative view underspecifies the difference between older scientific worldviews and present ones. Kuhn's account of science, therefore, was developed as an attempt to accurately reflect the major differences between older and present scientific worldviews and, thereby, keep philosophy of science in line with the history of science. The key features are the emphasis placed upon the revolutionary nature of scientific progress and the role played by the sociological characteristics of scientific communities during what he terms 'normal science'. Kuhn's picture of the way science progresses can be summarised by Figure 7.4 below.

The disorganised and diverse activity that precedes the formation of a science (pre-science) eventually becomes structured and directed when a single paradigm becomes adopted by the scientific community. A paradigm is composed of the general theoretical assumptions that the members of a particular scientific community adopt. Included within these assumptions are an array of problems considered important to investigate, theories about underlying structure and causal relations, theories that interpret the underlying structure empirically and methodological guidelines. Workers within a paradigm practice what Kuhn terms normal science. These workers will articulate and develop the paradigm in their attempt to account for the behaviour of the phenomena relevant to their discipline. In doing so, they will inevitably experience difficulties and encounter apparent falsifications, however they will usually deflect these falsifications onto some other aspect of the study rather than the prevailing paradigm. One reason for this is that the paradigm continues to yield theoretical understanding and empirical insights despite anomalies. However, there are always a number of non-intellectual factors that buttress the stability of paradigms to the point where they become 'doctrine'. One is professional training (standard textbooks are written assuming a particular paradigm). Another is professional authority

7. EPISTEMIC HUMILITY

(recognised leaders in a field espouse the paradigm).[16] Furthermore, proposals for research funding must be acceptable within a paradigmatic framework and professional publications largely screen out papers that breach the pattern of normal science.[17]

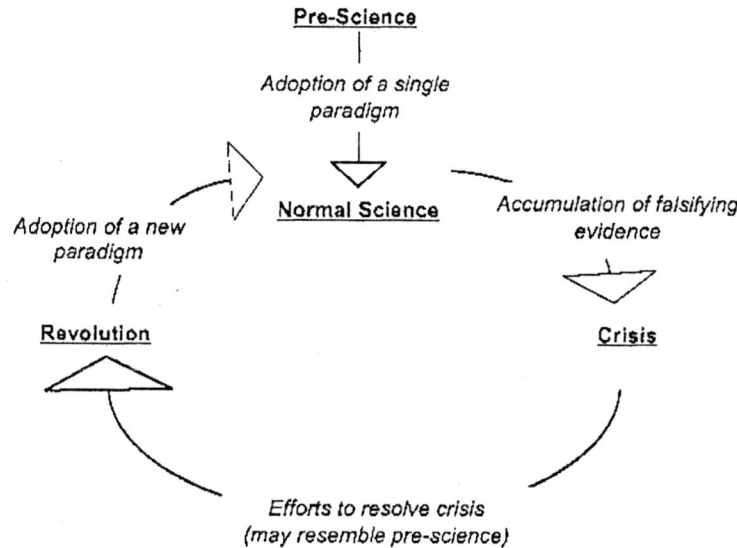

Figure 7-4. Scientific Progression According to Thomas Kuhn's Historical View

However, with the accumulation of falsifying evidence and the realisation that the prevailing paradigm is exhausting its potential for new discovery, the community eventually becomes aware of a crisis emerging within the paradigm. The crisis is resolved when an entirely new paradigm emerges and attracts the allegiance of more and more practitioners within the discipline. Eventually, the original paradigm is abandoned and a 'scientific revolution' takes place. The new paradigm, full of promise and not beset by any apparent difficulties now guides normal science until it too runs into

---

[16] A good example of authoritative knowledge is the hypothesised existence of black holes. When Vikram Chandrasekhar was a young scientist he performed a set of calculations suggesting that massive stars end their lives in gravitational collapse (i.e. a black hole). The foremost authority in astrophysics at the time, Sir Arthur Eddington, could not accept such an absurd sounding conclusion and thought that something (never specified) must have been wrong with the theory – and he said so. Although Chandrasekhar had the better argument, Eddington's conclusion was almost universally accepted for a number of decades.

[17] The concept of **peer review** within the sciences thus becomes a powerful tool for perpetuating the dominant paradigm. Burke (2000) discusses this under the heading **cultural inertia**.

serious trouble. Thus, science progresses not by becoming more objective or mathematically rigorous (as the positivist would have) but by becoming more imaginative.

The defining contribution of Kuhn's work is the revelation that science is fundamentally a socio-cultural undertaking (Kuhn, 1970b). Thus, the development of scientific texts, institutions, methodologies and most importantly, theories of explanation, is a social process subject to all the political, sociological, literary and anthropological influences as any other social process. In contrast to Popper and the logical positivists, Kuhn is not much interested in partitioning science from non-science. However, when pressed, he has claimed that: "the existence of a paradigm capable of supporting a normal science tradition is the characteristic that distinguishes science from non-science" and justifies the statement by the observation that 'normal scientists' are usually highly uncritical of the paradigm in which they work (Kuhn, 1962). Indeed, according to Kuhn, it is only by being uncritical of the paradigm that scientists are able to concentrate their efforts on the detailed articulation of their discipline (as seen through the paradigm). It is this lack of disagreement over the fundamentals, which distinguishes normal science from pre-science, which, according to Kuhn, is characterised by the fact that "there will be almost as many theories as there are workers in the field" (Kuhn, 1962)[18].

Following Wittgenstein (1953), Kuhn suggests that there is more to a paradigm than what can be tabled in the form of specific rules and directions. If one tries to give a precise characterisation of a paradigm in the history of science, it always turns out that some work within the paradigm violates the characterisation (Kuhn, 2000). However, Kuhn insists that this state of affairs does not render the concept of a paradigm untenable. Even though there is no complete characterisation, individual scientists acquire knowledge of a paradigm through their scientific education by solving standard problems, standard experiments and eventually completing a piece of research under a supervisor who is already a skilled 'paradigm practitioner'. The aspiring scientist will not necessarily be able to give an explicit account of the methods and skills s/he has acquired because much of the knowledge will be tacit, in the sense developed by Michael Polanyi (1973).

Kuhn's conception of a paradigm may be seen as a reformulation of the 'theory-ladenness of observation'. Paradigms are like 'reference frames'. Normally, our paradigm (reference frame) is taken for granted and mistaken for reality. Indeed, most people are not aware they are walking around

---

[18] Kuhn offers optics before Newton as an example of pre-science.

7. EPISTEMIC HUMILITY

carrying a frame of reference at all[19]. The reinforcing nature of such paradigms is discussed by de'Bono in his book *I am Right, You are Wrong* (de'Bono, 1990). In it de'Bono states:

> "Any [science] sets out a scaffold for perception which permits us to seek data which will reinforce [the paradigm]. In all these cases we see a broad type of circularity taking effect" (de'Bono, 1990).

The mere existence of unsolved puzzles within a paradigm does not constitute a crisis. Indeed, Kuhn recognises that paradigms will always encounter anomalies and apparent falsifications. It is only under special sets of socio-cultural conditions that these difficulties can develop in such a way as to undermine confidence in the entire paradigm. According to Kuhn, an analysis of the characteristics of a crisis period in science demands the competence of a psychologist and sociologist as well as that of a historian (Kuhn, 1970b). A crisis becomes serious when a rival paradigm appears. The new paradigm will be incompatible with the old one. Each paradigm will usually regard the world as being composed of different kinds of things and thus regard different kinds of questions as being meaningful. Kuhn argues that there is a sense in which proponents of rival paradigms "are living in different worlds" (Kuhn, 1977). In this sense, he claims that rival paradigms are incommensurable (Kuhn, 1977). A scientific revolution occurs when the relevant scientific community, as a whole, abandons the old paradigm in favour of the new one.[20]

Some aspects of Kuhn's writings give the impression that his account of the nature of science is a purely descriptive one, that is, he aims to do

---

[19] And those that are aware that they carry a frame of reference (or paradigm) through which they understand the world are often unable to completely understand its effects due to its tacit nature.

[20] The notion of 'paradigm', despite its usefulness in explaining the manner in which scientists learn and practice, exaggerates the extent to which a paradigm constitutes a single, coherent, closed system. Since Kuhn defines paradigms as closed systems (employing incommensurable languages), change can only ever be revolutionary – the entire paradigm must be totally accepted or totally rejected. Popper criticises the idea of incommensurability in his essay *The Myth of the Framework*, where he defines the myth as follows: *"a rational and fruitful discussion is impossible unless the participants share a common framework of basic assumptions, or, at least, unless they have agreed on such a framework for the purposes of discussion"* (Popper, 1994). Whilst I am in general agreement with Popper's critique of radical incommensurability, I am sceptical of Popper's characterisations of 'rational' and 'fruitful'. According to Popper, some frameworks are intrinsically superior to others and the 'fruits' of a 'rational' discussion between parties will be the discovery by one party of the hitherto unknown superiority of the alternative framework. According to Popper, paradigms can be compared (and hence ordered) on purely rational grounds. It is this sort of rationality that Wittgenstein (1953) critiqued in *Philosophical Investigations*, Rorty (1979) in *Philosophy and the Mirror of Nature* and Feyerabend (1975) in *Against Method*.

nothing more than to describe the practice of science. However, Kuhn insists that his account constitutes a theory of science because it includes an explanation of the function of the various components (Kuhn, 1963). Certainly, there is something descriptively correct in his idea that scientific work involves solving problems within a framework that is, in the main, unquestioned. A discipline in which fundamentals are constantly brought into question (as characterised by Popper's *Conjectures and Refutations* (Popper, 1969)) is unlikely to make significant progress simply because principles do not remain unchallenged long enough for scientific work to be done. Indeed, it has been suggested that it is philosophy, not science, which comes closest to being adequately characterised by *Conjectures and Refutations* (Chalmers, 2000).

After the publication of *The Structure of Scientific Revolutions*, Kuhn was charged with having put forward a relativist view of scientific progress (Hoyningen-Huene, 1993). That is, Kuhn proposed an account of progress according to which the question of whether a new paradigm is better than the one it has replaced does not have a definitive (or absolute) answer, but depends on the values of the individual, group or culture that makes the judgement. Kuhn was clearly not comfortable with that label and in the postscript to the second edition he tried to distance himself from it (Kuhn, 1970a). However his insistence that science is intrinsically sociological and that understanding science involves "examining the nature of the scientific group, discovering what it values, what it tolerates, and what it disdains" inevitably leads to relativism if it transpires that different groups value, tolerate and disdain different things. This, indeed, is how the constructivist school interpret Kuhn, developing his views into an explicit relativism[21].

Perhaps the most influential exponent of the relativist-constructivist school is Paul Feyerabend, whose controversial account is outlined in his book *Against Method: Outline of an Anarchistic Theory of Knowledge*

---

[21] Social constructivism is the position that affirms that the prevailing scientific paradigm is a construction of the socio-cultural processes at work within the relevant scientific community (as opposed to a representation that has closer correspondence to reality than its predecessor). The beginnings of constructivism can be found in Kant's Copernican tactic of implicating the 'self' in all seemingly objective representations of the world. However, whereas Kant assumed that all human kind wielded the same set of 'cookie cutters' on the dough, the constructivists think that different groups (sometimes linguistic, sometimes social, sometimes disciplinary) will wield different a priori concepts. Given that the mind's a priori categories differ from group to group, each group 'constructs' slightly different worldviews. Furthermore, each worldview is relative to the a priori concepts imposed and not absolute. Applied to science, social constructivists typically claim that electrons, muons, curved space-time etc all exist relative to a particular theory (or discipline) but do not exist relative to past theories (or do not exist relative to other disciplines) and can never be said to exist in any absolute sense.

7. EPISTEMIC HUMILITY

(Feyerabend, 1975).[22] According to Feyerabend, the goal of science is to expand knowledge. Movement toward this goal is facilitated by the rapid generation of theories of all types. Constraints that hinder the generation and consideration of theories must be confronted and removed. Methodological standards are one such source of constraint, they function to (falsely) legitimate some theories and inhibit the consideration of others. The culmination of Feyerabend's case against method is that there is no scientific method and that scientists should embrace methodological pluralism. According to Feyerabend, history reveals that if there is a single, unchanging principle of scientific method, it is the principle that "anything goes" and the only appropriate response is the Maoist one: "let a thousand flowers blossom".

Feyerabend's thesis that there is no scientific method throws much of the work of all who have come before him in the philosophy of science into serious question. As we have seen, one of the principle problematics within the discipline has been the issue of demarcation: the idea that science is demarked from other forms of knowledge by certain methodological criteria. The verificationists and positivists partitioned science from non-science by the existence of confirmatory empirical evidence (verifiability), the critical rationalists, by the possibility of refuting evidence (falsifiability) and Kuhn by the existence of a paradigm able to support a 'normal science' tradition (agreement). In opposition to all who have come before, Feyerabend denies that science is a privileged form of knowing and sees it simply as the sacred superstitions of recent Western culture (Laudan, 1996). The truth, he suggests, is that:

> "Science is much closer to myth than a scientific philosophy is prepared to admit. It is one of the many forms of thought that have been developed by man, and not necessarily the best. It is conspicuous, noisy, and impudent, but it is inherently superior only for those who have already decided in favour of a certain ideology, or who have accepted it without ever having examined its advantages and its limits" (Feyerabend, 1975).

Moreover, if the aim is to progress science through Kuhnian revolution, then according to Feyerabend, we must be willing and able to generate competitor theories to the ruling one. Scientific progress is thus enhanced by theoretical pluralism as well as methodological pluralism. According to

---

[22] *Against Method* was dedicated to Imre Lakatos as "friend and fellow anarchist". The implication being that Lakatos' attempts to find a rational basis for scientific progress through his methodology of scientific research programmes had failed and, accordingly, he had no other recourse but to adopt Feyerabend's 'epistemological anarchism'. Lakatos, for his part, was set to reply to Feyerabend's critique in the same publication (his part entitled For Method), however, his untimely death meant that the proposed joint work, For and Against Method, never eventuated.

Feyerabend, by defining science as an activity that is able to support a 'normal science' tradition, Kuhn seems to be legitimating uncritical dogmatism. Feyerabend labels this the principle of tenacity – the idea that paradigm practitioners tend to tenaciously defend the dominant theories of their disciplines in the face of seemingly falsifying evidence. In contrast, Feyerabend suggests that science adopt the principle of proliferation (the idea that scientists should be encouraged to continuously generate new theories) at the same time as the principle of tenacity. Whereas Kuhn suggests that tenacity and proliferation should govern different phases of research (normal science and pre-science), Feyerabend recommends that they operate at the same time. Thus, theoretical pluralism becomes a feature of normal science.

Whilst Feyerabend is unashamedly relativist-constructivist in regard to theory choice, Kuhn's writings contain two incompatible strands, one relativist-constructivist and the other not. This opens up two possibilities within Kuhnian interpretation:

1. To ignore the relativism and rewrite Kuhn in a way that is compatible with some overarching sense in which a paradigm can be said to constitute progress over the one it replaces.
2. To embrace the relativism within Kuhn's writings and develop a view of scientific progress that looks beyond the cumulative perspective on theory choice.

The first path, which may be termed the logical interpretation of Kuhn, suggests that a purely rational philosophy of science is possible. Paradigms can, therefore, be compared using strictly logical criteria (i.e. it is possible to generate an algorithm for theory choice) and, as such, the prevailing paradigm is always an improvement on its predecessor (in that it is closer to how the world truly is).[23] This is the path of Imre Lakatos and other critical rationalists who associate science with a neutral search for truth. According to the critical rationalists, the products of science are legitimated with respect to objective criteria such as their degree of correspondence with the absolute truth of how the world really is. Such an interpretation must, therefore, legitimate its claims to truth (as correspondence) with respect to ontology (the nature of the 'real', mind-independent, world).

The second path, which may be termed the sociological interpretation of Kuhn, questions the entire attempt by philosophy of science to construct an algorithm for theory choice. Rather, it suggests that a purely logical account of scientific progress is impossible. According to the sociological interpretation, if the replacement of one theory by another requires some

---

[23] According to the logical interpretation of Kuhn, the replacement of one theory by another is always progressive in the sense that the new theory can explain everything that the old theory could and more besides.

sense of cumulative progression, then the bulk of scientific practice must be considered irrational because scientific communities routinely accept new theories that could not explain everything that the old theory could. Moreover, the sociological interpretation suggests that a cumulative view of progress cannot be sustained because different paradigms are either wholly, or partially, incommensurable. The prevailing paradigm, therefore, is as much an emergent property (or social construction) of the disciplinary socio-culture as it is an obvious enhancement over its predecessor. This is the path taken by Paul Feyerabend and other sociologists of science who associate science with a community of inquirers. According to the sociologists of science, the products of science are legitimated with respect to subjective criteria (such as their usefulness, degree of acceptance, simplicity, *etc*) or with respect to relative criteria (such as their degree of coherence with other accepted theories).[24] Such an interpretation is agnostic about ontology (the nature of the 'real', mind-independent, world).

7.6 CONCLUSION: THE DRIFT TOWARDS EPISTEMIC HUMILITY

"Science plays its own game; it is incapable of legitimating the other language games ... But above all, it is incapable of legitimating itself, as speculation assumed it could ... [this] is changing the meaning of the word knowledge. A Game Theory specialist whose work is moving in this same direction said it well: 'wherein, then, does the usefulness of Game Theory lie? Game Theory, we think, is useful in the same sense that any sophisticated theory is useful, namely as a generator of ideas'."
- *Jean Francois Lyotard*

The reader will note that this essay has reached its conclusions without yet suggesting a precise characterisation of 'science' capable of overcoming the problems associated with the characterisations that have thus far been presented. As such, the description of science that seems to be emerging may be objected to on the grounds that it is too vague. Part of the response to that charge is to admit that it is vague, but to argue with Chalmers (2000) that a lack of a precise definition of science (and associated demarcation between

---

[24] David Bloor and the Sociology of Knowledge group at the University of Edinburgh have explicitly pursued the sociological interpretation of Kuhn. The sociologists of knowledge claim that science, as a whole, does not have logical grounds for theory choice. The research from Edinburgh has included personal, professional, social and political interests as part of the set of factors used for theory choice. Indeed, they have produced an imposing set of historical case studies, which they claim illustrates this (Bloor, 1991).

science and non-science) is not a weakness but one of the real strengths of the emerging picture. As Midgley (2000) has stated "we cannot know the exact relationship between knowledge, the language we use to frame knowledge and reality". As such, any account of the relationship between scientific theories, the methodologies we use to generate those theories and the world that those theories are intended to be about must contain a degree of uncertainty.

One of the weaknesses of early accounts of science is that they assumed that there is a single category 'science', and that various domains of knowledge (e.g. physics, biology, psychology, management, history, etc) or that various methods of inquiry (e.g. controlled experimentation, mathematical modelling, critical theory, hermeneutical studies, etc) either come under that category, or do not. As we have seen, the project of philosophy of science has, to date, been unable to establish a single categorisation that can come to terms with the various methods of investigation and socio-cultural processes at work within even the most unambiguous 'scientific' disciplines. It seems, therefore, that no one (not even the most advanced scientistic apologist) can achieve consensus on how to separate the scientific sheep from the non-scientific goats. As Chalmers (2000) points out:

> "There is no general account of science and scientific method to be had that applies to all sciences at all historical stages in their development."

Science, it seems, is a mixed bag and could possibly only ever be seen through Wittgensteinian lenses – as a family of activities with various similarities. Many features are common to many disciplines, but no set is definitive. Moreover, no process of inquiry, thus far suggested as the criterion of scientificity, has been able to withstand serious criticism. As such, it seems that the boundary between 'science' and 'non-science' is vague, subjective and value-laden.

Despite this state of affairs, however, science (as opposed to non-science) has achieved considerable prestige in contemporary society. Indeed, the term 'scientific' is commonly used as an epistemic honorific. However, if science cannot be partitioned from non-science, then the use of such an honorific is hard to justify. Indeed, the very idea of there being a thing called 'science' comes into question.

So where then does legitimacy reside?

Certainly not in acclaiming (or denigrating) items of knowledge because they conform (or don't conform) to some homespun criterion of scientificity. As Lyotard (1979) has argued:

"Science plays its own game; it is incapable of legitimating the other language games ... But above all, it is incapable of legitimating itself, as speculation assumed it could".

Each area of knowledge must, therefore, be analysed separately by investigating its aims, the methods used to accomplish these aims and the degree of success achieved. It is contended here that any area of knowledge (whether traditionally categorised as scientific, or not) should stand, or fall, by the degree to which its methods have been able to achieve the aims it has set for itself and whether these aims are in anyway useful or interesting. Such a position associates science with questions of value and puts it on the road to epistemic humility. Indeed, epistemic humility is virtually implied by the theory ladenness of observation. If our knowledge of the world is always filtered, interpreted and (in important ways) 'constructed' by our *a priori* faculties then we can never know things as they truly are and we are forced to accept a degree of humility with respect to our 'scientific' pronouncements.

This realisation, as readers of this present book would be aware, is contiguous with West Churchman's realisation that our knowledge of systems is always filtered, interpreted and (in important ways) 'constructed' by our *a priori* boundary judgements and, as such, we are forced to accept a degree of humility with respect to our 'systemic' improvements.

REFERENCES

Andersson, G., 1994, *Criticism and the History of Science: Kuhn's, Lakatos's and Feyerabend's Criticisms of Critical Rationality*, Leiden, New York, USA.
Ayer, A., 1959, *Logical Positivism*, Free Press, New York, USA.
Bayes, T., 1763, "Essay towards solving a problem in the doctrine of chances", *Philosophical Transactions of the Royal Society of London*, **53**:370-418. Reprinted in *Biometrika* 45: 293-315, 1958.
Bell, J., 1994, *Reconstructing Prehistory: Scientific Method in Archaeology*, Temple University Press, Philadelphia, USA.
Bhaskar, R., 1986, *Scientific Realism and Human Emancipation*, Verso, London, UK.
Bloor, D., 1991, Knowledge and Social Imagery. University of Chicago Press, Chicago, USA.
Bohr, N. (1928) "The quantum postulate and the recent development of atomic theory", *Nature*, **121**:580-590.
Bohr, N., 1935, "Can quantum mechanical description of physical reality be considered complete?", *Physical Review*, **48**:696-702.
Burke, M.M., 2000, *Thinking Together: New Forms of Thought Systems for a Revolution in Military Affairs*, DSTO Research Report (DSTO-RR-0173), Edinburgh, Australia.
Carnap, R., 1928, *The Logical Structure of the World*, University of California Press, Berkeley, USA. This edition first published, 1967.
Cassirer, E., 1923, *The Philosophy of Symbolic Forms Volume 1: Language*. Translated by Manheim, R. Yale University Press, New Haven, USA. This edition first published, 1953.

Cassirer, E., 1925, *The Philosophy of Symbolic Forms Volume 2: Mythical Thought*. Translated by Manheim, R. Yale University Press, New Haven, USA. This edition first published, 1955.

Cassirer, E., 1929, *The Philosophy of Symbolic Forms Volume 3: The Phenomenology of Knowledge*. Translated by Manheim, R. Yale University Press, New Haven, USA. This edition first published, 1957.

Cassirer, E., 1942, *The Logic of the Humanities*. Translated by Manheim, R. Yale University Press, New Haven, USA. This edition first published, 1961.

Chalmers, A., 2000, *What Is This Thing Called Science?* 3rd ed., University of Queensland Press, St Lucia, Australia.

Checkland, P., 1981, *Systems Thinking, Systems Practice*, Wiley, Chichester, UK.

De Bono, E., 1990, *I am Right, You are Wrong: From this to the New Renaissance, From Rock Logic to Water Logic*, Penguin, Harmondsworth, UK.

Einstein, A., 1916, "Die grundlagen der allgemeinen relativitätstheorie" (The foundation of the general theory of relativity), *Annalen der Physik*, **49**:769-822.

Feyerabend, P., 1975, *Against Method: Outline of an Anarchistic Theory of Knowledge*, New Left Books, London, UK.

Friedman, M., 1992, "Philosophy and the exact sciences: Logical positivism as a case study". in: *Inference, Explanation, and other Frustrations: Essays in the Philosophy of Science*, J. Earman, J., ed., University of California Press, Berkeley, USA.

Friedman, M., 2000, *A Parting of the Ways: Carnap, Cassirer and Heidegger*, Open Court Publishing Company, Chicago, USA.

Fumerton, R., 1992, "Phenomenalism", in: *A Companion to Epistemology*, J. Dancy & E. Sosa, eds., Blackwell Companions to Philosophy, Oxford, UK.

Goodman, N., 1954, *Fact, Fiction and Forecast*, Harvard University Press, Cambridge, USA. This edition first published, 1983.

Hazelrigg, L., 1989, *Social Science and the Challenge of Relativism Volume 1: A Wilderness of Mirrors – On Practices of Theory in a Gray Age*, University of Florida Press, Gainsville, USA.

Heisenberg, W., 1927, "Uber die grundprinzipien der quantenmechanik", *Forschungen und Fortschritte*, **3**:11-83.

Hempel, C., 1945, "Studies in the logic of confirmation", *Mind*, **54**:1-26.

Hempel, C., 1965, *Aspects of Scientific Explanation*.

Hempel, C., 1966, *Philosophy of Natural Science*, Prentice Hall, Englewood Cliffs, USA.

Hollway, W., 1989, *Subjectivity and Method in Psychology: Gender, Meaning and Science*, Sage, London, UK.

Hoyningen-Huene, P., 1993, *Reconstructing Scientific Revolutions: Thomas S. Kuhn's Philosophy of Science*, University of Chicago Press, Chicago, USA.

Hume, D., 1748, *An Essay Concerning Human Understanding*, Reprint Hackett Publishers, Indianapolis, USA. 1977.

Kant, I., 1781, *Critique of Pure Reason*, translated by Kemp Smith, N. Macmillan, London, UK. This edition first published in 1929.

Kuhn, T., 1962, *The Structure of Scientific Revolutions*, University of Chicago Press, Chicago, USA.

Kuhn, T., 1963, "The function of dogma in scientific research", in: *Scientific Change: Historical Studies in the Intellectual, Social and Technical Conditions for Scientific Discovery and Technical Invention, from Antiquity to the Present*, A. Crombie, ed., Heinemann Educational Books, London, UK.

Kuhn, T., 1970a, *The Structure of Scientific Revolutions*, 2nd ed., University of Chicago Press, Chicago, USA.

Kuhn, T., 1970b, "Logic of discovery or psychology of research?", in: *Criticism and the Growth of Knowledge*, I. Lakatos, & A. Musgrave, eds., Cambridge University Press, Cambridge, UK.

Kuhn, T., 1977, *The Essential Tension: Selected Studies in Scientific Tradition and Change*, University of Chicago Press, Chicago, USA.

Kuhn, T., 1998, "Objectivity, value judgement and theory choice", in: *Scientific Knowledge*, J. Kourany, ed., Wadsworth, Belmont, USA.

Kuhn, T., 2000, *The Road Since Structure*, University of Chicago Press, Chicago, USA.

Lakatos, I., 1970, "Falsification and the methodology of scientific research programmes", in: *Criticism and the Growth of Knowledge*, I. Lakatos & A. Musgrave, eds., Cambridge University Press, Cambridge, UK.

Lakatos, I., 1977, *Philosophical Papers, Volume 1*, Cambridge University Press, Cambridge, UK.

Laudan, L., 1996, *Beyond Positivism and Relativism: Theory, Method and Evidence*, Westview-Harper Collins, Boulder, USA.

Midgley, G., 2000, *Systemic Intervention: Philosophy, Methodology and Practice*, Contemporary Systems Thinking Series, Kluwer Academic/Plenum Publishers, New York, USA.

Newton, I., 1704, *Opticks*, B. Cohen, A., Einstein & E. Whittaker, eds., Dover Publications, New York, USA. This edition first published, 1952.

Peirce, C., 1905, "What pragmatism is", *The Monist*, **15**:161-181.

Polanyi, M., 1973, *Personal Knowledge*, Routledge, London, UK.

Popper, K.R., 1969, *Conjectures and Refutations*, Routledge, London, UK.

Popper, K.R., 1972, *The Logic of Scientific Discovery*, Hutchinson, London, UK.

Popper, K.R., 1974, "Darwinism as a metaphysical research programme", in: *The Philosophy of Karl Popper*, P. Schlipp, ed., Open Court, La Salle, USA.

Popper, K.R., 1979, *Objective Knowledge: An Evolutionary Approach*, revised edition, Oxford University Press, Oxford, UK.

Popper, K.R., 1994, *The Myth of the Framework: In Defence of Science and Rationality*, Routledge, London, UK.

Quine, W.V.O., 1953, "Two Dogmas of Empiricism", in: *From a Logical Point of View*, Harvard University Press, Cambridge, USA.

Ray, C., 2000, "Logical positivism". in: *A Companion to the Philosophy of Science*, W.H. Newton-Smith, ed., Blackwell Publishers, Oxford, UK.

Reichenbach, H., 1920, *The Theory of Relativity and A-Priori Knowledge*, University of California Press, Berkeley, USA.

Rorty, R., 1979, *Philosophy and the Mirror of Nature*, Princeton University Press, Princeton, USA.

Russell, B., 1912, *The Problems of Philosophy*, Oxford University Press, Oxford, UK.

Salmon, W., 1970, "Bayes theorem and the history of science", in: *Historical and Philosophical Perspectives of Science*, R. Stuewer, ed., University of Minnesota Press, Minneapolis, USA.

Schlick, M., 1915, "Die philosophische bedeutung des relativitatsprinzips", *Zeit Fur Phil. Und Phil. Kritik*, **159**:129-175. Translated by P. Heath, 1978, in "Moritz Schlick", *Philosophical Papers*.

Schlick, M., 1918, *General Theory of Knowledge*, Translated by A. Blumberg, 1974.

Trout, J., 2000, "Paradoxes of confirmation", in: *A Companion to the Philosophy of Science*, W. Newton-Smith, ed., Blackwell (Blackwell Companions in Philosophy), Oxford, UK.

Wittgenstein, L., 1953, *Philosophical Investigations*, Translated by G. Anscombe, Blackwell Publishers, Oxford, UK.

SECTION C

SETTING THE TONE: THE IMPORTANCE OF ETHICS

Chapter 8

IN SEARCH OF AN ETHICAL SCIENCE
An Interview with C. West Churchman An 80th Birthday Celebration

VAN GIGCH, J. KOENIGSBERG, E. DEAN,B.

Abstract: In 1993, we celebrated C.W. Churchman's 80th birthday. On this occasion, we asked Dr. Churchman to tape an interview, the transcription of which is printed below.

In this article we will not review Dr. Churchman's many accomplishments. Another journal (*Interfaces*, a TIMS/ORSA Journal, 24 (4), July-August 1994) is presenting a list of his publications. Suffice it to say that Dr. Churchman's ideas are of fundamental importance to the management community. I would characterize Dr. Churchman as a philosopher and an epistemologist. His thinking reflects a deep understanding of the sources of knowledge for the management discipline. Often, his colleagues have considered his writing to be esoteric and difficult to understand. However, once you overcome the" churchmanalia", you discover very rich ideas which, without any doubt, will become classic reading. Right at present, he is involved in a project to formulate the outline of an Ethical Science. This subject is not new; he has always stood for the conscience of management and for the morality of systems. We hope that this small tribute is a demonstration of our deep affection for a friend/colleague and our unbounded admiration for a great thinker. He has many admirers throughout the world and has been awarded honorary degrees from two Swedish University, at Umea and at Lund.

This interview was conducted on the campus of the University of California at Berkeley on April 30, 1993. The interviewers were: **John P van Gigch** (JvG), California State University, Sacramento, CA. (Also convenor of interview and editor of proceedings); **Ernest Koenigsberg** (EK), University of California, Berkeley, CA; **Burton Dean** (BD) San Jose State University, San Jose, CA. [1]

---

[1] This article appeared in *Journal of Business Ethics*, 16: 731-744, 1997. Copyright by Kluwer Academic Publishers, The Netherlands and UK.

We acknowledge the contribution of the W.A. Haas School of Business Administration, UC Berkeley which provided the venue for the interview as well as funds for incidental expenses. Ms. Chris Otis carried out the transcription.

Key words: Churchman C. West (1913-2004), interview; ethical issues, values; career achievements

INTERVIEW WITH C. WEST CHURCHMAN (CWC)

Participants: John P. van Gigch (JvG), Koenigsberg (EK), and B. Dean (BD).

JvG: Can you tell us what you consider is the biggest achievement of your career?
CWC: I've been thinking about the answer to that question since I received a copy of the questions for this interview. It is difficult to pick out a date and say that on that March date of such and such a year, I discovered something of horrendous importance. I've had a single purpose life.

I attended a Quaker school in Philadelphia. What I got from the Quakers was the knowledge that you can have a life dedicated to humanity. That was the best thing one could do. At the age of seventeen, I began keeping a journal and in that journal it says what I would do. At the time, I was a freshman at the University of Pennsylvania. The question was not, to what should I devote my life, but what course would be important to my major.

It has been my life's ambition, trying to figure out the nature of the human species and why it leads such a miserable life. We are endowed with intelligence, a sense of humor, caring, love, and al the rest of it, yet, I would estimate that more than 90% of human beings live lives that could easily be described as miserable. And I haven't changed on that ... I'm still struggling with this issue at the age of eighty. What I have come to realize, in the last five years, is that there is a lot of literature on ethics of humanity and lots of scientific literature on humanities and lots of literature on how to reduce the misery of human beings, but one of the characteristics of the human species is that it does not have the capability of transferring the written word into action. That's amazing. Here are all these bright ideas and no clear suggestion on how we go from sensible arguments to any kind of action.

When I got into Operations Research and Management Science, I was naive. I thought that the precision those two areas promised in looking at problems and trying to understand management, would carry with it, an acceptance on the part of management. I really did believe that we could

8. IN SEARCH OF AN ETHICAL SCIENCE

trace the problem of implementation of research findings, by undergoing a really drastic change, even in the language of management, and introduce a mathematical, more precise way and then use measurements.

In the early '60s, the students and I, here at UCB, wrote to the authors of 13 articles in journals of Operations Research and Management Science and asked them, "Dear author, you wrote a brilliant article on inventory (or queuing or whatever the topic was) and we're so interested, we want to know what you did about it? Did management accept your research and could you then see how successful you were in practice?" There was only one author who even had an idea about what happened. We were very anxious about implementation. The other 12 did not. So implementation was not taken seriously.

Then, I decided to find out how general this was and ran some experiments. I got five 5 MBAs running a little business with 3 products. They had to decide on the price of each product, the scheduling of each process, the production schedule, and the way in which they would respond to demand. There was an ideal solution, so there was no excuse. The MBAs should've been able to work out the mathematics. They could determine the mathematical model and derive the optimal solution.

We ran the experiment 40 times and, with two 2 exceptions, not one individual implemented the solution which was *told* to them. We primed one member and told him the solution and he tried to tell it to the others.

JvG: You even had some stooges there, didn't you?

EK: It says something about students ...

BD: It could also say something about the difference between the "real world" and the MBA world.

CWC: Well, we came into the MBA world. The MBA world is made up of unreal problems that have solutions and that is what we gave them. But, that was not what was blocking the managers.

Our result was the same as that which you'd find in the real world of practice of Operations Research and Management Science. Nobody was using the solution, yet companies were using a lot of money to find these solutions.

JvG: I was really moved by your intent when you said that "I had a single purpose life", and then ... I find that there's some contradiction between the purposes of managers and the ideals of academicians. We don't teach students in management to pursue the lofty purpose of saving humanity. In other words, management doesn't have this lofty purpose as its goal. Therefore, I don't understand how you came to a school of business administration, and pursued your life-long ambition to inculcate your philosophy in this setting.

CWC: Well, when I began my journal at seventeen, I put down a list of maybe 5 to 7 possible majors and ended up with philosophy, because that seemed to me to be the discipline that was most interested in the broadest possible view of humanity.

JvG: *Yes, I can understand that.*

CWC: I could understand that myself, but it was a lie. Today's philosophy departments are not interested in the misery of humanity; they're interested in the misery of philosophy.

Anyway, my journals didn't address that question. They were particularly interested in ethics from the point of view of its accuracy of its definitions and how you verify ethical imperatives, but they did not have any interest whatsoever in applying it. I didn't even see the world of application. I knew nothing about management. It gradually dawned on me that it's the world of management that I need to be in. If this is what I want to do, to improve the human condition, where am I going to do it, but in the world of management?

JvG: *So really ... within this word "management" you encompass much more than is encompassed in, probably, the mission of a school of business. You're really talking about bringing to society and applying to society everything you know and you can do about social ills.*

CWC: It was really World War II that saved me from philosophy departments. At that time, we academics really had a choice of trying to stick to the university or go out and do something and be a volunteer for military service or do research. I elected to go into a laboratory and do research. I was performing mathematical statistics. My boss said, "What I would like you to do is to be as sure as possible that the ammunition we produce here will fire when the GI pulls the trigger of his gun anywhere in the world". That was my first introduction to a true management problem.

JvG: *How old were you then?*

CWC: Twenty-eight.

BD: *Following up on that, isn't this a problem of quality as well as saving soldier's lives? Quality being that the firing mechanism would work properly. Did you have some contact with people who were concerned with quality problems?*

CWC: The originator of the whole notion of quality is not Dr. Deming, but Walter Shewhart, at Bell Telephone Labs, who was the developer of statistical quality control in manufacturing.

I went to Bell Labs several times. We used to explore together whether or not the issue of quality could or could not be applied to the kind of management I was hunting for. "Total Quality" has only fairly recently come to the attention of management schools, but it goes back to Schewhart before World War II. He was the founder of Statistical Quality Control. Part of my

8. IN SEARCH OF AN ETHICAL SCIENCE

job during World War II was to go around to the plants and introduce Statistical Quality Control so that the machines that were manufacturing were kept in statistical control.

But, then, of course, that's only part of my boss' question. The question involved, not only to make sure that soldiers were trained and that ammunition was shipped, but also the reason for the war.

EK: I found it very interesting that when defining quality control in a plant we refer to controlling the machine and seeing that the machine is in conformance. The whole new modern phase of statistical control says that you don't control the product, you control the process. But, you and Schewhart saw that in the '40s. Somewhere along the way we lost it, and, somewhere along the way again, we've re-discovered it.

CWC: I found out that management has a type of problem that no academic would even dream of taking on. This applies to shirts, to automobiles, to anything else. I found out that what you want to do is to make sure that the product, once it gets in the hands of the consumer, is used safely and ethically. And ethics was the main thing I cared about. I sometimes felt that I was the only one in my time who cared about that at all. There's a final justification that when the GI sees a sniper up in the tree and pulls the trigger, that the GI knows that he is doing an ethical act. Now, nobody told me to do that, but that's what I understood. And that's management. I then discovered where I needed to be.

After the war was over, I went back into the philosophy department and found that a majority of the department had no interest in the action part of philosophy. The response I got from the philosophy department was that they turned down my Ph.D. candidates. That's where you hit a professor hardest. You can't get to his salary very well. You can't overload him with courses, but you can kill his Ph.D. candidates.

JvG: I'd like to go back to the concept of ethics because I think that listeners and readers will see a contradiction here. You said that to encompass, in the concept of ethics, the act of the soldiers pulling the trigger is in conflict with your concept of ethics. Maybe you can explain what you mean.

EK: I think it's your (indicating JvG) concept of ethics that conflicts with his.

JvG: Of course ... the accepted concept of ethics.

CWC: Where were we pacifists in the 1940s? There was a Hitler who was on the rampage, taking over countries, suppressing Jews ... a figure that was a threat to the ethics of the world ... a dangerous man. Now, we're witnessing Dunkirk. The English Army was crossing the English Channel in rowboats. Nothing was going to stop them.

When Paris fell, we thought it was the end of Europe. What's a pacifist going to do? Above all, what is a pacifist? Above all: "The world must not fight". We couldn't say that that year. Above all, we had an evil of tremendous magnitude. My response was to go and work on bullets, steel, and all the other things, from a statistician's point of view.

JvG: Is it safe to ask you what is different today, in Bosnia?

CWC: We've learned a lot since 1940. In the Yugoslavian states, things are different. We admit that. Bosnia is not Hitler. It is something very different.

BD: You're an advocate of science. And using your example of Hitler and World War II, the German science used in concentration camps was not ethical. We used science to counter the German threat. My question is: Do you think science can and should be used ethically? Is it possible?

CWC: That's the biggest question of the interview. Can science improve the human condition, ethically? So far it hasn't done so. Science in the form of technology, in all its forms, has never been ethical, if, by ethical, you mean that all human beings, who have the ethical need for the technology, are served. Take agriculture, for example, which has some of the oldest technologies. Look at the state of agriculture today, where, even with impressive technology, in excess of 35,000 children die from starvation or starvation-related diseases every 24 hours. Compare that with World War II, where, every day, 7,500 military were killed or declared missing. Our war against the children, through the technology of agriculture, is four or five times more severe than the military war was. That's true of all the technologies that we have.

Kirk Smith, who conducts social research on the Pacific Rim, tells me that, by 1999, 25% of the homes will be electrified; 75% won't. Is that serious? You bet it is. Non-electrified homes get heart and energy through inadequate wooden stoves that contaminate the air which then creates something worse than the worst kind of air pollution Los Angeles ever had.

That's characteristic of all the technologies, including medical technologies. To day's science does not serve humanity. But there is such a thing as a science that could, which is what I am trying to promote today. What would that science be like? For one thing, I would tear down all the walls between the disciplines. For example, organization theory was developed in an institution that is badly disorganized. If we were operating sensibly, we would make sure that all the disciplines are working together. We all share the same kinds of problem.

EK: There's a line in <u>Oklahoma,</u> "I'd like to say a word for the farmer," and that's what I'd like to do. You're placing the blame on agriculture. The fault isn't on agriculture, but how we distribute agricultural products and that's part of organization theory. That's where we fail. We know how to do

8. IN SEARCH OF AN ETHICAL SCIENCE 147

a lot in farming, but we don't know how to get it from the farmer to the mouth.

JvG: *Isn't this true for every field? We know we have technologies, but ...*

EK: *We say it is not the farmer. It's the system we use to go from the ground, from the farmer to the mouth.*

CWC: I didn't say "farmer;" I said "agriculture". Agriculture goes all the way to ingesting and digesting the food. I do not separate production from consumption.

EK: *You need to have a definition of the agricultural system that includes the full cycle. If you include all the parts, like consumption, that's okay.*

CWC: You're a systems guy. How can you separate the farming from the consumption?

EK: *Many people do.*

BD: *In your models of science, application and methods, what is the role of government in solving problems? Is there a separate role of government beyond science?*

CWC: When I was in the Quaker school, they impressed on us that the need of humanity was for a world government. They wanted to get rid of the nations. The nations would be like states of the U.S.A. That impressed me a lot as a young man. One of the big problems was the national government. Since I have been in the Business School on this campus and other places, I have come to realize that what they were saying had the potential to be a solution to the management of the world, but not necessarily the right one. The problem is with implementation. What would it take to make one world government? Would you call for the obliteration of unsatisfactory human beings? Hitler's aim was the same thing; they're doing the same thing in Bosnia now.

If that's the aim, and you would do it through the implementation of a world government, "the hell with it." One of the things I've found about implementation is not that it fails, for it can be put into laws, but the failure is with the way the law is implemented, which is unethical. The implementation destroys the ethical idea. That was the way it was with Hitler. I've seen it over and over. The U.S. government is not democratic; it does not even approach democracy.

BD: *What would it take, what changes would we need in order to make government ethical and, therefore, make science ethical?*

CWC: There's already lots of it around. Russ Ackoff believes in interactive management. He's working to test, in practice, his idea of a democracy in a corporate environment, of all places. That's the last place that you would expect to find democracy.

JvG: *After reviewing your proposal, it occurred to me that you would have to introduce some kind of popular socializing democracy to replace the capitalistic style that currently exists. Can you explain your concept of this democracy that you visualize?*

CWC: I can explain some of the characteristics but not how to do it. That would be stupid and I have no idea how to do it. I'd be afraid that if I made up an idea somebody would implement it.

JvG: You have taught us about these wonderful ideas, including implementation. If you have these characteristics of a world government, we should start asking ourselves is it an ideal that can be implemented or a pipe dream or something only in your head? I'm moved by your intentions, but then I ask myself, "is this possible?"

CWC: I began as an idealist. My teacher was Edgar Singer. He said: "First state your ideal, what you think is the ideal world". We broke it down that way in the business schools and taught the problems of consumer goods and services, then marketing, etc. We think of the ideal way, with all the technology and how it could be used and managed, how to work towards it. Cooperation brought it about. That's what my first books were about - ideals. Then you need to study the problem, how to work towards the ideal and measure it. Measurement was the ideal.

For example, look at history in the' 18th century. When the physicists first started measuring the velocity of light in a vacuum, they could measure it to within several kilometers per second. Now they can come within 0.1 kilometer per second. That's progress. There is no way to get to zero but we could begin to approximate zero. Singer thought that approximation could be true of every idea.

The trouble with idealism is it's not very good. It fails to describe what needs to be done to take one more step. In a laboratory setting it's different. It's easy to describe what to do next, but with human beings, it's different. I have no idea how.

EK: In the early days of the Industrial Revolution the Quakers were idealistic managers of industry. For instance, they were the owners of Roundtree and Cadbury, the chocolate business. They saw their role as bettering the life of people. They improved the lives of their workers. For the times, they built comfortable homes for their workers; paid wages the people could live well on. At the same time, there were large profits to be made by doing good. They were doing well. But all that disappeared. Even Quaker places are not owned by Quakers anymore. How do you feel that the Quakers missed something in passing on this tradition?

CWC: It is not only the Quakers. In the 19th century, my idea of improving the human condition was prevalent, especially in this country. The notion of going out and setting up a commune, where you wouldn't be interfered with, was a popular one. But, not a single one of these communes lasted. This same spirit was present at the beginning of the kibbutz. In this

8. IN SEARCH OF AN ETHICAL SCIENCE

situation, you are your own isolated world and you make it the best you can. Then, you start allowing the outside world to impact on it. When that happens, you have to wonder how long the original community will keep its own characteristics. Within five to ten years most of these communities disappeared.

Why? Why have these ideal communes and the Quaker community disappeared? Now that's an interesting and workable scientific problem. "Why?", I ask you. The answer I get is "greed" sets in.

For the scientists, that's the end of the question. If we were working in a laboratory and I said, "Something about the lighting is influencing our experiment," you'd say, "Let's test it out," and we'd get excited about it. But, if you say "greed," you say ugh, and that's the end, you have popped the balloon.

JvG: Not popped ... it's just that the whole system that we have here is based on greed.

CWC: Why don't we study human greed? Greed is an addiction. One characteristic of all addicts is that there is never enough, E-N-O-U-G-H. Never enough. Like eating, do you need to eat everything? No, you say stop. But an addict cannot stop eating. He needs to consume all of it.

BD: You've indicated several attributes of misery, the latest now being greed. Have you thought that if you could change things, what would you change? What choices would you have made differently?

CWC: Practically nothing. There are lots of incidents where I shouldn't have been engaged in this or that. I shouldn't have been the first editor of *Management Science* because it was a waste of time. In the first volume you can see what I had hoped would happen. I hoped there would be a science of management. What I got as the editor, more and more, was mathematical model-building. I would have liked that to have been different. I would have liked to publish a true science of management journal that says we don't know what management is about. Let's start with what we've got and make that question our study and try to use our results, and not make it all mathematical. Management theory was turned over to mathematicians.

If I could change anything, I would like to be born now because I have an exceptional idea, mainly, how do you create a <u>science that will help the human condition.</u>

I want to start with logic. That was my first discipline in my thesis. I want to start with logic. Most of symbolic logic is of no particular use in management except set theory. Set theory begins with, "If all As are. Bs and all Bs are Cs, then all As are Cs." The question is, when is it the case that all As are Bs? Under what conditions does an item belong to a class? I taught logic for years and students never asked the right question. You, as a manager, want to know whether to hire this guy or this girl That's the question. Does that guy or girl belong to the class of people who should be hired? You want to know whether they have certain characteristics that will help you as a manager. I want to redo the foundation of mathematics.

BD: An emerging field is 'fuzzy logic." It seems to have an increasing number of applications. Do you feel that fuzzy logic can be applied to some human misery problems?

CWC: What Zadeh did was to become precise about fuzziness, which is something that no manager is able to do.

JvG: He fell into the same trap as operations researchers. He became precise about fuzziness. The error of operations researchers is to not recognize that, as you enlarge the slice of reality with which you are working, the problem that you are solving becomes more complex and, hence, less precise. We attempt to be more precise and very exact about problems, when the world isn't precise. It's exactly the opposite.

CWC: I have lived a life with all these theories. Chaos Theory, for example, is an attempt to be precise about chaos - How can you be precise about chaos?

EK: What they attempt to do is to define chaos so that they can be precise about it.

JvG: But, is that the way to go?

CWC: No.

EK: For some things.

CWC: Have you lived through a serious earthquake yet? You can't be precise about what it feels like to live through an earthquake.

EK: To know what you don't know ... when you can't be precise ... don't pretend to know that it's a rough approximation. But, when you know that you have found a way to put human feelings in a numerical scale, and that's more important than anything else, you've missed the whole point of every research.

CWC: To do good science you must have an essential qualification and that's humility. A lot of what's happened in this area has been numerical or mathematical programming where there's a great deal of competition to be successful. There is a lot of self pride. The most humble scientist I know of was Einstein. He wrote in a book, "I can't understand why they thought I was any good. I was just like any other scientist who had to puzzle things out. I happened to have had suggestions to make."

I don't have it (humility). I have lived a life struggling to get the Nobel Prize. This has been a real blockage for me. This place has disappointed me because I didn't get the promotions I thought I should. I didn't get as many honorary degrees as I thought I should. It's all part of a characteristic I had, in carrying out my work, that showed a lack of humility.

JvG: ... which, in a way, is a little bit like greed.

8. IN SEARCH OF AN ETHICAL SCIENCE 151

CWC: Greed is characteristic of fame. How do you get to be famous? Win the Nobel Prize.

JvG: But the whole world revolves around the notion of greed. We have to succeed, we have to achieve, get there.

BD: But is that true of both the western and eastern worlds?

EK: Maybe what we can observe of the eastern, maybe not.

CWC: Look at the influx of the gurus in California. They were not modest. Experts on spiritual life came here, made up Transcendental Meditation for which you had to pay $75. That's hardly humility.

EK: There is a lot of folklore about Japan and how they look at things differently. Their way of life may or may not be more like what they've written about it.

BD: Are there any managers that you admire?

CWC: The head of ARCO, the President or CEO, had the right idea. Part of what you're talking about when you talk about human misery is the notion of hierarchy. Hierarchy is part of the misery. People who get high on hierarchy begin to lose their sense of humanity. They think they are superior.

I have a friend who was head of a firm in Stockholm. He figured he shouldn't be the CEO, the CEO idea was wrong. In fact the whole building where he worked was symbolic of it all. It had the lower-level employees on the lower floors and increasingly higher levels of management on the higher floors - which was wrong. It was not his idea of an ethical firm. Later, he wrote a book about a company that operates ethically.

JvG: Are there any organizations, worldwide, that represent a different paradigm?

CWC: That was the original idea of science. Alcoholic Anonymous (AA) - there's no hierarchy there. It's an upside down pyramid. There is no central office. Other self-help and non-profit organizations are like this. Non-profit organizations haven't received nearly enough attention in business schools.

JvG: Some of them became "successful" promoting their goals, but then they became greedy, and more like big business. Fundraising became a big business. They lost their altruistic nature.

EK: There is a saying, "business is business." Maybe it can't be ethical.

JvG: The way business is organized today, by definition, it cannot be ethical, at least not in the way that West conceives of ethics.

EK: It can in some areas, where they show humanness. But that is more than likely a private company, where they help their workers. You won't find that in public companies.

JvG: That's true, when the private companies have to compete for a share of the market. Then the battle starts.

EK: That happens because the competition plays dirty, is doing illegal things. I don't know how an "ethical" company can stay in business. At least in the way that West conceives of ethics.

CWC: That's where you get carried into another area. If you go back to the Quaker companies, they were trying to help workers. They were not helping consumers. They were not helping those consumers that would never be able to afford their products.

EK: By ensuring that they had quality goods, they were.

CWC: That's another admirable quality, market sense. Look back at Thomas Edison. In the 1870s, Edison had a great idea. An invention was not an individual matter, it was a marketing matter. You have to go searching for the market for the invention. You have to go looking for those that can afford to pay for the product. That's very much the U.S. sentiment.

My ethics are different, Quaker-like. You have to watch out not only for those that can pay. You have to be concerned about those that don't have the ability to pay. How are you going to fill that gap? And that's a big gap. The Edison model is "the ability to pay model".

JvG: This company you're thinking of does not have the same economic goal that drives other companies. They can't. This ethical goal is contradictory to the economic goal.

CWC: What you could say to many producers today, is that they're missing a huge market, those people that want the product, but don't have the ability to pay for it.

BD: What is the role of government with companies that are not satisfying customers? Does government have a role beyond business to ensure that consumers get what they need? Is there a role for government, beyond business?

CWC: You could create that kind of service or role and call it government, but it's nothing like the government I live under. Government has the concept of greed. We elect the most greedy individuals we can find. It's the great American tradition. Otherwise it's communism. Serving the people with the greatest need, that's Lenin or Marxism. The whole spirit of Lenin was the practical man. He wanted a government to implement this idea. This same spirit of communism, serving the needy man, underlies what I'm talking about.

Marx was a revolutionary and was shown to be incorrect. We've seen what happens to that kind of communism. They went to a dictatorship which may be the only way to get anywhere. But is that the way it has to go for all idealists?

You have to hope that there is such a thing as a better human species. This whole idea of hope is not scientific in today's language, but it's a terrific theory. Without hope you have nothing.

8. IN SEARCH OF AN ETHICAL SCIENCE 153

EK: Do you think we'll find a better human species through genetic experimentation?

CWC: That's another fear I have. When I was growing up, pure science was just that, pure. We were looking to discover the truth. I was curious about the origin of the universe. There was no application. It was just pure science in 1945. I later discovered that someone was able to use that.

JvG: In the wrong way.

CWC: Yes. Now look at our world. We lived through Hitler, who claimed he would improve humanity. If he'd known much about genealogy, he wouldn't have obliterated the Jews, he might have tried to make much better humans. Who should produce what children? And I think there are guys out there now who plan to do just that. I don't know what they're up to. Again, under the notion of purity, science is threatening to destroy us. It's frightening.

EK: I have a question about March's work in organizational theory. Are you familiar with it and how do you feel about that work?

CWC: That work is part of an effort, on the part of the business schools, that came when there was an increased interest in the structure of the organization. We knew the meaning, okay, but what about the structure? What parts produce useful organizations? Like most, we never referred to ethics, or when an organization was ethical, which would be, namely, when and how do you reduce the inner disturbances and chaos that are present in so many organizations.

March and Simon's effort was an attempt to put it all together. And it did not, like Peter Druker, bring out problems of management per se. It was more descriptive and it was theoretical. It didn't interest me. It didn't raise crucial issues about life in an organization.

Our courses in the MBA program are centered around March and Simon's work. This is a school of management, giving an MBA. We're teaching people to manage through organizational theory which is not helping them know how to manage an organization. They become good at knowing the facts. But there is a difference between *factual science* and *ethical science*. It seems to me that it is all an attempt to use different ways of *describing* and *classifying* organizations, but it does not address *how to manage* them.

It was in the business school, where I first came across accounting. And I asked, what was its purpose? I was told it was to keep accurate data on the status of a firm for purposes of determining its financial position. Why? Why should we be accurate? What is the point of double-entry accounting: to be accurate. Why didn't we take it one more step? What's the point?

Later, when I was working with the railroads, I discovered that as the accounts were settled every month, there was a way they could save billions of dollars if they used random stratified sampling. It was accurate to 1%, not to a penny! Wow! They were spending millions of dollars on penny accuracy, but no one was asking why accuracy was important. I was stupid and earnest. Where were the savings going to occur? On accountants. They didn't need to hire so many accountants. I got strange reactions to my suggestions of stratified sampling. Some people asked, "We're not going to turn into a gambling outfit are we?"

BD: On the issue of Quality Control and Deming's Theory of Total Quality Control, would you contrast and compare your philosophy with Deming's?

CWC: Ethics. The issue is the same. The center of Total Quality Control is ethics. With Total Quality Control, ethical management is not reached, until the product being produced or service being provided has reached everyone who has a true need for it.

BD: Deming considers the entire environment. He does not advocate evaluating person/lei on their productivity, but is very much an advocate of team production.

CWC: Total Quality Control is mainly oriented toward the product and not toward the ethical nature of the market. I may be unjust about the whole idea of Deming's fit with what I am looking at, but there is nothing about ethics in it.

BD: There are strong similarities between what you advocate and Deming.

CWC: I have just seen one of the latest textbooks on it. I don't see anything about ethics. It's not in the index. Of course, maybe that's the wrong place to be looking for it.

ED: Deming is very concerned with the customer.

CWC: I want the customer who isn't served.

BD: Deming is very concerned about satisfying the customer who has the money to pay Jar the product. He is working with both governments and major corporations. What about the conflict between science and society? Science is interested in producing facts and answers. Government really wants immediate results. Government can't wait as long as science. Is there a conflict between government and science?

CWC: Let's be even more general and look at management and science. Management and science look at things very differently. You can tell because science thinks of solutions, how to formulate problems and solutions. That's the scientific way. No manager would expect to find solutions to problems if he carries the problem out. If he puts boundaries on it, he only comes to an approximate solution. But, it is not even approximate because he does not even know if it is accurate or not.

8. IN SEARCH OF AN ETHICAL SCIENCE 155

I want a science for management, something like the Institute of Management Science (TIMS) which is pursuing an effort to invent a science of management. Look at my boss. He asked me to look at ammunition to guarantee that the ammunition would work and the soldier is justified to pull the trigger. It is a management question but it is also a scientific question. I want a science for management. "Management Science" was a great label TIMS had to invent a science of management which would also be a management of science. I wanted an "X of X". I wanted to see it both ways. As it is, neither modern science nor management can satisfy the ethical demand.

BD: In your book, you call for an ethical management science. Is this possible in a non-ethical society? It seems like an ethical society would be a necessary condition for this to occur.

CWC: We would have a different meaning of science, a different meaning of management.

BD: But the society in which both operate is non-ethical. You have greed, war, hostility, etc. The environment if science is not ethical. How can science itself be ethical?

CWC: Here's a suggestion. Greed is a disease just like any other disease. It comes about because of [dysfunctional] interrelationships between human beings. Can the disease be lessened?

I'm an alcoholic, so I know that there is a way to overcome an addiction. For me, I never seemed to have enough to drink. The word E-N-O-U-G-H meant a lot to me. That's also characteristic of wealthy people, there's not enough wealth to be saved. It seems like an addition. Look at it this way, we may be able to get over it, not see people getting as wealthy as possible, and not have a society of greed, greed for fame, for political power, etc. That's a conquerable disease.

I want a science that's going to help me find out about that disease. But it won't look like today's science. It will not come from a collaboration of different disciplines. It will become a coordinated science where the experts are not today's experts of rigorous science but of a science whose main concern is the service of humanity. It won't look like today's science.

JvG: How can this science exist in the context of this society? Will we have to revolutionize our organizations and the institution if science to implement this program?

CWC: It's not that difficult. What if you were to set up a research organization in the middle of Los Angeles? Students go out to work on city problems and become part of the course. Some of the prime researchers are inhabitants of that city. They know things others don't, things that no urban planner knows. It's the same principle as AA. The main research is done by alcoholics to research their own addiction. They are the researchers.

I'm not talking about our current hierarchy of knowledge and human science. This new effort is starting already. Like AA, there are over 200 different kinds of twelve-step programs around that are all based on the same principle of the twelve steps of the AA approach to the problem.

The new science is coming, and we can get it. It will be in existence. The universities will be competitors with that science. That's what I want, but it will be institutionalized and then overrun with bureaucracy and then ruined!

BD: Let me pick up on your point of the research organization in Los Angeles, using the Total Quality Management (TQM) idea as a basis for the question. One of the basics of Total Quality Management is that you have teams of workers who know the process. It's not the traditional scientific approach. Workers are multifunctional in that they know the production process, accounting, marketing, etc. They have access to data and attempt to solve problems.

The Graduate School of Business of Chicago is organized around the Total Quality Management (TQM) theory. Faculty and students work together with administration to develop curriculum and the pedagogy. And I am sure there are other examples of where teams of workers are working together to solve questions and improve their organizations.

My question is: Is there some way to capture what is going on within the organization in order to study it systematically rather than in an ad hoc Fashion, which is what's currently happening? We have examples but not a theory about how these things operate. We know they are successes after the fact, but we are not clear on how they do it.

CWC: You don't know whether that's a good way of doing it or not because that has to do with the management of inquiry. In the study of management, I do want to create a science where each human being is a scientist and move away from the distinction between being a human and being a scientist. Russ Ackoff is doing the same thing that you described at Chicago. There are lots of examples around. If I didn't find examples, I'd be worried, but I find them everywhere. The only thing that's not emphasized enough is ethics. I want to see the improvement of the human condition. We need to bring about joy in the human condition. Maybe that's what we need, a Department of Human Joy. But to what department does that really belong?

EK: Be careful about that ... Hitler used the phrase "strength through joy".

JvG: It's interesting that we're talking about ethics in the context of the School if Business. Within the academic community and the accreditation community they want us to teach ethics, but that would go against the concept of greed, which is also taught and greed takes precedence. An undergraduate has no idea what you mean by ethics. Therefore, when he graduates, he goes out and sets up his business in a very unethical way.

EK: *The business schools teach ethics but they don't understand West's concept. Their concept of ethics in a corporation means to be legal and to operate just within the boundaries of the law but to never get caught.*

CWC: Take, for instance, our courses on Conflict Resolution. These courses are designed to teach how to influence workers so they don't fight with management and what management wants. It is about getting the rest of the organization to go along with management. It is not about ethics. You resolve the conflict when you make the proposal and you know beforehand that the rest of the organization will agree beforehand, rather than reach a situation with conflicts afterwards. That is not ethics.

BD: *Is there a hierarchy of ethics? Are you more concerned with satisfying the higher aspect of ethics, e.g. social responsibility?*

CWC: There isn't any good hierarchy. My mind works that way. Perhaps if I had to select a higher ethic, I would say "kindness." I have the same idea as Immanuel Kant. Kant's main idea was Ethics by a Moral Law. He wanted to reach a kingdom dictated by the highest principles. If you're motivated by kindness, true human kindness, you've got it. But most people aren't so motivated. You can't feel kind about starvation or killing 50 000 people on our highways. Laws do not operate kindly. In the courtroom, lawyers are not kind and neither is the medical profession. If you spend any time in the hospital you'll find that medicine is not kind. They treat you like a machine.

JvG: *Let's talk about greed and kindness. if greed is a disease that we must eradicate, how do we use kindness to do it? Do we encourage people to be kind? Our society is not like that. We're not kind, because of greed, competition, pressure and the like.*

CWC: There are more books being written now on this idea of "care" than anything else. I'm sure of that. What this means is that there's a bigger push for the human species. How do you care? Not like the good Samaritan who sees a mugging victim and takes him to a house down the road to get help. That's not how we care today. I try to care for those who drink too much. I can't tell them to stop because they don't know what I'm talking about. You have to learn how to care. Learning is what we professors are all about. Why isn't there a course on that. There should be a number of courses on caring. Human Care. What Department would you put that course in?

EK: *Nursing . . no, not really. Stafford Beer, in his writing, quotes from the Bhagavad-Gita and this is what he ways, "In reality, action is entirely the outcome if all the modes of nature's attributes; moreover, only he whose intellect is deluded by egotism is so ignorant that he presumes. I am doing this." He gets to the greed aspect of the managers: I did this, so, therefore, I deserve everything I can get. He introduces the Creed of Greed. But Stafford*

Beer quoting Bhagavad-Gita is coming close to what you're trying to do in your book.

CWC: With one notable exception. The Gita makes no mention of implementation.

EK: He says that all action goes toward implementation.

CWC: His is not a book of implementation. "If you listen to me then you will do what I tell you". That's what I get from Gita enthusiasts. His work is really about taking control. We're famous for control devices. We have beautiful technologies. They are control devices. What they don't tell us, however, is how to control desire. That's the Gita message. You haven't got anywhere near control if you can't control the desire function.

EK: We know how to control influence, through advertising, etc. But we can't control desire. We know how to enhance the desire Junction.

JvG: Is desire a purpose?

CWC: It's one purpose. We don't have to have a desire function. A lot of biological functions are carried out without a desire function. Gita's point is that where humans go off the track is because they can't control their desire. For example, just ask Mr. Perot, "How many more billions of dollars do you need to be satisfied?" If he is honest, his answer will be that he will never reach that point.

The Gita is one of my best examples. Some of my students introduced me to the Gita. It's a marvelous book. It starts with a young man who is confused, wanting to know why he should go out in the field and kill his family that's at war with itself. He is talking to his character. What comes out of this is that we are all in a state of confusion because our ethics are in a state of confusion. We don't know how to control the desire in our lives. The Gita is a control book.

Stafford is exactly right. It says the message clearly. I don't get it out of Kant. Plato is the first management book about the management of cities. He tried to implement it. He wanted to have the dictators implement his ideas in his Republic. And he almost got killed trying.

BD: Wouldn't you say that the Ten Commandments are a fair representation of ethical management principles?

CWC: Yes, it's an example. But, in the analysis of its content and justification, it just doesn't work for me. From the translation that I have, I don't get it. It doesn't help me understand. They are a good guide for some, but I do not understand them. Plato, who was really one of the first great Western management theorists, for each step in the Republic, he gives a justification. That is what I call science. I didn't get that from the interpretation of the Ten Commandments. Also, they come from Yahweh through Moses and his conversation with Him on a mountain top. I can appreciate that. I think most of us listen to an inner voice. But, if you believe

8. IN SEARCH OF AN ETHICAL SCIENCE

that the Ten Commandments come from Yahweh, there is less of a tendency to analyze them, to be reflective.

BD: But aren't the Ten Commandments obvious statements about how we should ethically manage ourselves and society?

CWC: Not for me. To go back to World War II, "Thou shalt not kill", wasn't very obvious even to this pacifist. If we didn't oppose Hitler I could see a world of terrible tragedy. I had a strong feeling. Not a solid belief. I felt frightened, we all did. But we believed we had to defend ourselves and our children, and to do so, we had to kill the enemy.

JvG: I have a little bit of a problem - maybe others do too. If others were to listen to our tape, they would hear the concept of war talked about and hear that war is ethical. Can you temper your statement with an explanation of what you mean by that? I think it can be misinterpreted.

CWC: We make generalizations all the time. For instance, all drinking, all alcohol is bad. Well, that's not true. We say the same thing about drugs and there are some drugs that are good. The problem is to identify when the ethics are different. That's a systems approach, straightforward systems approach. There are no true generalities for action. That does not mean you can go backward and find a standard basis. I am not a Kantian. I don't try to find a moral law. You can try to find a basis, but they are always changing. So the rules change, but not the overall ethical purpose: service to humanity. That's an invariant.

JvG: We keep coming back to how do we decide and who decides, in this case, what's ethical?

CWC: Who decides what kind of mathematics is appropriate? We didn't do that at the beginning of the Institute of Management Science (TIMS), and the Operations Research Society of America (ORSA), and we should have.

JvG: Do you really have this faith in mathematics? Really?

CWC: What is math?

JvG: Logical thinking, but...

BD: No, it's applied psychiatry.

CWC: We still don't know. Mathematics is described as rigorous thinking. Described as deductive thinking as opposed to inductive thinking. Mathematics is described in Principles of Mathematics by Whitehead as pure logic, applied. We don't even know what the number system is. How can it be the final foundation when we don't know what it is.

JvG: Maybe what went wrong is how mathematics is applied. It seems synonymous with making a field or a discipline more scientific. This is what befell management science.

CWC: I don't think it's scientific at all. I don't think modern science is scientific.

JvG: This was the purpose of TIMS, trying to make decision-making more scientific.

CWC: Maybe we should've said, back in the 1950s, "We're starting something that is going to call into question what science means". Because the current state of knowledge gathering is terribly deficient in terms of its management.

BD: Do you think you're giving mathematics too much credit? Mathematicians had an axiom for parallelism and then later found there were some non-Euclidean geometries. Do we need a variety of mathematics or geometries to describe human science or science applied to the human condition?

CWC: Consider Plato. The Pythagoreans were the strongest mathematicians in Plato's time. They didn't think of the deductive stuff. They would not have said that set theory or propositional calculus are the foundations of math. They said that numbers had a spiritual nature. Saying five plus seven equals twelve symbolizes something far beyond just the manipulation of symbols. But eventually, the numbers were stripped of their spiritual meaning.

EK: The concept of calculus is spiritual.

CWC: Numbers mean to us, human beings, incredible things. Just look at the mystery of April 15. That's a spiritual holiday in this country.

JvG: I'd like to ask you to conclude our interview by giving us your definition of hope. I know it is important to you. Why don't you tell us about hope and with this we will close out the interview.

CWC: I used to think, because I'm a logician, that definitions had to be rigorous. It's a paradox. You can explain A by B, but B is not clear enough so you do C, but C is not clear enough because it doesn't define rigor. So you have to define rigor. You can't do that except by D. It's a lot of nonsense. Definitions should be meaningful And meaning goes deep to the spiritual side of me. It can't be made rigorous at the present time.

HOPE is the spiritual belief in an ethical future.

JvG: On behalf of all of us, thank you, Dr. Churchman, for kindly agreeing to take part in this interview. We appreciate the time you have spent with us. May the years to come be as fruitful and as creative as all those earlier ones. We wish you continued health and happiness.

Chapter 9

ETHICS AND ENLIGHTENED PERSONAL RESPONSIBILITY

FRANÇOIS,C.
GESI- Asociación Agentina de Teoría General de Sistemas y Cibernética

library@iafe.uba.ar

Abstract: Three decades after the publication of C.West Churchman's main works, it seems necessary to reevaluate his perspective on ethics and to reframe it within the much modified and ever more complex interrelations between the widely different cultures in the world. In fact, his views on ethics are, like everyone's own, culturo-centrated.

On the other hand, it seems obvious - taking particularly in account Magoroh Maruyama's work on "mindscapes" - that different personal psychologies lead to also quite different Weltanschauungen and ethical stands.

Anyhow, in the present dangerously unstable situation of mankind as a whole, only a genuine transcultural ethics can save us from a variety of possible global mega-catastrophes, mostly engineered by man himself.

The crucial questions are: Can we hope that such a globally recognized and accepted ethics will emerge in time? And what should be its prescriptions?

9.1 MUSINGS ABOUT CHURCHMAN'S VIEWS ON ETHICS

9.1.1 Ethics?

We should start with the following quotation from *The Design of Inquiring Systems* (p.238);

> "The religious Weltanschauung, ... describes a certain kind of relationship – such as love, adoration and obedience – between men and other men, or between men and some superior being, or between men and 'Nature'".

To begin with, the religious Weltanschauung does in fact prescribe (not merely describe) certain kinds of relationships. This is specially true in the case of the three great monotheistic religions: Judaism, Christianism and Islam.

All three prescribe more or less the same type of behavior, in most cases.

But there is a ploy: The "faithful" should only adore the "true" God ... and moreover strictly in the specific ways established by the sacred scriptures.

And, in most cases, this love, adoration and obedience is not due to god. When and as described by any other creed, but exclusively to the only "proper" one. Those people who are not regarded as true believers in the "true" God are most generally not loved. No obedience is due to "alien" scriptures and "heretics" are excommunicated.

Indeed, the "infidel", the "pagan" or even simply the non-believer can find himself in dire straits and is frequently marginated or hostigated. He can even be killed, under the most sanctimonious reasons, be it during a cruzade, or by an inquisitor, or by some Jihad fighter.

All this is ominous in view of the religious origins of ethics in the cultures that emerged from monotheistic faiths. The case of Eastern creeds, as Hinduism and Buddhism would need a different approach. The first mainly prescribes a rigid and highly fragmented social order, while Buddhism is oriented to personal introspection and betterment.

Ethics, as derived from different religious beliefs, has thus in each case a specific flavor, specially in the Western World.

9.1.2 Why, how and what do we believe?

While we are generally unconscious of the deep sources of our individual behavior, it is however evident that our personal ethics largely depends on

9. ETHICS AND ENLIGHTENED PERSONAL RESPONSIBILITY

our own deep-seated – and also very, or more or less, personal Weltanschauung. This german term is used here because it is much more significant than the the english "worldview". In effect, it conveys the individual attitude oriented toward the observation of our environment (anschauung) and generally includes an intentionality, in the meaning of Brentano and Merleau-Ponty. Our outlook is not, and cannot be, neutral or "objective". We depend of what, for the lack of a better term, we should call our "temperament", and moreover, each of us is located within the particular culture he/she was reared in.

M. Maruyama widely explored the resulting behavioral inclinations of different people in varied settings and introduced a classification of so-called individual "mindscapes", after protracted research in many countries and different cultures. The following summary surely does not give full justice - far from it – to the wide embracing views of Maruyama. But it will – hopefully – give us a better understanding of our own psychological and cultural conditioning as observers and – mostly – believers.

Maruyama distinguishes four main types of personal "mindscapes", with the following predominant characteristics:

H type: Homogeneistic – hierarchical - classificational
(The world is organized in a very orderly way, that should be discovered, understood and respected)

I type: Heterogeneistic – independent - random
(Everybody does, or should do, what he/she likes – No very well defined order or organization can be recognized, nor should be imposed, and the general results are mostly random)

S type: Heterogeneistic – interactive – pattern maintaining
(People are different, but they are interacting and the more or less repeated interactions tend to create and maintain specific behavioral patterns)

G type: Heterogeneistic – interactive, pattern generating
(People being different and interactive in different ways, tend to generate new behavioral patterns)

Maruyama freely admits that his classification may not be perfect and that only few individuals may respond to a totally dominant mindscape, absolutely exclusive of some characteristics of another type.

However, admitting these limitations, it seems possible to hypothetize a classification of ethical feelings in correspondance with each type. What follows is of course in no way Maruyama's responsability.

H type: We should be respectful of the existing order. This implies mostly rigorous respect of the ethical commandments in force

I type: in our society, which leads in many cases to authoritarian attitudes.

I type: Society should not impose us our ethics. Such views liberate creativity. But they can also lead to socially destructive anarchy.

S type: We should find our place, integrating ourselves in social groups through working interactions. Socially oriented ethics seem to become paramount. But this could lead to obdured conservatism and stagnation, once patterns become dominant and self-reproducing.

G type: The need for adaptivity and evolution – individual as well as social – should be recognized in ethical terms. However, it could be difficult to evaluate the results, also in ethical terms.

9.1.3 ... And what should we accept as ethics?

It seems obvious that the different attitudes thus described did and do exist in every society at every level. Even Socrates trial, and self-accepted death sentence could be interpreted in terms of ethical conflict between irreconciliable mindscapes.

As we are all reared and educated in some specific culture, we acquire mainly the ethical hereditary Weltanschauung of our forefathers.

This fact leads us to some very awkward question marks: If your culture prescribes you some specific behavior, it remains to be seen if:

- Such behavior should be considered ethical in another cultural setting? (For example, what should you feel about Cruzades, or Jihad, in accordance – or not – with the traditional views of your culture ... the culture of the "others". Or, what should be in ethical terms, the social status and rights of women ... in all cultures?)
- You will personally accept fully, partly, or not at all such or such prescriptions? (This is for example the case of the conscious objector to military service, or opposition to blood transfusions, in accordance with some heterodox religious beliefs ... but in contradiction to orthodox social obligations or prescriptions)

Still further on, we are confronted with some very uncomfortable options.

Can we all admit eventual universal ethical principles? We all generally admit that human life should be respected. But what should we do if we are mandated for patriotical reasons to kill "ennemies" in war. And what would these "ennemies" do in turn?

And can we hope that a planetary ethics could emerge, and what should it prescribe? Which agreements and obligations could be stated and enforced, for example about the individual and nationally collective plundering of

9. ETHICS AND ENLIGHTENED PERSONAL RESPONSIBILITY

natural ressources in non-sustainable conditions? What should we make, in ethical terms of Garett Hardin's "Tragedy of the Commons" or John Lovelock's "Gaia"?

It seems obvious that these intricated questions cannot be resolved as long as individuals in all the different cultures of this world remain blind to any view seemingly incompatible with the dominant prescriptions of their own culture.

This implies one of the most critical jump in mankind's history. We cannot anymore afford the mere existence of six billions blind cultural parrots in a world burdened with nuclear weapons and tinkering with genetic engineering. Massive and collective irresponsability – or even simply non-responsability – could easily drive mankind to global suicide.

Considering the limits of multiple and many times divergent traditional ethics based on particular cultural Weltanschauungen, we need a massive mutation of all individuals toward a vision of ethics in terms of personal responsability. This is for sure a tall order ... which brings us back to Churchman.

In a review of Churchman's (1979) *The Systems Approach and its Enemies*, John van Gigch concluded: "The visionary must at one and the same time live his vision and the reality of the collective conscious".

This is in good accordance with a Japanese saying: "Vision without action is a dream, action without vision is a nightmare".

The main point seems finally to be in the quality of the "vision": action based on an incorrect vision could lead to a still more horrendous nightmare.

Possibly we would have to replace the "visionary" by a helmsman ("Kybernetes"!) who should be first of all a very able watchman. The ideal watchman tries to see what other people do not (or not yet) perceive. To be good at this, he should have a correct (i.e. working) model of what he is witnessing ... and a clear awareness of his own way of perception.

He should moreover correct his model every time some discrepancy emerges between his model and the perceived reality (as proposed by Stafford Beer in his famous lecture on "The surrogate world we manage"),

Wishful thinking, self-righteousness or an illuminist stand would not help. On the contrary, it could lead the "visionary" very much astray.

In relation to such views, two different currents of thought do run through West Churchman's work.

The first starts from a deep preoccupation with morality and ethics and the second one defines ways for constructing systems models and designing inquiring systems, obviously as a springboard for sound management.

As to morality and ethics, to begin with, we should not confuse both, as they are very different sources of "oughtness" (A. Bahm), i.e. what "should" be, or "should" be done.

Morality would merely be the acceptance and respect of a code of rights and duties as prescribed in a specific culture by religious belief, by law or by custom.

These prescriptions - and their practical effects (as we are now more and more aware of) - can be widely different from one culture to another. Such is the case for example in matter of social status (aristocracies, castes, women's rights, etc.); access to priesthood and exercise of the same; money lending and many other matters and behaviors.

Some examples and comments may be useful.

An extreme case are the situations in which one human being is supposed to have the right - or even the duty - to kill other human beings, acting for example as a public executioner, or in war as a soldier, or even maybe when murdering a "tyrant"

Any such behavior that does not fly in the face of culturally acceptable values and norms can thus be condoned.

Ethics prescribes a different kind of oughtness in which the individual is supposed to assume personally the responsability of his behavior, in accordance with his own informed judgement.

This is of course much more tricky, as for example in the border case of the murder of the "tyrant". The quote marks imply the personal and decisive judgement about the perverse nature of the "tyrant" ... Such an opinion is quite subjective and would possibly not be admitted by many other people ... whose views are of course subjective as well.

This means that the base for ethics is not merely obedience to the law, custom, etc., but necessarily "well informed judgement". And it is, of course, quite difficult to reach such judgement. Moreover, it is still generally much more difficult to convince everybody else that this judgement is really sound and fair - not to speak about the liars, scoundrels or simply stupid people who would not or could not accept it, for their own biased reasons or unreasons (...just as we ourselves and our ways may seem questionable to them!)

This problem - already quite unconfortable in itself - can be hopelessly complicated in specific present or future socio-cultural situations. This fact is duly acknowledged by West Churchman, for example in *Thought and Wisdom*.

Mass killlings, as in WW1 and WW2, were widely admitted until recently as being of course very unfortunate, but anyhow an unavoidable aspect of international relations.

Only that the virtuously named "rules of war" should be respected (and if not, "that's really too bad, but what could be done!?")

Such collective and horrendous mass human sacrifices of innocents by other innocents were waged from all sides under the invocations of "patriotic

9. ETHICS AND ENLIGHTENED PERSONAL RESPONSIBILITY 167

defense of one's country, of civilization, freedom, democracy, true faith, culture, etc."

We should meditate the recent comment of a german veteran of WW2 at the 60th anniversary of the 6th of June 1944 D.Day, when he met an American veteran, also about 80 years old. Both discovered that they were probably firing at each other in the same spot on D.Day. He said "We were praying, praying ... but still relentlessly trying to kill each other".

What a pathetic confession and how significant for one like this author, who was not so far away from these Norman beaches at that time.

As a consequence, I am obliged to confess that I feel quite uneasy with preachers of any denomination, as well as with so-called "practical" people. We should be suspicious about qualifying (or disqualifying) adjectives as "acceptable, satisfying, or negative, unsustainable" and unfortunately "moral" or "ethical" or their opposites, so frequently used in dubious ways by dubious people.

All of these judgements, implicit in the wordings, are ultimately based on criteria. And criteria are subjectively accepted (or rejected) by individuals whose information and judgements are poor, even when they are honest and well-intentioned. As noted by Churchman, thought is not enough; wisdom should be sought.

Consequently, it must be admitted that religious and ethical values and norms, imperfectly as they are invoked at time, are still the best behavioral rules currently available, as the result of a millenary experience of social life.

They should not be simply thrown overboard. In fact, social exercise of rationality is still a quite new endeavor, whose results have not always been very encouraging. But we should now look farther away.

West Churchman saw the problem. He wrote for example in *Thought and Wisdom* (p.19): "Is it possible to secure improvement in the human condition by means of the human intellect?"

Of course, he himself widely contributed to enlighten this intellect. But the answer to his query depends widely of the understanding and diffusion of his work, as done by his disciples.

He described in *Though and Wisdom* how unilateral approaches to human problems in their merely technical aspects, lead to lopsided illusory "solutions".

Even brilliant physicists, biologists or engineers, able to solve difficult theoretical or practical problems related to matter, energy and physiology, are short in many cases of a minimum understanding of individual and collective human conundrums and quandaries.

As men are not quarks, nor electric motors, nor even rabbits, the quality and type of intellect and knowledge needed for the understanding and management of human affairs, are definitely different from any specific technical knowledge.

It is the honor of Churchman to have searched and deepened the "Systems Approach", explaining clearly to anybody who would care to listen, how coherent (and frequently ignored) interrelations arise and how important they are.

He would, and possibly did agree with Thomas S. Eliot's comment: "Where is the knowledge we lost with information? And where is the wisdom we lost with knowledge?"

All his long career has been dedicated to elaborate better structured ways to knowledge and wisdom.

Let us all be grateful.

But let us also admit that ethics is a permanent and responsable construct for each of us and not only a collection of stale recipes.

REFERENCES

Churchman, C. West, 1968, *The Systems Approach*, Dell Publ. Co., New York.
Churchman, C. West, 1979, *The Design of Inquiring Systems*, Basic Books, New York.
Churchman, C. West, 1979, *The Systems Approach and its Enemies*, Basic Books, New York.
Churchman, C. West, 1982, *Thought and Wisdom*, Intersystems, Seaside, Ca.
Maruyama, Magoroh, 1979, *Mindscapes*, World Futuree Society Bull. Nr.13.
Maruyama, Magoroh, 1994, *Mindscapes in Management*, Cambridge Univ. Press, Dartmouth, U.K.
Maruyama, Magoroh, 1995, "Individual epistemological heterogeneity across cultures", *Cybernetica*, **37**.
Merleau-Ponty, Maurice, 1945, *Phénoménologie de la Perception*, NRF, Paris.

Chapter 10

MAY THE WHOLE EARTH BE HAPPY [1]
Loka Samastat Sukhino Bhavantu

BEER, S.

Abstract: Appreciation of different viewpoints is the starting point for better management

Key words: happiness, systemic and harmony

10.1 INVOCATION

West Churchman and I have been friends since we met at the first international conference on operational research in Oxford in 1957. About twenty-five years ago, he was staying at my house in Surrey outside London, and roamed around my library. After he had left, I found that he had written on the fly-leaves of my copies of his own books.

In his influential book *Prediction and Optimal Decision*[2] he thanked me touchingly for my critical review essay, which had appeared in the American journal *Philosophy of Science*.[3] Let us recall that his subtitle was *Philosophical Issues of a Philosophy of Values*. This is a far cry from the technique oriented aridity that we have later come to expect under titles concerned with either forecasting or optimality. Churchman is surely the outstanding modern philosopher of systems theory; and his concern with

---

[1] This essay appeared in *Systems Practice Action Research*, 7(4): 439-450. Copyright Kluwer Academic/Plenum Publishers, UK and USA.
[2] Churchman, C. West, 1961, *Prediction and optimal Decision: Philosophical Issues of a Science of Values*, Prentice-Hall, Englewood Cliffs, N.J.
[3] Beer, Stafford, 1963, "Prediction and optimal decision", *Philosophy of Science*, **30**(1) January.

ethical questions has never been an academic preoccupation. His book appeared as the turbulent sixties began - a decade of grave moral disquiet: civil rights, Vietnam, and "free speech" were all matters that split the American nation. Churchman the activist emerged from behind the dreaming spire of the Berkeley campanile to run personal risks which others will catalogue but I remember.

His (then) new book was *The Systems Approach*,[4] I dare say that it became equally influential. Churchman wrote in my copy of this a brief testimonial of friendship "without the necessity of frequent 'feedback' because of love". And so it has worked out ever since. What do we know about any such mode of communication? It has nothing to do with E-mail, for certain: psi-mail, maybe. What do we know about the brotherhood of love itself? Very little, it would seem, as we look around the devastation of our world today. But Churchman has devoted much of his practical effort to the causes of love and peace.

As the Churchman oeuvre attests, West is a great expert - appropriately enough - on western philosophy. In this essay I put forward some recent thoughts that are based on ancient Vedantic philosophy. They are transmitted from East to West in the hope that he may enjoy them, in the spirit of his messages to me. It is appropriate to take as title an ancient invocation[5] of which he would surely approve: "*May the whole earth be happy*". In the Sanskrit: "*Loka samastat sukhino bhavantu*".

Lokham refers to the world - to Gaia herself, as we might say, and taken to include all living species whose habitat is here. Sukham, "happiness" in brief, is close to what Aristotle later called eudemonia, well-being; but it encompasses the satisfaction of needs that are not only physical and mental, but aspire to spiritual truth. It seems to me that the dictum offers a comprehensive invocation of what we sanctimoniously call the Quality of Life. We may even link these ancient Eastern ideals to our (less) ancient Greek heritage through the words satyam (truth) sundaram (beauty) and sivam (goodness), upheld in both cultures as cynosures of civilization. I apologize for knowing neither the Sanskrit nor the Greek for "gross national product per capita".

10.2 MANAGEMENT AS ACTION

Each of the three approach lanes to my remote cottage in the Welsh hills runs through a small farm, each of which in turn is guarded by sheep dogs.

---

[4] Churchman, C. West, 1968, *The Systems Approach*, Delacorte Press, New York.
[5] Murthy, P.N., 1988, "Towards 2000 and Beyond", Thirtieth Sri Mokshagundam Visvesvaraya Memorial Lecture, Institute of Engineers (India).

As my battered old truck comes near, dogs rush upon it, barking, and chase it off the territory on the further side. They drop off the chase after a hundred yards or so with their tails in the air - quite evidently satisfied with a job well done. I was just going home anyway, fellows, but never mind.

There was a successful company once whose staff constituted a real team. The workforce was hard-working and well-organized. They had a wonderful relationship with both suppliers and clients, and made steady profits. When the annual general meeting approached each year, there was a big party for the press: the company had done very well, and would be doing even better next year, announced the man who organized the party. The published accounts confirmed it all, and revealed that the party organizer had collected a large bonus. This was in order because he was the CEO.

The Bhagavad-Gita[6] says:

> "In reality, action is entirely the outcome of all the modes of nature's attributes; moreover, only he whose intellect is deluded by egotism is so ignorant that he presumes 'I am doing this'."

This is the philosophy of holism writ large in terms of management action. Managers often claim credit for the details of outcomes that are actually systemic products, a credit from which they are disqualified by Ashby's Law of Requisite Variety.

When first offering the Gita quotation in that context thirty years ago, I wrote[7] that it reveals "an insight into the nature of system, secondarily into the proliferation of variety, and thirdly into the generation of spontaneous control activity". The first clause, translated directly into cybernetic jargon instead of English might read: "Output is a self-regulating black box function of input variety". But the sting is in the second clause. It speaks of the "self-determination of the system from our own nature, of the implicit control which cybernetics purports to discover in nature 5,000 years too late to count as original".

There is much more than this quotation that recommends the Gita for management studies, because this scripture is wholly concerned with the nature of right action and is a monumental rebuke to the Creed of Greed which has lately characterized our era. The world is imprisoned in its own activity, says the Gita: yes - that is how we lose sight of the concept of social good, and are latter-day Philistines into the bargain. "There is no such thing as society" said Prime Minister Thatcher's notorious dictum. So all that is left is the individual and selfish desire.' S/he is the slave of his/her own activity, says the Gita - whose authors had never even seen an executive

---

[6] The *Bhagavad Gita*, own translation. There are many different versions available in English, for example 14 below.
[7] Beer, Stafford, 1966, *Decision and Control*, John Wiley, Chichester.

wrestling with the laptop computer's appointments and jobs-to-be-done program.

You might think from this that the recommendation might be to take less action, or to remove from the fray altogether, in adumbration of workshops on Executive Coddle, and "death is nature's way of telling you to slow down" slogan-mongering. On the absolute contrary, the Gita teaches that there is a positive duty to act. But the action must be in harmony with the whole situation, not an imposition. Above all, it must be disinterested. The duty to act does not carry rights to the fruits of action, and "only the petty-minded work for reward". So much for the Eighties Ethic of the First World.

Consider a manager who seeks to perform right action for its own sake, and (as the Gita says) is like the leaf of a water-lily that rests unwetted on water: "s/he rests *on* action, untouched by action". Such a person will try to refuse the credit for success which is difficult; what is more difficult still is to refuse the blame for failure - so long as the action was right. Our manager will find it wearisome to operate on so lofty a plane, and might well consider withdrawing his effort. Interestingly enough, this was exactly the predicament of Prince Arjuna before the battle of Kurukshetra, which is the scene of the Bhagavad-Gita. He was not at all happy about fighting in a battle against his own kith and kin. However, Arjuna was fortunate to have the divine incarnation, Krishna, as his charioteer: and thence flows the philosophic discussion of the ethics of right action. Not only is there a duty to act, but personal disinterest in outcome must raise one above it; not only must one not run away from distasteful action, but "only action can free you from the obligation to act". That last sentence can stand a second thought.

These arguments are of a subtlety and nobility unknown to the Creed of Greed. Recently there has been a spate of conferences under the title of "Business Ethics". It is sad to turn from the Gita to confront the fact that these important meetings concern simply the prevention of fraud. There should be more to business ethics than that. It is sad too that the Gita has sometimes been used among Hindu extremists to justify violence - just because the argument draws its high dramatic effect from a scene of battle. But Krishna is not alone in this misrepresentation. The teachings of another Prince of Peace have with equivalent perversity been manipulated in the cause of a multitude of heinous crimes.

The four-thousand years old Sankhya philosophy, part of the Vedantic tradition, is the basis of Raja Yoga. It teaches that belief is nonsense. Only direct experience counts. This would make a powerful aphorism for modern management, and one that might find approval in the Business Schools. But they would need to follow through. They would find that the ultimate meaning of their new motto lies in the direct experience of self - which few people ever confront. They would also have to face the lesson or direct experience in the implementation of the techniques of socio-economic management that they teach. This is that they don't work.

A School of Vedantic Management would teach the eight-fold path to wisdom of Raja Yoga through direct experience of self. In managerial terms it would come out as the Gita says: wisdom is right action - 'the two are one'.

10.3 THE CREATION, RECURSION AND CLOSURE OF SYSTEM

All great cultures have supported creation myths. Our own is sponsored by modern physics: the Big Bang Theory. This, of course, is no more a statement of "objective truth" than any of the others. Its most well-known exponent, Stephen Hawking, himself says[8]: "We see the universe the way it is because we exist'. And this is a view taken even further by some physicists in the anthropic cosmological principle,[9] which see humankind as the centre of or indeed the reason for the universe. Others have not hesitated to advance to 'The Theory of Everything" that emerged from the ten-dimensional constructs of Superstrings[10].

All the myths, including that of the Big Bang, are fundamentally interested in the origin and nature of time itself. While our own culture uses concepts and measures such as Hawking's we still have fundamentalists among us who accept the seven days and seven nights of Genesis; although some of those concede that an Old Testament day might be longer than the terrestrial day. Otherwise, the two myths have much in common - except for the order in which the earth and moon appeared. Both contemporary myths, let us note, involve not only a primary Big Bang, but an eventual Big Crunch.

So does the oldest creation myth in the book. It is the Vedantic theory of yugas, the world ages. There are four world ages, and between them they last for 4,320,000 years. It takes a thousand such cycles to account for a day in the life of God (Genesis enthusiasts please note). At the end of the day (thus defined) "all matter in the universe is reabsorbed into the universal spirit".[11] During the night, which is also some four trillion years long, "matter persists only as potential for reappearance" (ibid). All this refers to one day in a year of 360 days in the Life of Brahma; and he lives for a hundred years before

---

[8] Hawking, Stephen W., 1988, *A Brief History of Time*, Bantam Press, London.
[9] Barrow, John D. and Tippler, Frank J., 1986, *The Anthropic Cosmolqgical Principle*, Clarendon Press, Oxford.
[10] Peat, F. David, 1991, *Superstrings*, Scribners, New York.
[11] Hope, Murray, 1991, *Time: The Ultimate Energy*, Element Books, Dorset, England.

the Big Crunch. The whole of this immense cycle then renews. So time is not itself an episode suspended in nothing.

In short, the Vedantic philosophy of the yugas is structurally recursive in its treatment of time, matter and energy. It exhibits closure, both within its recursivity, and also metasystemically beyond it. Modern systems theory is better equipped to reflect and foment such Vedantic notions in our times than is contemporary physics - and to more satisfying philosophic outcomes. A full treatment of this cybernetic contention would consider especially the constructivist theory of Heinz van Foerster and his associates,[12] and the theory of autopoiesis advanced by Maturana and Varela.[13] But in a short writing perhaps I might be permitted to use illustrations from my own work, in this and the following sections.

The problems encountered in discussing creation, and thereafter of its contained created systems, begin with the logical incompleteness of the language in which they are cast. They are in their nature rambling Goedelian sentences in which some propositions can neither be proved nor refuted. This is exemplified in physics by asking questions about chance and necessity in the universe, and is epitomized in Einstein's agonizing over his belief that "God does not play dice" - held in the face of new discoveries in quantum theory that required a random basis, and seemed to imply that he did. It was exemplified in the context of management by a discussion of production planning in my first book.[14] To deal with this, I offered the principle of "Completion from Without", which proposed to give closure to a Goedelian sentence by a device which could not be understood - except insofar as it did indeed make the closure. The: managerial cybernetics later propounded (6) was based upon this principle. But instead of using my inelegant form of words, Einsteins's spirit might be quieted by ostensive definition: "God is the dice". The Black Box is what it does.

Returning to the most revered of the Vedantic scriptures, we find the device in frequent use in the Gita. Among the examples comes a dictum of the Lord Krishna: "I am the discussion among disputants", or in a poetic translation:[15] "I am the logic of those who debate". The principle has been available for thousands of years, had we or Einstein but recognized it.

---

[12] Siegel, Lynn, 1986, The Dream of Reality: Heinz van Foerster's Constructivism, W.W. Norton, New York.

[13] Maturana, Humberto R. and Varela, Francisco J., 1980, *Autopoiesis and Cognition*, D. Reidel, Dortrecht, Holland.

[14] Beer, Stafford, 1959, *Cybernetics and Management*, English Universities Press, London.

[15] Prabhavananda, Swami and Isherwood, Christopher (translators), 1944, *Bhadavad Gita*, Chapter X, Vedanta Society of Southern California.

10.4 VEDANTA AND THE VIABLE SYSTEM

During the fifties and sixties I was preoccupied with the development[16] of a cybernetic model of any viable system: the created component of a more inclusive viable system - identically and therefore recursively, defined. This is a holistic model involving the intricate interaction of five identifiable but not separable subsystems. Only recently have I made comparisons between the Viable Systems Model, VSM, and the Vedantic Creation Model, SRISTI. This tabulates the results:

Table 10.1.

SRISTI THE VEDANTIC CREATION MODEL	DESCRIPTION	VSM THE CYBERNETIC VIABLE SYSTEM MODEL
MAHA SHAKTI	is the power behind creation and the establishment of identity	SYSTEM FIVE
MAHAS WART	lays down the large lines of development for the whole	SYSTEM FOUR
	drives the energies of the autonomous components	SYSTEM THREE
MAHAKALI	and guards them with a powerful vigilance	SYSTEM THREE STAR
MAHAI.AKSMI	highlights and guides the rhythms of autonomous components with grace and harmony	SYSTEM TWO
MAHARASWATI	presides over the detailed organization and execution of autonomous components	SYSTEM ONE

The account of the five subsystems that holistically elucidate the VSM is spread over three books,[17] two of them very thick, and it is difficult to give a brief description of them. There are profound reasons for this. In the first place, a viable system is not a hierarchical structure: all components are mutually interdependent, and reflect each other (as it were) holographically. Any diagrammatic layout, any listing of components, has the appearance of

---

[16] Beer, Stafford, 1962, "Towards the cybernetic factory", in: *Principles of Self-Organization*, von Foerster and Zopf (eds.), Pergamon.

[17] Beer, Stafford, 1972, *Brain of the Firm*, Allen Lane The Peguin Press, London, 2nd Ed. John Wiley and Sons, Chichester, 1981.
Beer, Stafford, 1979, *The Heart of Enterprise*, John Wiley and Sons, Chichester.
Beer, Stafford, 1985, *Diagnosing the System for Organizations*, John Wiley and Sons, Chichester.

hierarchy: something must be mentioned first, or be drawn above something else. Secondly the characteristics that underwrite viability are the managerial phenomena connoted by such words as regulation, control, communication, coordination - not to mention the word management itself. But the essence of the cybernetic discoveries embodied in the VSM is that these phenomena are generated by the holistic performance of the entire system. They do not reside anywhere at all; least of all can anyone of them be identified with anyone subsystem. The nearest that I have come to such specificity is to identify various "regulatory loops".

The force of such a view derives from the neurocybernetic studies[18] in which the VSM has its origins: the localization of function in the brain is an intractable topic, because of the redundancy of potential command - and the impediment applies to all viable systems. Thus I have consistently evaded demands to "explain" the five VSM roles, to make them "more understandable", because the cybernetic insight is denatured in the process. This is also the reason why I gave the subsystems numbers (and paid the price among those who insist this infers hierarchy), instead of simple but misleading names. We should not do well to nominate the loudspeaker as "the voice" in a radio, nor to label the moving piston of the engine as a vehicle's "speed".

Users of the VSM may wish to consider the descriptions given in the above tabulation against this background. They are (at least) satisfactory in that they are highly informative for their length, do not fall into the localization trap, and are recursively effective via the component systems. The interesting thing is that I did not write them. They are Vedantic statements, drawn from the ancient creation model called Sristi. According to this, the Great Power, Maha Shakti, is attended by other "magnificent powers", all having the prefix meaning great. Each has a responsibility, rather than a technique or an office, which qbviously interacts with the other responsibilities. I particularly like Mahalaksmi's wording, which in the inelegant parlance of VSM system Two, is "anti-oscillatory". 1 have taken only one liberty with the Vedantic descriptions,[19] in using the term "autonomous components" where my original says "created forms"; but 1 should be surprised if the (not available) Sanskrit does not equally support that reading.

According to the VSM, every viable system contains and is contained in other viable systems: this is why the model is called cecursive. Exactly the

---

[18] Beer, Stafford, 1962, "Towards the cybernetic factory", in: *Principles of Self-Organization*, von Foerster and Zopf (eds.), Pergamon.

[19] Murthy, P.N., 1989, "Some philosophical perceptions about design as a creative process", address to the Fourth Indian Engineering Congress at Bhubaneswar, 31st December (transcript).

10. MAY THE WHOLE EARTH BE HAPPY

same is true of the Sristi model. Mahashakti is conceptualized as a goddess, the divine mother; and she is in herself recursive. The great sage Sri Aurobindo writes[20] of three modes of her being:

> "the transcendent supreme, original Shakti, who is above the worlds and serves as a link between creation and the still unmanifested mystery of the supreme; the universal, cosmic Mahashakti, who creates all beings and contains, penetrates, supports and directs the millions of processes and forces; and, lastly, the individual, who personifies the power of the two most vast aspects of her existence, makes them alive and close to us, and interposes herself between human personality and divine nature."

As Larousse[21] comments, she is:

> "at the same time, the whole of divine power, yet this does not prevent her from being entirely incarnate in the mother of the family of each household, and in the Kundalini at the base of the vertebral column in the case of each human being."

Similarity, the VSM has been mapped onto body cells, the central nervous system, the family, townships and the country as well as to all kinds of industrial and service enterprise worldwide.

10.5 OF FORM AND MEASURE

Results, of the foregoiing analysis were deeply surprising to me, despite a lifetime's conviction that knowledge is perennial. Often its expression is contingent on current technology - and this provokes confusion between the two. Worse still, it may brand the notion of transcendental wisdom, which is called prajna in Veaanta, as superstition rather than knowledge at all. Perhaps this applies to ideas about underlying forms and energies that are not directly susceptible to empirical investigation. Has our own culture lost sight of the "mixture" of Empedocles, and of Aristotle's "pneuma"; and was the "ether" of a quondam physics so readily to be dismissed? At any rate, the Chinese "chi" has recently been admitted to Western society via acupuncture, and the moving meditation of tai chi chuan. It is hardly surprising. that there is a vedantic version of all of these in the form 'of prana, a word used for breath in breathing control - but having more profound and immaterial connotations in ontology too.

---

[20] Aurobindo, Sri, 1970, *The Nother*, Sri Aurobindo Ashram, India.
[21] Herbert, J., 1965, "Hindu Mythology" in: *Larousse World Mythology*, Pierre Grimal, ed., Hamlyn, London.

Bridging thought from East to West needs to remember that the word science means knowledge; and that knowledge is ageless, not an instant reification of the latest technological gimmickry. For example, our comtemporary approach to quantification is governed by "'state art the art" computers, and the brands of mathematics that are most useful in their service - beginning with binary logic and passing on to the inversion of matrices and all kinds of iterative processes. It is algorithmic (dare we loosely say "left-brain") at root. What of the origins of number in Vedantic mathematics? It is seriously suggested in India that "VM", as its protagonits refer to it, might change approaches to the teaching of mathematics to children.

No-one can deny that the Western approach is frequently counterproductive; and if many children shy away, it could have to do with the denial of intuitive (let's say "right brain") entry. Perhaps the sudden emergence, and subsequent success, of set-theoretic pedagogy owed much to an intuitive component, as well as to its Open University pioneers in person. A story is told of the great mathematician Gauss at the age of about six. The teacher, seeking respite from a rowdy classroom, told the children to sit down and add up the numbers from 1 to 100 – "that will keep them quiet for a bit". The juvenile Gauss raised his hand and said 5050. It may not be immediately obvious how he did it, but it was not by the pedestrian route that most "properly trained" people adopt.

Earlier this century, Jagadguru Shankaracharya created quite a sensation by writing a book called *Vedic Mathematics*. There is of course a technology involved, and there is no room to explain it here; but essentially results were obtained by meditation and visualization rather than by reckoning as we normally think of it. But see Gauss. See also the extraordinary history[22] of Srinivaisa Pamanujan. He was an unschooled boy from outside Madras - born in 1887, and dead at the age of thirty-two. He was installed in Cambridge by the preeminent mathematician G.H. Hardy. -Ramanujan used Vedic visualization, and declared that "an equation has no meaning for me unless it expresses a thought of God": this, then, had to do with the *pranas*. And although his language was not (shall we call it) "academically correct", this man was called "the one superlatively great mathematician whom India has produced in the last thousand years'.

Although he usually hears it said that the Arabs invented calculation modulo ten, together with the absolute value of zero per coluumn, it seems[23] that they had learned the system' from ancient India in the first place. The *Taittiriya-Samhita*, dating from before 1000 BC gives a Sanskrit name to

---

[22] Kanigel, Robert, 1991, *The Man Who Knew Infinity*, Washington Square Press, New York.
[23] Pramhans, S.A., 1991, "Mathematical heritige of ancient India and its t:ransmission", in: *Vedic Mathematics*, H.C. Kbare, ed., Rastriya Veda Vidya Pratishthan, Delhi.

10. MAY THE WHOLE EARTH BE HAPPY

each order of magnitude up to ten to the nineteen. Other Vedantic sources provide names in groups of five orders of magnitude up to ten to the sixty. Numbers were classified[24] into even and odd, yugma = pair and ayugma = not-pair; and the same classification appears in distinguishing between even and odd arithmetic series. We can hardly fail to note that modern physics divides particles according to discrete, quantized spin values, which arise because of their quantum-mechanical nature. The bosons (such as photons) have full spin, while the fermions (such as electrons) have half-spin. As I have suggested before: coincidence is the inability to see what really matters.

As to the possible relevance of all this to perennial knowledge, mention has already been made of potentially new teaching methods for children. Sri Abhay Kashyap has gone so far as to suggest a new type of computer based on Vedic mathematical insights, which would go to very high speeds of number processing, and would thereby improve on number-intenslve research in AI and pattern recognition. His proposal appears on page sata of the cited work, pleasingly enough; that is to say page ten to the power of two. Has the computer industry made such a huge investment in established equipment and operating systems that it cannot change? Has AI itself too great an investment in its own paradigms to countenance it? Throughout its existence management science has bemoaned the entrenched positions of management; today it is itself vulnerable to its own favorite criticism.

10.6   PREPARATION FOR ACTION

We began by considering management as right action, the importance of direct experience over unfounded belief, and the focus in yoga on direct experience of the self. People prepare themselves for most forms of action, from playing squash to climbing mountains, by toning-up their bodies, by "getting into shape" - and such preparations may even extend to tutored autosuggestion that prepares the mind to succeed. We hear nothing, however, about preparation for right action in management.

Thus I mentioned in passing the eightfold path, which is founded on the Vedantic aphorisms of Patanjali, and should be taught as a necessary aspect of any programme in Vedantic management. The Gita descrihes "closing the gates of the body, drawing the forces of the mind into the heart, and by the power of meditation concentrating vital energy in the brain". This is evidently formidable preparation for right action.

Attention is drawn once more to what is profound. Already medical people have started to recommend meditation as a palliative for executives

---

[24] Skukla, K.S., "Vedic Mathematics: the deceptive title of Sawamiji's Book", op. cit.

who have survived their first infarct. But doing that is not profound. It is simple enough advice, and likely to help in controlling high blood pressure. But it is not preparation for right action. Raja yoga means the royal road to Union - with the cosmos and with oneself.

Again, the modern executive is admonished to be abstinent: s/he frets about weight and cholesterol, calories and salt, goes jogging, pumps iron. Says the Gita: "the abstinent run away from what they desire, but carry their desires with them". Contemporary attitudes are a royal road to frustration, anxiety and probable collapse.

There is no space here to evolve an integrated plan for teaching Vedantic management as an intellectual discipline, and yogic practice as the discipline of self. In all religions prescriptions exist that strive for that integral result; Tantric yoga integrates sexuality too. So a great deal remains to be said, if there is interest in these preliminary thoughts. At least an image of a balanced manager is evoked.

Such a person, please note, is calm rather than frenetic, happy amid all the adversities of our age. If meditation generates that state of being, it is not on account of mystical claptrap, but for good scientific reasons that were understood in the Vedas. Today they can be expressed by saying that the eighth step on the yogic path, called samadhi, floods the brain with endorphins. You may observe the same characteristic smile on the painted and sculpted faces of sages and saints in every culture in the world as a result.

Loka Samastat Sukhino Bhavantu
May the Whole Earth be Happy

NOTE ON GAUSS: In case the form of the answer failed to indicate the infant Gauss's visualization of the problem, the pairs 1 and 100, 2 and 99, 3 and 98, and so on, of which there are fifty, each equals 101. Thus the total is 5050.

ACKNOWLEDGEMENTS: I am much indebted to Professor P.N. Murthy, who heads the Systems Engineering and Cybernetic Centre of Tata Consultancy Services in Hydrabad, for rewarding conversations and his gift of Lecture Transcripts; also to Dr. Ashok Jain, Director of the National Institute of Science, Technology and Development Studies in Delhi for inviting me to discuss these topics at a staff seminar in January 1993.

SECTION D

ENTREPRENEURSHIP

Chapter 11

AN OUTLINE OF A DESCRIPTIVE THEORY OF THE ENTERPRISE
A Systemic Contribution

ERIKSSON,D.M.

Abstract: C. West Churchman seems to have been wrestling with this dilemma of Systems Thinking. On one hand we need a comprehensive understanding of the system under consideration if we are to secure the opportunity to generate ethically sound design. On the other hand, it is not really feasible to secure comprehensive understanding of a system. This essay is a small contribution to the second challenge regarding the design and management of enterprises. The question posed here is: in what terms may an enterprise be conceived if a reasonably comprehensive understanding is to be achieved? The proposed outline of descriptive theory of an enterprise is derived from a French School of Systemics, denoted here as Le Moigne's Systemics, and is supported by several empirical case studies. Its main constituents are the Projective Constructivist Epistemology and the General Systems Theory conceived as a theory of modeling. This derivation has resulted in a set of enterprise concepts (i.e. categories, or constructs) that support the conceptualization of an enterprise. The value of the outlined theory of the enterprise is advanced here in two ways. The proposed theory is compared from a theoretical viewpoint with several other current theories for enterprise description. This is followed by a presentation of an empirical real-life case study and its value for problem identification. Both of these demonstrate that the proposed theory provides a rich and comprehensive understanding of the enterprise in question, which is its very purpose.

The parable of a drunkard looking for his key
It is past midnight and the avenue is very dark,
illuminated only by some clear spots of light,
projected by sparsely placed lamp posts.
At the foot of one of them,
a drunkard with hiccups had squatted down,
attempting to look for a key to his house,

which was some fifteen meters away.

A sympathetic passer-by stopped to help and asked:
"Do you remember where you lost this key?"
"Yes", answered the drunkard,
"in front of the door of my house, when I tried to unlock it".
"But why then, do you not look in front of the door?"
wonders the passer-by.
"Because here at least, under the street light,
there is light and I can see..."
Cited in J.L. Le Moigne (1990:6)

11.1 INTRODUCTION

This essay introduces an outline of a Systemic Theory of the Enterprise, based on the systems theory advanced by Jean-Louis Le Moigne, a French systems thinker. It is a descriptive theory aimed to support the manager by supplying a comprehensive conceptualization of his enterprise, whether as an analysis of a current enterprise or as the design and development of a desired enterprise. This Systemic Theory of the Enterprise is made up of a system of enterprise categories, or constructs, where each construct represents some distinct properties of an enterprise. The remaining part of this Introduction presents the relation between the theoretical outline proposed here and C. West Churchman's texts.

In our view, Churchman's life-long inquiry into the human condition may be perceived as a quest for a solution to a central systemic dilemma of all human activities. The dilemma, perhaps most clearly expressed by Churchman (1979), in his *Systems Approach and Its Enemies*, concerns the inherent conflict between the ethical and the epistemological insight into systemic thinking. From an ethical viewpoint we have the imperative for comprehensive or holistic understanding of an inquired system enabling derivation of an ethically desirable solution to a faced system problem, i.e. our aim to avoid sub-optimization. From an epistemological viewpoint on the other hand, we have the inevitable lack of comprehensiveness in any non-trivial systems inquiry, i.e. it is not possible to acquire a totally comprehensive understanding of the whole system.[1]

---

[1] Following this important insight, Churchman argues in his *Systems Approach and Its Enemies* (1979) that any attempt to logically incorporate or coordinate other rationalities – i.e. ethical, religious, political, and aesthetic - into a systems approach or any other so-called analytical-rational approach, however well conceived it might be, is unconditionally anti-systemic. Rather, these rationalities are incoherent and the only rationally sound and

11. AN OUTLINE OF A DESCRIPTIVE THEORY OF THE ENTERPRISE

This dilemma provides us with a means of conceptualizing Churchman's contributions into two areas. One area regards the actual process or procedure executed in order to generate knowledge of a social system. Churchman deals with these concerns, probably most eloquently, in his *The Design of Inquiring Systems* (1971), where he presents five different inquiring systems. The central questions of this include: "how to design the inquiring process that generates knowledge of a social system", and "what kind of validity may such a process aspire to for the generated knowledge?" The second area of Churchman's contributions relates to the content or characteristics of the social system itself. Churchman developed a set of categories for use when conceptualizing a social system. Briefly, these are: Client, Purpose, Measure of Performance, Decision Makers, Components, Environment, Planner, Implementation, Guarantor, Systems Philosophers, Enemies of the Systems Approach, Significance (Churchman 1971:43; 1979:79-80,106f). Hence, metaphorically speaking, these categories serve as a pair of spectacles that enable the conceiver to clearly see the key features of the social system conceived.

While we have earlier contributed to Churchman's area of the process of inquiry into social systems (see Eriksson 2003), we aim here to contribute to the area of substance or characteristics of a social system, and particularly those social systems understood as an enterprise.[2] Given the challenge offered by Systems Thinking – i.e. to generate holistic understanding of an inquired object – the outline of a Systemic Theory of Enterprise presented here aims to support the generation of comprehensive conceptions in an enterprise, rather than limited and detailed ones. Finally, the theory assumes a phenomenological understanding of an enterprise (in the Kantian sense of the distinction between the phenomenon and the thing-in-itself which may give rise to several phenomena). This means that the validity of the proposed theory shall arise from its feasibility or viability in its actual use rather than whether or not it provides an accurate description of an independent reality.[3]

The next section gives a brief presentation of some of the methodological features employed in the development of the theory presented here, while the

---

operationally feasible solution is to employ a truly critical reflection upon the very lack of comprehensiveness of any system inquiry or proposition. This is a last attempt to guard us against deception, dogma and manipulation. It seems to us that Churchman's student Werner Ulrich knew how to read him and also conceptualize an approach to deal with this dilemma in a heuristic manner, rather than a theoretical one (Ulrich 1983).

[2] In preliminary terms, an enterprise is understood here to be a collection of resources – human and non-human – that conduct activities, which in turn construct processes generating output for consumers, all in relation to goals and governed by rules.

[3] This distinction between phenomenological and ontic status of a theory is further discussed in Eriksson (2004: Chap. 3).

section after that presents the theory itself in terms of its constituting enterprise constructs. The following section briefly positions the proposed theory in relation to Systems Thinking, Organization Theory and Enterprise Modeling Languages in order to articulate the theoretical value of the theory. The next section consists of a case study that shows the empirical value of the theory suggested. The final section summarizes our message.

11.2　A NOTE ON METHOD

While a detailed description of the research process employed in the generation of the Systemic Theory of the Enterprise can be found in Eriksson (2004:Chap.2), we will present here only some of the key features of this research process.

The research process followed the guidelines of Action Research arising from the action research program developed over the last three decades at Lancaster University in the UK (Checkland 1991; Checkland & Holwell 1998). The research has been conducted in the form of numerous case studies, executed over a period of approx. 8 years, in a business, public and military context, within some 50 projects of varying scope and length.

The generation of the theory was made by an iterative interaction between theoretical-deduction of proposals and empirical-induction from actual experiences in the conception of an enterprise. The enterprise constructs constituting the Systemic Theory of the Enterprise were initially derived from a particular systemic theory, labeled here as Le Moigne's Systemics. The latter's axioms are inherited from J. Piaget's developmental psychology (1937), which subsequently gave rise to its distinct Projective Constructivist Epistemology (Le Moigne 1994b, 1995a, 1995b; Eriksson 1997). This epistemology in turn provides foundation for a General System Theory, understood as a theory of modeling. It includes a set of general modeling principles, such as the general system, the general process, and the Decision-Information-Operations System Model (Le Moigne 1990, 1993, 1994a; Eriksson 1997, 1999).

The enterprise modeling constructs that constitute the proposed Systemic Theory of the Enterprise are presented in the next section. These constructs are of two types: elementary and composite. While the majority of the elementary constructs were generated through theoretical derivation from Le Moigne's Systemics, most of the composite enterprise constructs were generated by means of empirical induction (see Eriksson (2004, Chap. 10-11), for all details of both the deduction and induction of the proposed enterprise constructs).

11.3 THE PROPOSAL: A DESCRIPTIVE THEORY OF THE ENTERPRISE

The Systemic Theory of the Enterprise outlined here is constituted by a system of enterprise constructs, each representing some specific aspects of an enterprise. This includes elementary enterprise modeling constructs and composite enterprise modeling constructs. As an analogy, the elementary constructs may be regarded as single letters in the alphabet, while the composite constructs as a composition of two or more letters into one distinct unit. As with natural language, where various single letters and combinations of letters are combined into sentences and paragraphs, the proposed enterprise constructs should be used in specific manner to combine combinations that represent a particular enterprise.

11.3.1 Elementary Enterprise Constructs

There are currently thirteen distinct elementary enterprise constructs as set out below (see Eriksson (2004, Chap. 10) for a detailed specification).
- *Consumer:* an external actor that consumes an Output of the enterprise.
- *Output:* something generated by an enterprise, whether product or services, tangible or not, and ultimately constituted by information and/or matter-energy.
- *Out-Channel:* something that forwards an *Output*, from an enterprise's Process to the Consumer.
- *Activity*: an event conducted by one or several Resources which is part of a Process, and which transforms a Transformation-Object as governed by Rules.
- *In-Channel*: something that forwards an Input from a Supplier to one of the Processes of the enterprise.
- *Input*: something that is provided by a Supplier, via an In-Channel to a Process of the enterprise, whether product or services, tangible or not, and is ultimately constituted by information and/or matter-energy.
- *Transformation-Object*: an entity whose initial state is the Input, final state the Output, which is transformed by Processes and its constituting Activities; it is ultimately constituted by information and/or matter-energy.
- *Supplier:* an external actor that provides *Input*, via *In-Channel*, to the enterprise's *Processes*.
- *Resource*: an entity that contributes to the execution of Activities and thereby *Processes;* may be human or non-human. There are two distinct types of Resource within an enterprise.

- *Active Resource:* a Resource that is an active actor, or processor, in the particular context, and which drives the execution of Activities. It may be human, such as an employee, or non-human, such as a computer or a robot.
- *Catalyst Resource:* a Resource that is passive in the particular context yet is in some way needed to support the execution of Activities by the Active Resources, such as a building which is needed for production.
- *Rule:* regulations, policies, directives, norms – whether formal and explicit or informal and implicit – that stipulate or govern an Actor's execution of an Activity; can typically be expressed in the form of "IF x = 1 THEN y = 2".
- *Goal:* an expression of a desired state, of some property – typically a variable – of the enterprise, in terms of the constructs presented above.

The elementary enterprise constructs presented above can as such be used to support the conception of an enterprise. Figure 11.1 is a principal diagram illustrating all these constructs and how they should be related to each other, as the "grammar" of this system of constructs, to continue our analogy of the alphabet.

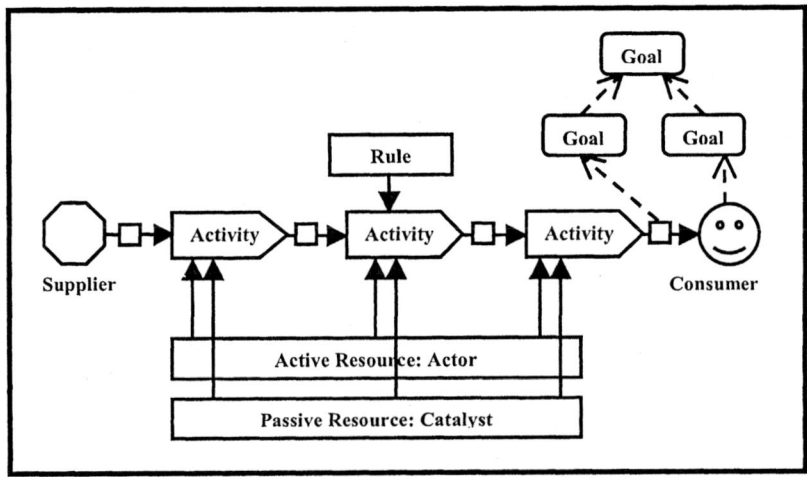

Figure 11-1. Illustrates a principal schema of all the elementary enterprise constructs, including the rules by which they should be related to each other. Starting from the right, Consumer receives an Output via an Out-Channel from an Activity, which receives a Transformational-Object from a previous Activity, which is directed by a Rule, and which in turn receives a Transformational-Object from a previous Activity that receives an Input via an In-Channel from a Supplier. The three mentioned Activities are executed by an Actor, and with the support of a Catalyst. The Consumer has an assigned Goal as has the Output, and there is another Goal to be fulfilled by the two mentioned goals. The symbolism employed is optional.

11.3.2 Composite Enterprise Constructs

Ten composite enterprise constructs are presented below with the aim of providing an additional arsenal of constructs to support the conception of an enterprise (see Eriksson (2004: Chap. 10), for a more detailed presentation).
- *Process:* two or more Activities that transform an Input, provided by a Supplier, into an Output, to a Consumer; executed by Resources, and governed by Rules.
- *Sub-Process*: a Process that is a part of another Process, hence a constituent of the latter.
- *Supra-Process:* a Process that contains two or more other Processes, and is thus constituted by them.
- *Core-Support Process:* a Process that generates an Output, which constitutes a resource for another Process; the first is then a Support-Process while the second is a Core-Process.
- *Support-of-Support Process:* a Process that generates an Output that constitutes a Resource for a Support-Process, where the latter's Output, in turn, constitutes a Resource for a Core-Process.
- *Core-Policy Process*: a Process that generates an Output that is made up of Rules governing the execution of another Process, the first is then a Policy-Process while the second is then a Core-Process.
- *Human-Actor Resource*: an Active Resource constituted by a human being, that typically constitutes some professional Role in the enterprise; the latter may be defined in terms of the Activities to be conducted, Outputs to be generated, or by inherent properties, which may be physical and mental, where the latter may be cognitive and emotional.
- *Machine-Actor Resource:* an Active Resource constituted by a machine (i.e. not a human being), which may be defined in terms of the Activities it is conducting, Outputs it is generating along with Inputs it is receiving, and its various inherent properties.
- *Organization-Structure Resource*: Organization Units, constituted by roles, together with the relation between these Units, that create a structure, which typically constitutes formal command and control streams, which in turn manifest the formal decision-making authority.
- *Organization-Culture Resource:* the various values (shared and diverging) held by human actors in the organization of the enterprise; these values, together with cognitive contents constitute attitudes which in turn give rise to human behavior – both formal and informal – hence Activities and Processes, that in turn, affect the Output of Activities

190   *Chapter 11*

conducted and thus performance of an enterprise. Figure 11.2 illustrates several of these composite enterprise constructs.

11.4 POSITIONING: POTENTIAL VALUE OF THE PROPOSAL

One way of assessing the value of the Systemic Theory of the Enterprise proposed here is to conceptually compare it with other approaches aimed to support the understanding of an enterprise, and then identify the differences. In the following text a comparison is made with key approaches within Systems Thinking and within Organization Theory.

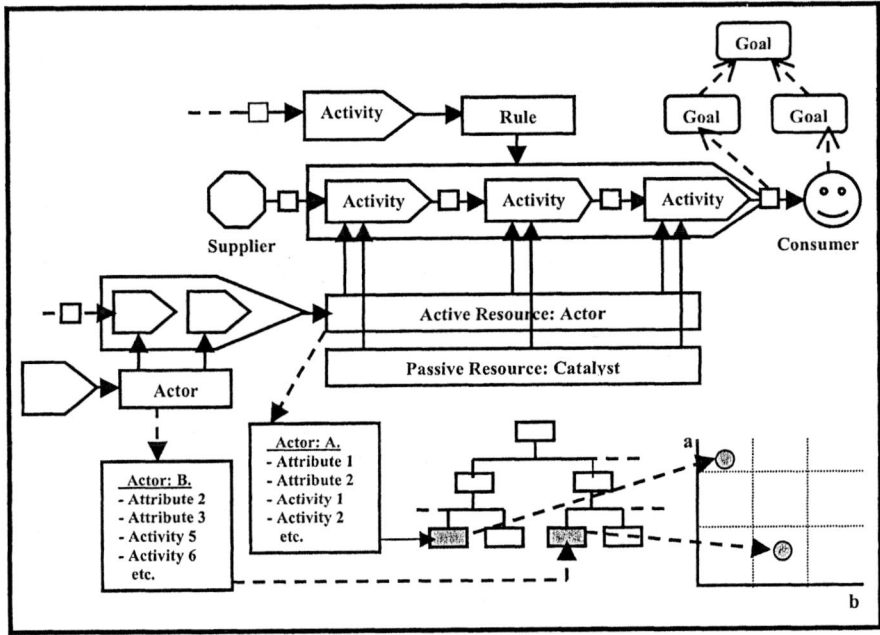

Figure 11-2. Illustrates several of the composite enterprise constructs discussed earlier. Beside the elementary enterprise constructs illustrated in Figure 11-1 above, the Figure also illustrates: the Process composed of the three Activities mentioned, the Core-Policy Process which formulates the Rule, then the Core-Support Process that generates properties of the Active Resource Actor mentioned above, and then the Support-Support Process that generates the properties of the Actor executing the Core-Support Process, then there are two Actors articulated in terms of their inherent Attributes and Activities; these Actors are then allocated within an Organization Structure,

and also within an Organization Culture. The symbolism employed is optional.

11.4.1 Comparison with Systems Thinking

Systems Thinking may be understood to be the intellectual domain that employs the idea of a system as a conceptual tool for managing human affairs, whether public, business or any other kind (Checkland 1981). A somewhat established distinction classifies the various theoretical and practical contributions within Systems Thinking into three main strands, the so-called Hard Systems Thinking, Soft Systems Thinking, and the Critical Systems Thinking (Flood & Jackson, 1991). It is symptomatic and relevant to note here that C. West Churchman contributed to the birth of all of these, in various ways. In the following text, the Systemic Theory of the Enterprise outlined here will be compared briefly with these three strands of Systems Thinking (see Eriksson (2004: Chap. 4) for more details).

11.4.1.1 Regarding Hard Systems Thinking

Hard Systems Thinking is typically associated with systems approaches such as Operations Research, Systems Analysis, Systems Engineering, and Systems Dynamics (Eriksson 1998). Whichever approach is regarded, they have common properties: these approaches include methods of problem solving, typically comprising the phases (i) problem formulation, (ii) problem modeling, and (iii) derivation of a problem solution. Hard Systems Thinking approaches provide quantitative methods for problem solution. For example, Operations Research includes methods such as various linear programming methods, network models, goal programming, integer linear programming, deterministic and probabilistic dynamic modeling, forecasting models, decision analysis, queuing systems, simulation modeling, etc.

These approaches are typically either general, as with Linear Programming which is a mathematical modeling technique designed to optimize the usage of limited resources, or more limited in scope and specialized, as is the case with transportation models, which are a specific sub-class of Linear Programming, dealing with situations in which a commodity is shipped from a source to a destination.

The following three properties at least serve to differentiate the various Hard Systems Thinking (HST) approaches from the proposed theory, see Table 11.1.

1. HST tends to model a limited part of an enterprise – e.g. a network model determining the shortest route between two cities on an existing network of roads – while the proposed framework provides support for a comprehensive articulation of an enterprise, where and if needed – i.e. from consumers and outputs, throughout channels, processes, activities, transformation objects, supplies and rules, to the various resources, including humans, organization culture and machines.
2. HST focuses on numeric representation of a situation, with a limited and often very generic type of enterprise constructs, i.e. "resources", "source", "destination", while the proposed approach focuses on a rich set of enterprise constructs, with no numeric representation necessary.
3. HST provides strong support for problem *solving*, e.g. the how-to-do things, while being at the same time rather weak in its support of the actual problem *formulation*, e.g. what-to-do. The proposed approach has the opposite emphasis, i.e. it supports formulation and articulation of enterprises, including their problems and opportunities, but does not support their solutions. In this way, the proposed approach is complementary to the various HST approaches.

11.4.1.2 Regarding Soft Systems Thinking

Soft Systems Thinking is associated with systems approaches supporting formulation and structuring of managerial problems, such as Soft Systems Methodology (SSM), Interactive Planning (IP), Strategic Options Development and Analysis (SODA), Metagame and Hypergame approaches (Eriksson 1998). These are typically (i) non-quantitative approaches with emphasis on support of (ii) the articulation of a messy situation followed by structure of a formulation for the problem or issue. However, as is the case with Hard Systems Thinking approaches, these methodologies tend to model some (iii) limited part of an enterprise. Further, they often provide (iv) methodological stipulation about activities that should be pursued if the problem is to be formulated, which is not the case with the proposed Systemic Theory of the Enterprise. A comparison of Soft Systems Thinking approaches and the proposed approach shows that these may be complementary in that the latter provides a rich language for comprehensive articulation of an enterprise, while the former provides a procedure for identifying managerial problems, yet typically providing a rather limited way of articulating an enterprise. Hence, for example, in the case of Soft Systems Methodology (SSM), a managerial situation is conceived in terms of its Customer, Actors, Transformation, Weltanschauung, Owner, and Environmental constrains. Even this most advanced approach of modeling within Soft Systems Thinking manifests only a limited part of an enterprise compared to the proposed approach (See Table 11.1 for comparison). Further, SSM offers also the so-called "Activity Modeling" of a desired system or situation, which is a chain of logically coherent activities. When comparing with the framework proposed here, the Activity Modeling approach is a rather primitive way of articulating some desired operations of an enterprise, or human affairs, as it focuses only on the activities, thus generating a highly reduced articulation of the enterprise.

11.4.1.3 Regarding Critical Systems Thinking

We select here only the original approach to problem solving associated with Critical Systems Thinking, namely Werner Ulrich's *Critical Systems Heuristics* (Ulrich 1983). This is focused on support of social system design. It provides theoretical and methodological support for critical reflection upon the assumption of observations and judgments involved in the definition of an 'ought-to', i.e. design of something. Hence, it provides support for dealing with the question of why-to-do something, as compared to the two previous questions: what-to-do, and how-to-do. As the proposed Systemic Theory of the Enterprise aims to provide support to articulation of an enterprise only and is not a methodology for generating the content of a design, i.e. an ought-to, the proposed framework may be regarded as complementary to Critical Systems Heuristics.

Table 11-1. Illustrates an overview of the three mainstream Systems Thinking approaches and the proposed Systemic Theory of the Enterprise

Approach	Systemic Theory of the Enterprise	Hard Systems Thinking	Soft Systems Thinking	Critical Systems Thinking
why-to-do?:	No	No	No	Yes
what-to-do?:	Yes	No	Yes	No
how-to-do?:	Partly	Yes	No	No
scope of modeling:	Comprehensive & Narrow: Enterprise	Narrow: Problem & Design	Narrow: Problem & Design	Narrow: Problem & Design
type of representations:	Non-quantitative: rich set of constructs	Quantitative: very limited set of constructs	Non-quantitative: limited set of constructs	Non-quantitative: limited set of constructs

11.4.2 Comparison with Organization Theory

Enterprises such as public schools, hospitals, manufacturing companies or professional service firms typically all include two or more human beings.

Organizational Theory has generated a vast amount of often-incompatible theories, aspiring to describe, explain and even prescribe organizations and their behaviors. In this account we have selected Morgan's (1986) seminal work, *Images of Organization,* on organization and its theories, which presents a set of "images", or notions, of an organization. Five central images of organizations are compared with the proposed approach to enterprise conception: (i) *organizations as machines* emphasizes parts, processes and regularities, (ii) *organizations as organisms* articulates an organization's adaptation to its environment, (iii) *organizations as brains* focuses on an organization's information processing capabilities, (iv), *organizations as cultures* accounts for the various differences in values, attitudes and perspectives within an organization, and (v) *organizations as political systems* refers to organizations as systems of power conflicts and coercion.

The proposed Systemic Theory of the Enterprise accounts for aspects related to the five notions or images of organizations, where the fifth notion, politics, is not developed in any detail here, see Table 11.2.[4]

The purpose of organization theory differs from the proposed theory; while the latter aims to provide cognitive tools to represent an enterprise the former attempts to describe, explain and prescribe an organization and its behavior. Hence, the proposed approach provides *descriptive* support of an enterprise, and therefore an organization while organization theory, considered in its various dialects, offers attempts to *describe, explain* and eventually *prescribe* the behavior of an organization.

The significance of the proposed approach is that it may provide an integrative approach to articulation and thus description of an enterprise and its organization: this does not limit itself to one notion or image of an organization but rather is an integrative means which may include aspects of all of the mentioned approaches.

Table 11-2. Illustrates an overview of the relation between the here-proposed Systemic Theory of the Enterprise and the Organization Theory, including its five notions of an organization. The proposed approach accounts for organization notions as a machine, an organism, a brain, a culture, and partly as a political struggle.

Image of Organization	Account by the Systemic Theory of the Enterprise	Example of the account
Machine	Yes	Processes, Activities, Structure,

---

[4] The comparison of the proposed theory with Systems Thinking and with Organization Theory should not be regarded as a complete comparison. We have elsewhere (Eriksson 2004; Chap. 10) presented comparison with various Enterprise Modeling Languages and with three types of the Theory of the Firm. Yet numerous other approaches to represent human activities and affairs are not considered here, for instance the rule-based representation as developed by Artificial Intelligence research.

11. AN OUTLINE OF A DESCRIPTIVE THEORY OF THE ENTERPRISE

Image of Organization	Account by the Systemic Theory of the Enterprise	Example of the account
Organism	Yes	Transformation Object Consumer, Supplier
Brain	Yes	Information, Information Flow & Processing
Culture	Yes	Organizational Culture in the enterprise
Politics	Partly	Formal decision making

11.4.3 Summing up the Potential Value

In summary, the proposed Systemic Theory of the Enterprise may be regarded as a complement to the various theoretical strands within Systems Thinking as well as within Organization Theory. This is because the proposed approach provides the richest system of enterprise constructs to support a practical articulation and representation of the diversity of an enterprise in a coherent manner. Hence, whether identifying an optimal solution to a defined problem, formulating the problem itself, or critically questioning the assumptions of the definition and solution to a problem, there is a need for constructs, or categories, to support the expression of the problem: this is provided by the Systemic Theory of the Enterprise. Continuing with the Organization Theory, the proposed theory may both support the articulation of an organization in terms of one of the various available notions of the organization, but also act as an integrative framework to combine the articulation of an organization in terms of several notions, or images, of the organization.

11.5 REALITY CHECK: A CASE STUDY

A short empirical case study is presented in this part of the essay, to manifest the practical managerial value of the proposed theory. An enterprise is described first in natural language, and then in terms of the proposed Systemic Theory of the Enterprise. Thereafter an analysis is made to identify the generated insight and the enterprise constructs employed to generate this insight.[5]

---

[5] Space limitations force us to present a simplified case study with associated problems. This may give the reader the impression that problems defined there can be identified just as well without the support of the proposed theory. This may naturally be so in very simple cases, however in more complex cases the value of the proposed language

11.5.1 Presentation of the Enterprise in Natural Language

This intervention was conducted in spring 2002, followed by post-implementation evaluations at the end of that year and in mid-2003. The object of investigation was the adverse event reporting (AER) operations (i.e. side effects reporting) at a Nordic branch of a leading pharmaceutical company. A few years earlier, this company had lost a major lawsuit brought about by negative side effects of one of its pharmaceutical products. The company's managers assumed that one of the reasons for this was malfunctioning internal control and follow-up routines for adverse events caused by the company's drugs, both during the various stages of clinical trials and when the products were on the market. In order to prevent occurrence of similar situations in the future, the company's headquarters issued a carefully designed adverse event reporting policy. Two key imperatives of this policy were: (i) all employees are responsible for reporting any known adverse event information caused by the company's products to the local adverse event reporting unit; (ii) adverse event information received by any employee must be reported within two working days.

11.5.2 Problem and Hypotheses

Managers of medical operations within the Nordic branch recognized that adverse event reporting imperatives were not being obeyed and that the adverse event reporting routines were inadequate for the task. Thus, the initial question was *why did the adverse event reporting routines not work according to the policy?* One of the working problem-hypotheses for solving this problem was to develop a new computerized information system to support and secure reporting of adverse event information. Hence, implicitly, *the problem-hypothesis for the cause of the malfunctioning operations was the lack of suitable tools for communicating known adverse events.* As will be shown below, the conducted enterprise investigation will give rise to alternative and more detailed problem-hypotheses.

The description of the enterprise below articulates its malfunctioning state. Once the cause of the malfunction had been identified, a design for the desired situation was formulated and explained in terms of the proposed enterprise constructs. However, for space limitation reasons, the first-mentioned situation only is presented here.

---

manifests itself typically by generating contra-intuitive findings, which may not be achieved by intuitive everyday understanding.

11.5.3 Articulation of the Enterprise with the Proposed Constructs

The following text explains the investigated enterprise in terms of the proposed enterprise constructs using alphanumeric symbolism.

We start with the *Consumer* of the AER; this is the Adverse Event Management Centre located at the company headquarters in the USA.

This *Consumer* receives an *Output*, an adverse event report, which includes specified information about the adverse event and its circumstances (i.e. symptoms, probable causes, patient, treatment, physician, etc.).

This *Output* is communicated to the *Consumer* via an *Out-Channel*, which is a dedicated computerized Information System, i.e. a *Machine Actor*.

The AER operations are executed through the AER process, *Core Process*, which includes four main *Sub-Processes*; which are:

"Receipt of adverse event information and forwarding", "Processing of received adverse event information", "Formulation of an adverse event report", and "Reporting of an adverse event report".

These *Sub-Processes* are further broken down into numerous constituting *Activities*.

The *Sub-Process* "Receipt of adverse event information and forwarding" receives an *Input* in the form of information about an adverse event that has occurred.

Hence, the adverse event report is the *Transformation Object*.

This *Input* is communicated via an *In-Channel*, which could be telephone, e-mail, mail or a face-to-face meeting between a *Supplier* and a *Human Actor*.

The *Supplier* can be executed by numerous external *Human Actors*, representing different *Roles* or *Organizational Units*.

The group of *Suppliers* includes the following *Roles*: Patients, Physicians, Nurses, Pharmacies, Health Insurances and The National Health Care Agency.

All these types of *Suppliers* use all types of the specified *In-Channels*, although some *Suppliers* prefer to employ one or two specific *In-Channels* only.

The first *Sub-Process*, i.e. "Receipt of adverse event information and forwarding" can be executed by two types of *Human Actors*, which are both part of the company.

One type of *Human Actor* within the Medical Department i.e. an *Organizational Unit* within the *Organizational Structure* is the *Role* of the Safety Officer.

The other type of *Human Actors* not included of the Medical Department, include the following *Roles*: Sales Representatives (approximate number: 70); Sales Managers (approx.10); Product Managers (approx. 15); Business Unit Managers (approx. 10); Medical Information Officer (approx.1); Medical Advisors (approx. 3); Clinical Managers (approx. 8), and Logistics Officers (approx. 5).

The *Core Process* of AER operations utilizes a *Machine Actor*, in other words a dedicated computerized Information System, for reporting, processing and storing the received adverse event information and reports.

The *Human Actors* constituting the *Role* of the Safety Officer also constitute the *Transformation Object* of a distinct *Core & Support Process*, namely the Safety Officer Development Routine.

The other *Human Actors* mentioned which are not part of the Medical Department e.g. the sales staff, are also *Transformation Objects* of a similar *Core & Support Process*, namely the education and training routine of adverse event reporting.

The first-mentioned *Core & Support Process* (Safety Officer Development Routine) will be conducted by the company's own Safety Officers, i.e. a *Role* constituted by an internal *Human Actor*.

The second-mentioned *Core & Support Process* (Education and Training Routine of Adverse Event Reporting) will be executed by Safety Trainers, an external *Human Actor*.

The AER operations are governed by an adverse event reporting Policy, which constitutes the *Rule* for the behavior of the *Human Actors*, and is formulated by a the medical policy formulation process, i.e. a *Core & Policy Process*.

The enterprise has an *Organizational Structure*, including its *Organizational Units* and these units constituting *Roles*. The model also shows their relation to the *Organizational Units* and the *Human Actors* and their *Roles*, and explains the relation between a *Human Actor* and its *Role* and the constitution of the *Role*.

Furthermore, a sales process is the *Core-Process* within the company's sales operations. This comprises four *Sub-Processes* which are:"Preparation of a sales visit", "Execution of a sales visit", "Reporting an executed sales visit", and "Receipt of adverse event information and forwarding", which is conditional in that it is executed only if a sales representative has received adverse event information.

The fourth *Sub-Process* "Receipt of adverse event information and forwarding" is simultaneously a constituting *Sub-Process* of the above presented AER *Core Process*.

The sales process is executed by the sales representative *Role*, which is part of the *Human Actor* construct also executing the initial *Sub-Process* of the AER Process.

The same sales process is also governed by a sales policy, i.e. a *Rule*. This sales policy stipulates the existence of operational *Goals*, motivating effective execution of the sales process.

The sales policy states: "IF: four sales visits are executed per field working day, on average, THEN: a financial bonus will be awarded to the sales representative".

In operational terms, this sales policy stipulates the two operational *Goals*: Goal 1: "To make an average of four visits per field working day," in order; Goal 2: "To receive financial bonus".

The sales process, or more specifically its potential final *Sub-Process* and its executing *Human Actors*, is governed by the medical policy, which is a *Rule*, and which includes the adverse event reporting imperative: "IF: adverse event information is received THEN: it must always be reported to the Medical Department within two working days."

Finally, the enterprise investigation provided an attitude index, illustrating the sales representatives' actual position in regard to the importance they assign to the sales policy and to the medical policy. This attitude index shows that the sales policy is significantly more important to the sales representatives than the medical policy.

11.5.4 Re-formulated Problem and its Hypotheses

The symptoms of the investigated problem were perceived in terms of adverse event reporting imperatives that *were not being obeyed* by the members of the organization, while the policy imposed that *all* adverse events known to the company's employees *must be reported*, and that these adverse event reports must be executed *within two working days*. Consequently, the question being investigated was: Why did the adverse event reporting routines not operate in accordance with the policy? A working problem-solution-hypothesis within the company was that *deployment of a new local computerized information system would secure reporting of adverse event reports*. Thus, the working problem-hypothesis for the cause of the malfunctioning operations was the *lack of suitable tools for communicating known adverse event information*. However, the description of the enterprise above articulates the investigated situation in a way that can support the following reasoning.

Finding #1: Receipt of adverse event information and forwarding it may be handled by a variety of *Human Actors* and their respective *Roles*, for example safety officers, sales representatives, clinical managers, etc. If this is to work properly these *Roles* need to know that adverse events, as such, exist, and that any received adverse event information must always be reported within two working days. The task of training staff in adverse event information reporting routines should be executed by the *Core & Support Processes* – i.e. the safety officer development routine and the adverse event reporting training routine. However, this investigation showed that the second process mentioned was never executed, and was scarcely even conceptualized within the company.

This finding acts as the foundation for an explanatory problem-hypothesis: *The various Human Actors and their Roles, for example the sales representatives, did not report received adverse event information as they had not been instructed and trained to do so.*

As a consequence, a problem-solution-hypothesis would be: *To operationally establish the adverse event reporting training routine (Core & Support Process) and to ensure that all staff are trained in and familiar with it.*

Finding #2: The model articulates a conflict of policies and associated goals. Two different *organizational units* – the medical unit and the business unit – had independently issued two different policies or *Rules*: the *medical policy* regulating the conditions for reporting received adverse event information, and the *sales policy* which defines the conditions for earning bonus (goal) based on number of conducted sales visits (sub-goal to the mentioned goal). The conflict is founded on the fact that the two policies attempt to govern behavior of one and the same type of *Sub-Process*, i.e. receipt of adverse event information and forwarding it – which belongs to two different *Processes* – the *adverse event reporting* process and the *sales process*.

Further, these two policies simultaneously govern one and the same *Role*: that of the sales representative. The result of this is that when the same sales representative receives adverse event information from a physician when conducting a sales visit, two conflicting policies govern his or her behavior.

On the one hand, the medical policy stipulates that he or she should take the time to report this information; on the other hand, the sales policy imposes a goal structure to make the sales process efficient, so that sales representatives spend their time conducting sales visits. In other words, two different interests compete for the same time slot with a sales representative.

Next, the two conflicting policies force the sales representative to make a choice, which is ultimately based on his or her attitude, convictions and values (part of *Organizational Culture*). As is proven by this investigation, a sales representative's attitude is to follow the sales policy rather than the medical policy.

11. AN OUTLINE OF A DESCRIPTIVE THEORY OF THE ENTERPRISE

These articulations of the investigated enterprise provide a foundation for formulating the following problem-explanation-hypothesis of the cause(s) of the malfunctioning adverse event reporting process:

Adverse event information is not reported by sales representatives as this may limit their achievement of sales goals and consequently their financial bonus, while sales representatives generally consider it more important to earn their bonus than to report adverse event information; the underlying cause of this is that the Medical Policy and the Sales Policy are conflicting, or not coordinated with each other.

11.5.5 Resolution of the Problems

This explanation provides a foundation for formulation of a problem-solution-hypothesis for the investigated problem in the following terms:

Measure #1: *The Medical Policy and the Sales Policy must be re-formulated to coordinate with each other, in such a way as not to expose sales representatives to a conflict of interest. An example of this could be to re-formulate the incentive structure to provide financial bonus for reporting adverse event information.*

Measure #2: *Both the staff recruitment process and the staff development process need to be reformulated to ensure that all staff possess the core attitudes and values underlying the policies governing the company's operations.*

This investigation has delivered two key causes explaining why the adverse event reporting policy was not followed. These – the lack of training and the staff's prioritization of economic incentives – are not necessarily two *alternative* causes but may very well *complement*, i.e. enforce each other.

The employment of the Systemic Theory of the Enterprise as proposed here has generated an articulation and understanding of the investigated enterprise, leading to the formulation of alternative problem-hypotheses for explanation and resolution of the problems. Hence, the benefit, or value, of the proposed enterprise constructs is manifested as it provides support for generation of a richer articulation of the enterprise, which in turn gives a greater insight into the problems perceived.

11.6 SUMMARY

C. West Churchman's life-long inquiry may be characterized as the quest for a systemic comprehension of social systems. His contributions may be conceived as attempting to characterize the social system as such and characterizing the inquiring process necessary for understanding the social system. In this sense, this essay proposes a contribution to the first mentioned area: the characterization of those social systems that are conceived as enterprises.

The Systemic Theory of the Enterprise outlined here provides a rich system of enterprise constructs or categories, aimed to support the manager in his or her conceptualization of the managed enterprise, whether in the analysis of current states or design of desired situation. The proposed theory has been generated from a combination of theoretical deduction arising from the so-called Le Moigne's Systemic theory, a theory of modeling, and empirically induced from numerous case studies executed in Action Research mode.

This essay suggests that the Systemic Theory of the Enterprise may well complement the main strands of Systems Thinking – i.e. Hard, Soft, and Critical Systems Thinking – by providing them with a richer system of enterprise constructs to articulate the enterprise in question. Further, the proposed theory also complements the typical strands of Organization Theory as it may both support them with a richer set of enterprise constructs to generate the conception of an organization, and also provide a framework for integrative conception of an organization in terms of several different organizational theories.

We have also presented a case study of an enterprise, described first in natural language together with the perceived problems, and articulated thereafter with the proposed system of enterprise constructs. The latter gave rise to a richer articulation of the enterprise, which in turn gave rise to a greater insight into the problems and the generation of alternative solutions. In summary, the proposed Systemic Theory of the Enterprise supports a richer expression of an enterprise than the other approaches discussed, yet further investigations do need to be conducted to further expand the ability to generate comprehensive understanding of social systems.

Finally, as one of C. West Churchman's key concerns was the moral dilemma of all system design – i.e. how can we rationally justify the normative foundation of all prescriptions formulated – the theory presented here may perhaps serve to help us somewhat as it supports the generation of a comprehensive understanding of the enterprise under investigation, which is necessary if we are to avoid sub-optimized solutions which may be harmful to the entire system...

REFERENCES

Checkland, P.B., 1981, *Systems Thinking, Systems Practice*, Wiley, New York.

11. AN OUTLINE OF A DESCRIPTIVE THEORY OF THE ENTERPRISE

Checkland, P.B., 1991, "From framework through experience to learning: The essential nature of action research", in: *Information Systems Research*, H-E. Nissen, H.K. Klein, R. Hirschheim, eds., Elsevier, Amsterdam.
Checkland, P.B. and Holwell, S., 1998, "Action research: its nature and validity", *Systemic Practice and Action Research*, 11(1):9-21.
Churchman, C.W., 1971, *The Design of Inquiring Systems*, Basic Books, N.Y.
Churchman, C.W., 1979, *The Systems Approach and Its Enemies*, Basic Books, New York.
Eriksson, D.M., 1997, "A principal exposition of Jean-Louis Le Moigne's systemic theory", *Cybernetics & Human Knowing*, 4(2-3):35-77.
Eriksson, D.M., 1998, *Managing Problems of Postmodernity: Some Heuristics for Evaluation of Systems Approaches,* Interim Report IR-98-060/August, International Institute for Applied Systems Analysis, A-2361, Laxenburg, Austria.
Eriksson, D.M., 1999, *An Application of the Decision-Information-Operation System Model*, Luleå University of Technology, Luleå, Sweden.
Eriksson, D.M., 2003, "Identification of normative sources for systems thinking: an inquiry into religious ground-motives for systems thinking paradigms", *Systems Research and Behavioral Science*, 20:475-487.
Eriksson, D.M., 2004, *Four Proposals for Enterprise Modeling*, Doc.Diss, Chalmers University of Technology, Department of Industrial Management and Economics, Göteborg, Sweden.
Flood, R.L. and Jackson, M.C., eds., 1991, *Critical Systems Thinking: Directed Readings*, Wiley, Chichester.
Le Moigne, J.L., 1990,. *La modélisation des systèmes complexes*, Dunod, Paris.
Le Moigne, J.L., 1993, "Formalism of systemic modeling", in: *Some Physicochemical and Mathematical Tools for Understanding of Living Systems*, H. Greppin, M. Bonzon, Degli and R. Agosti, eds., University of Geneva, Geneva.
Le Moigne, J.L., 1994a, *La théorie du système général. Théorie de la modélisation*, 4th ed., PUF, Paris.
Le Moigne, J.L., 1994b, *Le constructivisme, Tome 1: Des fondements*, ESF éditeur, Paris.
Le Moigne, J.L., 1995a, *Le constructivisme. Tome 2: Des épistémologies*, ESF éditeur, Paris.
Le Moigne, J.L., 1995b, *Que sais - je? Les épistémologies constructivistes*, PUF, Paris.
Morgan, G., 1986, *Images of Organization*, Sage Publ., CA.
Piaget, J., 1937, *La construction du réel chez l'enfant*, Neuchâtel, Delachaux et Niestlé. (Also: *The construction of reality in the child*, Translation M. Cook, Basic Books New York, 1971).
Ulrich, W., 1983, *Critical Heuristics of Social Planning: A New Approach to Practical Philosophy*, Haupt, Bern.

Chapter 12

ORGANIZATIONAL EFFICIENCY AND VALUES
A Tribute to West C. Churchman

PARRA LUNA, F.

Abstract: A brief look through West Churchman's works suffices to see his concern for values and ethics in the behaviour of social organizations, be they political or of any other nature. Values are, in fact, what Churchmann stresses when he addresses the question of the problems facing the world, viewed as a political-social organization. But, where to start - he asks - to eradicate hunger, if first what needs to be changed is political will, and even before that politicians' education, and prior to that, the education of the citizens who elect politicians, and so on? Where to break the vicious circle? What system of values would have to be adopted to make that possible? What organizational strategy can be followed? What approach to organizational or political efficiency should we take? There are, of course, no easy answers to these questions. What can be broached is the way to advance towards an operational definition of the ultimate goals of human organizations, without deviating too far from the needs (that values fulfil) of the universal human being, who, in addition to being free, must be able to share with others. The difficult compatibility between **Distributive Justice** and **Freedom** is the core of the problem as West Churchman posed it. In this regard, what I personally am going to try to put forward is an operational (axiological) definition of the concept of ORGANIZATIONAL EFFICIENCY following the lines of methodological and ethical concerns of C.W. Churchman.

Keywords: Distributive justice, freedom, organizational efficiency

But proposing an appropriate or scientifically valid reply to the questions set forth entails solving certain basic theoretical problems, the first but not the most important of which is to attempt to distinguish between the concepts "Company" and "Organization". The former is a social-legal entity whose primary aim is to earn the highest possible returns on capital, along with other implicit although secondary or media-oriented aims. The latter, while it may concur with businesses in the pursuit of such secondary/media-oriented aims, does not a seek financial profit, at least as its *raison d'être*. In much of the literature on organization theory, however, no such distinction between the two types of social systems is drawn (Mullins 1996, Kreitner and Kinicki 1996, Espejo et al. 1996, Fernandez-Rios and J. Sanchez, 1997), even though these systems (organizational and entrepreneurial) differ significantly in a number of other aspects worthy of note. To give a few – far from exhaustive – examples, whilst organizations are constrained by (often statutory) rules and regulations, companies are private ventures that may change course or line of business whenever they wish; by contrast, whilst the Damocles sword of competition hangs over the latter, organizations are not subject to such pressure; whilst companies are obliged to be creative to defend themselves from competition, organizations merely follow the rules; whilst companies that don't grow fall behind (mergers, acquisitions, etc.), organizations have no such concerns; and finally, whilst companies are implicitly subject to a certain ethical vulnerability deriving from the confrontation between capital and labour, organizational behaviour is regarded to be ethical as long as no rules are broken. Companies, in short, as Dithurbide (1999) points out, are "a void, with no constraining rules, a 'free trade zone' with their own legislator". Although such an understandably exaggerated assertion is not to be taken literally, it is nonetheless clear that companies enjoy degrees of freedom and face hazards that are lacking in most other organizations. For these reasons, all stemming from their participation on a free market, companies may be said to be constantly in a state of tension unknown to any other manner of organization.

A second theoretical problem arises because the term "Efficiency" is enormously polysemous and at the same time its meaning overlaps with that of certain homonyms. A number of authors (Cameron and, 1983, for instance) have substantiated the existence of this problem. It suffices, in this regard, to cite the confusion around the terms Efficacy (when a company reaches its proposed aims); Effectiveness (when its employees (and its stakeholders in general) accept its behaviour); "Efficiency" (when the company behaves rationally); "Profitability" (when the unit of measure is the capital gains generated); "Productivity" (when production, profitability, etc., are related to the number of company employees); "Yield" (referred to net financial profitability); "Success" (when the ultimate aims are reached); "Growth" (referred to turnover and/or staff); "Development" (when certain desirable levels are reached); "Excellence" (when financial profitability and system expansion prevail), and so on and so forth. As Zammuto (1982) sustains, the term Efficiency is subject to considerable semantic chaos

It may be deduced from the ideas associated with these terms that none fully expresses the notion of GLOBAL good governance, which would cover not only the attainment of a company's initial objectives (efficacy for many), but also their ability to do so at the lowest cost or with the least possible effort (what others call Efficiency), while at the same time generating consent or acceptance among workers and other stakeholders (usually referred to as "Effectiveness" in political science literature). What concept might include at least these three chief components of a company's desirable behaviour?

[In view of this need for a term that expresses the idea of company efficiency, efficacy, performance, etc., GLOBALLY (incorporating both economic and social aspects), a viable solution might be to redefine some one of such terms, such as EFFICIENCY. This would entail not restricting it to mean merely the ratio between "ends achieved/resources used" - not only a dilutive but a frankly dangerous definition - but supplementing it with the other meanings ascribed to the words discussed above. If the initial question is to be answered with any rigour, account must necessarily be taken of the most obvious and prominent complexities of the company as a social system. And this, in principle, is simply a question of adding to rather than subtracting from the perspectives adopted to depict the concept as a more integrated whole.

The first operation should consist in labelling "ends achieved/resources used" – the ratio normally called "efficiency" – differently to release the latter term it from its present content. What might this action be called, in view of the fact that it consists of achieving the result pursued at low cost, expense, effort, etc., i.e., with a lower consumption of energy? In principle the denomination ECOLOGICAL might be accepted insofar as consumption is minimized. Although provisionally, the adjective "ecological" might be adopted to mean business behaviour with a suitable output/input ratio, the classic expression of the transformation process taking place in any company, or even in any system (Von Bertalanffy, 1968; van Gigch, 1978).

Broadening the term Efficiency to mean more than exclusively that ratio paves the way for the development of a more complex as well as operational and quantitative definition of Business Efficiency (EE). There are two stages to this process: a) the definition of Efficiency in general in any type of social organization seen from the systemic standpoint (or as a series of interrelated elements targeting specific aims), whether or not the system is a company; and b) the extension of that definition to the Company as a privately owned entity and consequently relatively free of bureaucratic restrictions, but subject to the risk of loss of competitiveness and obliged to earn at least a minimum financial profit to survive.

The third theoretical distinction is related to the terms "system", "organization" and "company", inasmuch as all companies are organizations and all organizations are systems. In other words, system is the most general of the three concepts and the one of greatest theoretical interest. The wealth of literature produced since Bertalanffy's formulation in the nineteen fifties (from Henderson to Parsons, from Buckley to Luhmann, from Churchman to van Gigch, from Foerster to Jackson/Flood, to name a few) has given rise to the introduction of three fundamental concepts: Epistemological Totality, Relevance and hard (as in Klir) or soft (as in Checkland) Formalization, all of which have had a considerable impact on sociological theory in connection with the company as the fundamental socio-economic cell of today's society. The numerous and influential works of W. Churchman on measurement, values, goals of organizations and ethical behaviour, are seminal for this approach. In light of this, I'll be using the concept of system as a methodological approach to the notion of organizational efficiency.

I'll begin this discussion with a necessarily abridged introduction to an eminently sociological view of the business system. Table 12.1 illustrates the different "social domains" that make up the complex and turbulent world of business relationships. These domains may be INTERNAL or IN-HOUSE (Employees, Shareholders); EXTERNAL (Customers, Suppliers, Financial System, Labour Market, Government Entities, Society and Inter-company Agreements) or MIXED (Associates, Trade Unions, Parent Companies and Subsidiaries) which may be either inside or outside the company, cutting across the boundary line (solid line) that separates the company from its surrounding environment at the operational level.

Table 12.1 Social Domains vs. Values-Objectives Matrix

Social Domains	Pattern of Values
Society	Health, Conservation of Nature
Customers	Health, Security, Material wealth, Knowledge, Freedom, Distributive Justice, Prestige, Self-fulfilment
Public Authorities	Security, Material wealth, Distributive Justice
Shareholders	Security, Material wealth, Distributive Justice, Prestige
Employees	Health, Security, Material wealth, Knowledge, Freedom, Distributive Justice, Conservation of Nature, Prestige, Self-fulfilment
Associates	Security, Material wealth, Knowledge, Freedom, Distributive Justice, Prestige, Self-fulfilment
Suppliers	Security, Material wealth, Knowledge, Prestige, Self-fulfilment
I-C agreements	Security, Material wealth, Distributive Justice, Self-fulfilment
Financial system	Security, Material wealth, Prestige
Trade unions	Health, Security, Material wealth, Knowledge, Freedom, Distributive Justice, Self-fulfilment
Labour market	Prestige

Allow me to preview the basic needs/values model (which I'll develop later) both to strike a balance between and respond to the various social environments. The matrix in Table 12-1 crosses social domains against company Values-Aims to give some insight into the powerful interests that conform business activity and its functional dependence on such domains, all of which aspire to obtain something from the company: salaries, dividends, profits, jobs, taxes and so on. Like it or not, there's no denying the central role of private enterprise as the primary economic footing of modern society (upon whose private profits all of a society's structure depends), even when viewed from the most critical, counter-cultural or post-modern perspective.

Taking these premises as our theoretical point of departure, we must now reformulate the initial question in more general terms: When is a social system efficient? In principle, and as I announced, when it is "Efficacious", "Ecological" and "Effective" all at the same time. And not only that, but it must be all three in relation to other comparable systems within a common social/economic/political/geographic/ historical Environment. It is generally accepted that nothing can be described as good/bad, high/low, beautiful/ugly except in comparison to something else. It is entirely possible to have a highly efficacious, ecological and effective system that is nonetheless the least efficacious, ecological and effective of all similar or comparable systems in its surrounding region. The relative nature of Efficiency would appear, then, to be indisputable. Hence, yet another aspect must be introduced in our discussion of this new concept: the INTERNAL/EXTERNAL dimension involved in this spatial comparison.

A TIME-BASED criterion might also be introduced whereby a social system would only be regarded to be Efficient if it improves its performance with respect to its immediate past. However, this "Progressive" criterion would be cancelled out or engulfed by another, "Adaptability", since a system may be classified as Efficient even when it is less so than in preceding periods, providing that all other systems (or the average of all comparable systems in its Environment) prove to be even more regressive.

The conceptual model of Systemic Efficiency (ES) I'm suggesting would correspond, then (in principle), to the five propositions set out in Table 12-2.

Table 12-2. Requisites for Efficiency in social systems

1.	A social system is Efficient if and only if it is EFFICACIOUS (achieves its aims).
2.	A social system is Efficient if and only if it is ECOLOGICAL (reaches a desirable Input/Output ratio).
3.	A social system is Efficient if and only if it is EFFECTIVE (is accepted by social environments)
4.	A social system is efficient if and only if it is ADAPTED (is at least equally efficacious, ecological and effective as the average of comparable systems in its Environment).
5.	A social system is Efficient if and only if is axiologically BALANCED (if the variation between the levels reached in the Values is significantly small).

These five minimum requirements attempt to comply with the principle of "epistemological totality", which means that they must accommodate the nine Values-Aims listed in the matrix in Table 12.1. Thus, for instance, in connection with efficacy, the entire System of Values attained or fulfilled (from Health to Prestige) must be compared to the System of Values envisaged or expected as Churchman (1950, 1959, 1974 and 1989) put it. Indeed, it would be decidedly rash to define the concept of Business Efficiency or systemic-social efficiency in general without taking account of this prior epistemological concern.

Having outlined these prerequisites for the new ES construct, it is now time to draw attention to the axial problem facing studies on System Efficiency, specifically as regards the literature on organizations: the lack of a general theoretical model for space-time comparisons of organizational efficiency. The general consensus is that since there is no such model, the subject of Efficiency can hardly be addressed. Some authors, such as Goodman, Atkin and Schoorman (1983) have even recommended refraining from engaging in such studies altogether in view of the conceptual immaturity prevailing. Others (e.g., Kreitner and Kinick, 1996) sustain that there are any number of criteria for organizational efficacy. This is confirmed by Handy (1993), Baguley (1994), Dunderdale (1994) and Mullins (1996), who focus on the concept "Effectiveness" and whose definitions, not unexpectedly, do not concur. Similarly, Cameron (1986) argues that "there is no single approach to evaluating efficacy that is suitable under all circumstances or for all types of organizations". According to Y.W. Buckley (1999), one of the founding fathers of the Theory of Social Systems, "in complex multicultured societies one may ask if such a standardized value system is possible to define for the varying and conflicting values, wants and needs of the several ethnic, cultural and class subgroups. It will clearly be a difficult job to define adequately the key terms and justify the types of measures and mathematical techniques necessary to get at the global performance of social systems in a meaningful way". And this is despite the fact that, as Buckley himself notes, system and organizational efficiency is the subject of CONSIDERABLE discussion, as it must be, given the relevance and interest it holds for anyone involved in or dependent upon a social system. But precious few operational definitions have been proposed to measure the concept and, according to what these authors sustain, while much desired, no universally applicable theoretical model has yet been generated.

In other words, research and discussion have been severely crippled by a working hypothesis launched "urbi et orbi", to the effect that: business/organizational efficiency is unmeasurable for two main reasons: a) there is no acceptable universal theoretical model and b) no such model can be built because "efficiency" means something different to every individual. Since every individual is unique, the argument goes, unrepeatable in his/her specific system of values, ideology, interests, likes and circumstances, no common system of values can be validly formulated. In this regard, Arrow's well-known Impossibility Theorem would reinforce the hypothesis by attempting to prove that it's impossible to develop a Global Utility Function from individual utility functions. We would, therefore, have mere individual mental "constructs", not a sufficiently validated theoretical concept, to work with.

The logic of this discourse is so convincing that some authors (such as Goodman, Atkin and Schoorman, 1983) recommend postponing studies on Organizational Efficiency, while others (Hannan and Freeman, 1977) go even further and advocate the pure and simple renunciation of such studies. The hypothesis implies, then, two lines of action: a) desistance from the task of calibrating and space-time comparing the efficiency of organizations, which are created solely and exclusively to reach a certain degree of efficiency and whose very existence depends on achieving that aim; and b) questioning or openly rejecting the possibility of integrating individual opinions into a single global opinion that minimizes the distance from each. What this premise actually questions is the viability of democracy. As posed in most of the literature, the problem is as flexible and permissive as it is ethically and functionally disturbing, since it would appear to grant the persons in positions of responsibility in social systems licence to define the efficiency of organizations at their discretion at any given time.

Are other possibilities in sight, outside continuing to be immersed in the shadow of the semantic chaos around the concept of Efficiency described above, and in particular around the possibility of measuring it? Indeed, as Fernández-Ríos and Sánchez (1997, p. 45) sustain, there are three reasons to continue to pursue studies on Efficiency ("Effectiveness" – EFO – for these authors), namely:

"THEORETICAL REASONS: The studies on EFO are central to organizational theory (Goodman and Pennings, 1977; Pfeffer, 1977; Cameron and Wheten, 1983) and are associated with all organizational models.

"All conceptualizations on the nature of organizations implicitly or explicitly include notions on EFO and the differences between effective and ineffective organizations. For instance, Contingency Theory emphasizes the consistency between an organization and its environment as the key feature of effectiveness (Galbraith, 1977; Lawrence and Lorsch, 1969; Hall and Clark, 1980).

"EMPIRICAL REASONS: EFO is the last dependent variable in organizational research (Cameron, 1986). The need to prove that a given structure is, in some way, better than others makes the notion of effectiveness a central empirical issue.

"PRACTICAL REASONS: People continually pass judgement on the effectiveness of organizations and this is the ultimate standard for determining the validity of an organization's design and operation.

"There are always reasons to judge effectiveness: even in the absence of ideal and satisfactory indicators of importance on effectiveness there would be a tendency to resort to others of less scientific quality but which would provide for some comparison. Unfortunately, people's judgements are often based on criteria that are either unrelated to or inconsistent with organizational accomplishment."

The problems identified will obviously not be easily solved. Nonetheless, following Machado's recommendation, we should, "make our way as we go" and the scientific "way" is ultimately one of trial and error and try again in pursuit of the final goal. The discussion below addresses a few reasons why it would be feasible to make gradual progress in the development of such a necessary general model, or in any event, to intensify the need to continue research on SYSTEMIC EFFICIENCY (SE) (or whatever other term may be adopted to denote this concept, providing it reflects all organizational concerns as a whole, regardless of the differences in approach, perspectives and results). There are essentially four such reasons:

1. As we have just seen, the notion of Efficiency or some related term is continually used in everyday life in a wide variety of social systems, from companies distributing profits to sport teams electing new officers, not to mention democratic parliaments solemnly debating the "State of the Nation". The fact that this concept is inappropriately "calculated" for want of a theoretical model should serve as encouragement to develop a

more highly evolved theoretical-methodological approach, in keeping with scientific standards.
2. Moreover, how can explanatory structures and the transformation processes of complex systems be as profusely analyzed as they generally are, in pursuit of scientific explanations of their behaviour, unless this concept is first defined and understood? It would be illogical, and it would be appear to be methodologically impossible, to try to explain the variance of variables $X1, X2, \ldots Xn$ around a dependent variable Y without first measuring the latter. There is no explaining something that isn't understood. And yet, this type of analysis is often effected in depth and usually with enviable literary brilliance, although naturally to the detriment of the scientific merit of the result. It appears to be quite necessary, then, to continue to study the Efficiency of social systems, since the definition and measurement of this notion are requisite to explanatory analysis.
3. It's not true that there is no universal theoretical model that can be applied to social systems. If, as most authors sustain, the components of social systems are human beings, such systems can and should only pursue what such persons, as a single global human race, pursue. And in most situations that arise in time and space, that is tantamount to saying anything that is incumbent upon all human life on the planet Earth. The most obvious example is the Universal Declaration of Human Rights proclaimed by the United Nations on 28 June 1948. The general model of human needs that it explicitly seeks to fulfil can be gleaned from a simple analysis of its articles. Health, Security, Standard of Living, Education, Freedoms, Justice, Personal Fulfilment and Worldwide Solidarity are the eight essential – inasmuch as both universally accepted and desirable – Values that emerge as the objectives of all individuals in whatever type of social system, from the family to the Nation-State, from sports clubs to churches or from schools to workplaces. Such Values, with the same or different names, are also to be found as well in the models recommended by authors who have attempted to establish common patterns of human needs, such as Maslow (1970), Laswell (1936), Deutsch (1974, 1980), Terleckyi (1970) and Bunge (1999). In summary, at least three philosophies to support this general model can be established: a) PHILOSOPHICAL models, defended by authors such as E.H. Carr, B. Croce, M. Gandhi, A. Huxley, S. de Madariaga, J. Maritain and P. Theilard de Chardin in their collective work "Los derechos del hombre" (Carr et al. 1973); b) SOCIOLOGICAL models, such as developed on the basis of the works of Bauer, Biderman, Gross et al. (Bauer et al. 1966) by the host of authors comprising the social indicators movement; and c) POLITICAL models, such as those introduced by most historical

Constitutions based on the 1787 American and 1789 French texts and ultimately summarized and universally accepted in the aforementioned UN Declaration of Human Rights. These have been further supplemented by the more recent Stockholm 1972 Conference on the Human Environment and subsequent meetings on sustainable development, whose implicit result has been to add the value "Conservation of Nature" to the 1948 declaration of rights. Cortina (1994) called this set of key universal values "third generation rights", for they are a compilation of the rights issuing from three historic moments: 1787/89 (earliest constitutions); 1948 (UN) and 1972 (Stockholm). This model is summarized in Table 12.2 as the basis for a possible general model with multiple operational possibilities.

4. While it's true that this universal model is designed, in its implementation, to generate different systems or sets of indicators through which it is operationalized (and which might, therefore, appear to detract from its universal nature), it is likewise true that once the universal and topical nature of the ("third generation") values is accepted, the indicators that operationalize each Value may be identical for all human systems, even though such operation may not always be economic in nature. What I mean by this is that while an indicator such as "percentage of households with running water" certainly affords very little pertinent information in developed countries, this should not prevent it from being taken into account in inter-system comparisons of different degrees of socio-economic development. It would obviously be irrelevant in a comparison of the degree of economic development between the U.S.A. and Canada, for instance, but not between the U.S.A. and some less developed countries. The determining issue is, then, the General Model of Universal Values-Aims and not the indicators to be applied in each case, which must be in keeping with the systems compared and the research aims.

Based on these four arguments, the UN 1948 Model of Values is adopted as the basis for the general model, with the sole addition of the Value "Conservation of nature", which wasn't included by the founders of the UN in 1948 simply because ecological sensitivity was less developed then than it is today. Models therefore, no matter how general, are like science itself: dynamic, contingent and provisional. And the possibility of progress lies precisely in the provisional nature of scientific knowledge. The importance of the Model described below is that it could be adopted today for any definition of Efficiency in whatever type of social system, large or small, simple or complex, modern or primitive. The UN model is necessarily a system with transcultural aspirations deriving from its irrefutable universality.

12. ORGANIZATIONAL EFFICIENCY AND VALUES

Table 12-3. Reference pattern of values (PRV)

NEED (N)		FUNCTION	VALUE PURSUED (Y)	
Kind	Symbol		Kind	Symbol
1. Physical and mental well-being	N_1	Health-care related	HEALTH	Y_1
2. Material sufficiency	N_2	Economic	MATERIAL WEALTH	Y_2
3. Protection against contingencies	N_3	Security-related	SECURITY (LAW AND ORDER)	Y_3
4. Understanding of Nature	N_4	Research and educational	KNOWLEDGE	Y_4
5. Freedom of movement and thought	N_5	Liberating	FREEDOM	Y_5
6. Equity	N_6	Redistributive	JUSTICE	Y_6
7. Power and esteem for others	N_7	Prestige-building	PRESTIGE	Y_7
8. Harmony with Nature	N_8	Naturalist	ENVIRONMENTAL CONSERVATION	Y_8
9. Self-fulfilment	N_9	Humanistic	QUALITY OF ACTIVITIES	Y_9

The possible strength of the model as a theoretical basis for the concept of Systemic Efficiency (ES) would derive from its three most outstanding features, namely: 1) From the theoretical standpoint this is a model that summarizes the contributions of many authors from many different perspectives, as well as the theoretical-semantic structures (systems of socio-economic development indicators) used by the major national and international organizations in the production of the statistical data. 2) From the standpoint of political practice, the vast majority of the governments of today's Nation-States, Regions, Departments or Cities tend to name the ministries, regional and municipal departments, and so on, after the chief tasks and functions that such Values-Aims represent, affording empirical-political support for the aforementioned theoretical assumptions. And 3) finally, it would be no easy task to overturn a model proclaimed/accepted by the World's approximately 200 Nation-States, with such different and diverse cultural, mental, ideological and economic backgrounds. Its transcultural and universal nature (the UN model implicitly/politically represents the nearly six billion human beings that inhabit the Earth) is what would possibly lend it indisputable validity as a general theoretical model.

The Annex sets out a system of indicators deriving from the hypothetical application of the model to companies, with no other intention than to better illustrate the possible content of such Universal Values. But, I must reiterate that it is not specific indicators, but the axiological-universal skeleton of the model that is decisive. A more detailed description can be found in F. Parra Luna (1983, pp. 215-263). The empirical indicators that operationalize the model in each case must, naturally, meet the standard methodological requirements of soundness, significance, maximum independence, heurisitcity and unidirectionality.

Before proceeding to apply it, however, it may be useful to review some of the chief characteristics of the general model for social systems inspired by the theory of transforming entities (Von Bertalanffy, Lange, Piaget, Easton, Van Gigch):

- All social systems inevitably consist in a process that transforms "Inputs" into "Outputs". The former are the resources used, the latter the degree to which people's needs are met. The INPUT (X)-TRANSFORMATION (T)-OUTPUT (Y) process is formalized in such a way that $Y=f(X)=TX$, where (only "ecological") system Efficiency would be expressed as: $T = Y/X$.
- Values are taken to be the "flip side" of Needs (Kluckhohn, C., 1951) and since social systems are made up of people with recurring unfulfilled needs (humans are "needy beings"), the only "Outputs" that social systems can produce are VALUES and they wouldn't be able to produce anything else even if they wanted to. Social systems are fated/determined to over- or under-produce Universal Values-Aims as their *raison d'être*.
- The generation of such Values encounters difficulties deriving from the existence of geographic, biological, economic and similar ceilings (or floors) and the dialectic interrelationships between the Values themselves. Sometimes this interrelationship is positive (for instance, a higher degree of Knowledge usually goes hand-in-hand with greater Material Wealth and vice-versa), but at others it is dialectically negative (for instance, greater Freedom is associated with less Distributive Justice and vice-versa). Such axiological dialectics lie at the root of the different ideologies.
- Given the historically contrasted difficulty of balancing the "production" of Values-Aims at relatively equivalent levels, social systems inevitably constitute axiological profiles deriving from the different emphasis placed on or achieved with certain Values at the expense of others. Politically speaking, for instance, capitalistic systems are characterized by their emphasis on the Values FREEDOM and MATERIAL WEALTH at the expense of the Values LAW AND ORDER-SECURITY and DISTRIBUTIVE JUSTICE; communist regimes, in turn, emphasize the Values DISTRIBUTIVE JUSTICE and LAW AND ORDER-SECURITY at the expense of FREEDOM and MATERIAL WEALTH, whereas Fascist regimes generally emphasize the Value LAW AND ORDER-SECURITY primarily at the expense of the Values

12. ORGANIZATIONAL EFFICIENCY AND VALUES

DISTRIBUTIVE JUSTICE and FREEDOM. Since it may not be possible to attain the maximum levels in all Values simultaneously, the most highly developed political systems attempt to balance their axiological profiles as fairly as possible. In such political terms, perhaps Northern European social democracies may be taken as more or less successful examples of attempts to strike such a balance.

- Values-Aims appear both at the real or objective $Y(O)$ and the perceived or subjective $Y(S)$ levels. Global outputs or "Y" are the result of the integration of both, expressed as the average: $Y=(Y(O) + Y(S))/2$.
- Finally, it must be made very clear that the Reference Pattern of Values-Aims (the simple nominal list of the nine Values in the model) is one thing and the System of Values (axiological profile) that each social system pursues/attains in each case, quite something else.

Now that the theoretical-conceptual grounds have been established, I'll go on to describe the methodology to be used, which consists of the STANDARDIZED quantification of each indicator in a common range of variability, from "0" (minimum or worst case) and "100" (maximum or best case). Let's assume, for instance, that a company's management regards the indicator "Turnover" (which for the previous year was 1.2 billion monetary units) should range between 800 million (assumed WORST CASE) and 1.6bn (assumed BEST CASE). These two values represent the two extremes of a continuum designed in such a way that both are unattainable under normal circumstances and serve only as a comparative percentage reference between indicators. This is their purpose. And it is somewhere within that hypothetical range of variability that the company's managers will set a realistic goal to be reached, such as a turnover of 1.35bn, which should be expressed as a percentage within the established range. Hence, if the difference between Best and Worst Case (1.6bn-800m=800m) is equal to 100, the difference between the target and the Worst Case figures (1.35bn-800m) is equal to $X=68$. This means that the target figure for turnover is 68% of what was initially defined as the "ideal". If this simple proportional rule is routinely applied to each and every one of the indicators used, we'll have a whole series of standards or targets, all expressed in PERCENTAGES and therefore directly comparable. This yields AXIOLOGICAL PROFILES that are both readily visible and open to criticism and eliminates the diversity of units of measure in which indicators are generally expressed.

If the actual turnover reached in the period considered was 1.28bn, for instance, the percentage attained would be 60. The negative deviation (68-60=8) would be directly comparable to the results attained for the other indicators and simply overlaying the graphs of the (expected and actual) axiological profiles would afford an instantaneous view of the positive and negative deviations, along with the areas of responsibility concerned and the Reference Pattern of Values to which they pertain. The differences between the two profiles would translate into the deviations between the PROJECTED SYSTEM OF VALUES and the SYSTEM OF VALUES ATTAINED for the year. For the intents and purposes of an initial evaluation of Business Efficiency and particularly as a tool for management control, this particular company would be able to tell that it had worked to 68% of the "ideal" level and to 88% of its target.

This is usually generalized as follows: if we have "n" standardized indicators (y_1, y_2, ... y_n) we can express the Outputs or "Y" as the average of all "n" ($y_1 + y_2 + ... y_n$)/n ="Y". The Inputs or "X", in turn, follow the same principle, but bearing in mind the two differences discussed below:

- Unlike Outputs or "Y" (which is generally a complex index comprising several indicators), the Inputs or "X" usually refer to a single indicator: the cost or total expenditure budgeted/laid out to produce the Outputs or "Y". But there is no diversity or, consequently, any need to standardize.
- In contrast to the Output indicators, the BEST CASE in the range would always correspond to the lower amount and the WORST CASE to the higher, and the respective percentage is calculated in such a way that a small percentage signifies good management in terms of cost savings. For instance, to reach the turnover budgeted above (1.35bn) a total cost target of 1.05bn was set within a range of 800m (best case) and 1.3bn (worst case). Therefore, the respective percentage is: if the difference (1.3bn-800m=500m) is equal to 100, the difference (1.05bn-800m=250m) is equal to X=50%. If the actual expenditure amounted to 1.1bn, the percentage would be 60%, for a negative deviation of 50-60=-10%. If this information is computerized and routinely processed, it would also constitute a management control and deviation analysis tool (F. Parra Luna, 1989, 1993). Or, as J.A. Garmendia (1987) put it: "Ensuring the efficacy of the (organizational) apparatus calls for monitoring not only the engine (rules) but also the brakes (deviations)."

From the perspective of the Efficiency obtained with such indicators, we find that $T = Y/X = 68 / 50 = 1.36$ as the BUDGETED index in the Input/Output ratio, whereas the index actually attained would be $60 / 60 = 1$, for a negative deviation of -0.36. Since both numerator and denominator are expressed in percentages, the values of "T" may be interpreted as follows: if T=1 systemic transformation is NEUTRAL; if T is greater than "1" transformation is MULTIPLICATIVE and if T is less than "1" it is REDUCTIVE.

Finally, the existence of comparable axiological profiles (or numerical vectors), namely the Projected (Y_p) and Actual (Y_r) Outputs, makes propositions 4 and 5 in Table 12.1 operational.

With this initial theoretical-methodological background, the concept SYSTEMIC EFFICIENCY (ES) may be formally estimated under one of two variations: A) for ORGANIZATIONS in general and B) for profit-seeking COMPANIES. My apologies to the reader for the overdose of symbols in the following discussion, but I see no other way of inter-relating the various items to calculate a quantitative estimate:

12.1　FOR ORGANIZATIONS IN GENERAL

A1　INTERNAL DIMENSION

A11 Degree of LEVEL EFFICACY (Efn)
$$Efn = (Yr / vy)r) / (Yp/vy)p) \qquad (1)$$
where: Yr is the average of the percentages of the actual vector
Yp is the average of the percentages of the projected vector
(vy)r is a measure of the internal variability of the percentages of the actual vector
(vy)p is the variability of the projected vector

Therefore, "Efn" fluctuates around one. If Efn = 1, the projected value concurs exactly with the actual value. If Efn is less than "1", the expected efficacy has not been attained and if Efn is greater than "1", actual performance was better than expected.

A12 Degree of PROFILE EFFICACY (Efp)
$$Efp = r\ (yr, yp) \qquad (2)$$
Where "r" is a correlation coefficient that ranges from 0 to 1 (Mills, 1962) between the actual and projected vectors. The closer this coefficient is to "1", the higher is efficacy.

A13 Degree of ECOLOGICAL EFFICACY (Ec)
$$Ec = TR / Tp \qquad (3)$$
where:　Tr is Yr/Xr, actual Outputs/Inputs ratio
Tp is Yp/Xp ratio, projected or expected ratio.

"Ec" also fluctuates around one. If Ec = 1, the degree of Ecological performance (in the broadest sense) is neutral. If Ec is less than "1", more (human, financial, etc.) resources have been expended than necessary; and if Ec is greater than "1", there has been a net savings on resources.

A14 DEGREE OF EFFECTIVENESS
A141　LEVEL (Etn)
$$Etn = YSr/YSp \qquad (4)$$

where YSr is the vector of values perceived as real by staff and YSp the vector they expected. A result greater than "1" means conformity with system behaviour, whereas values of less than "1" mean frustration.

A142 PROFILE (Etp)

$$Etp = r(YSr, YSp) \qquad (5)$$

where "**r**" is the correlation coefficient between vectors YSr (subjectively perceived as real) and YSp (subjectively expected). The closer the value is to "1", the greater is the acceptance among system stakeholders.

A16 DEGREE OF AXIOLOGICAL BALANCE (Eq)

$$Eq = 1/1+v$$

Where "v" is the internal variability of the objective outputs vector, Y(O).

We find that INTERNAL EFFICIENCY (EI) can be formally summarized as follows:

$$EI = Efn \cdot Efp \cdot Ec \cdot Etn \cdot Etp \cdot Eq \quad (6)$$

This, of course, is a multiplicative relationship, which is subject to the disadvantage of possibly reducing EI to a value much lower than "1", due primarily to the effect of the correlation coefficients, "**r**". An additive relationship might be adopted instead, namely:

$$EI = (Efn+Efp+Ec+Etn+Etp+Eq)/5 \quad (7)$$

In both expressions the terms may be weighted or otherwise. The latter has the advantage of being centred more closely around one. Generally speaking, if EI is greater than "1" Internal Efficiency is positive. If it is less than "1", efficiency is negative.

A2   EXTERNAL DIMENSION

As I mentioned earlier, the assessment of any system entails a comparison with other systems in its environment, a comparison which is fairly problematic under highly competitive circumstances, since the necessary data are usually lacking and/or scantly reliable. The problem might be posed from a dual standpoint: IDEAL (or utopian) and OPERATIONAL or realistic.

A21 IDEAL ADAPTATION (AI)

$$AI = EI/\overline{EI}$$

where **$\overline{EI}$** would represent the Internal Efficiency of all the different comparable systems in the environment as a whole – information which, I insist, is unlikely to be available.

A22 OPERATIONAL ADAPTATION (AO)

In light of the difficulty involved in using the above expression **$\overline{EI}$**, in some organizations it may be feasibly replaced with some global indicator of the results obtained by the organizations in the environment. Thus, for instance, if the usual socio-economic indicators are not all available for a less developed Nation-State, the use of only one might be significant. In this case "Life Expectancy", for instance, could be a good choice, inasmuch as it is demographically simple to compute and is regarded to be positively correlated with other development indicators (per capita income, level of education, level of health care, and so on). Sports organizations would be another example: the difficulty in obtaining indicators such as y1, y2, ... yn might be offset by using teams' final classifications, information that is readily available, as the maximum expression of the organization's *raison d'être*.

In these cases, then, it would be feasible to replace the expressions EI and **EI** with some global indicator (G) such as mentioned, so that the expression:

$$AO = GE/GE \quad (9)$$

would be indicative of the degree of acceptance, where GE is the global indicator representing the system measured and GE the same indicator for the neighbouring systems.

The summary formula for ORGANIZATIONAL EFFICIENCY (EO) would be expressed as follows:

$$EO = (EI + AO)/2 \quad (10)$$

i.e., a simple arithmetic mean, which may be weighted or otherwise.

More explicitly, the formula would be:

$$EO = (E_{fn}+E_{fp}+E_c+E_{tn}+E_{tp}+E_q+AO)/7 \quad (11)$$

a numerical expression that likewise fluctuates around one: the higher the value of this expression, the higher is the degree of ORGANIZATIONAL EFFICIENCY.

12.2 FOR COMPANIES

Companies are slightly different insofar as the Outputs in a business system cannot be represented by $Y = (y1 + y2 + ... + yn)/n$, even where one or several of these indictors represent financial profit, and no matter how heavily they are weighted with respect to all the others. In this type of organization financial profit is too axial to consider it to be just one of several indicators. The primary aim of private enterprise is to obtain at least a standard return on the capital used, and if this requirement isn't fulfilled the company isn't viable. The financial profitability indicator often appears, then, as the sole criterion to determine Business Efficiency. Martín López (1997) put it this way: "Compliance with the teleological assumptions suffices to fulfil the necessary conditions for a company to exist and survive".

When it comes to evaluating a company's profitability, at least two dimensions can be taken into account: a) the difference between

PROJECTED and ACTUAL profitability, and b) profitability compared to the figures for other significant companies in the ENVIRONMENT.

B1 PROJECTED-ACTUAL profitability (Rpr)

$$Rpr = Rr/Rp \qquad (12)$$

where "Rr" is the ratio between net profit earned and the mass of capital used (or lacking that, turnover) during the respective period. "Rp" would be the same ratio but between profits and mass of capital as projected at the beginning of the period.

B2 COMPANY-ENVIRONMENT profitability (REE)

$$REE = Rr/RE \qquad (13)$$

where RE is the average net profit/capital used ratio for the companies that define the business Environment.

Consequently, company profitability (R) would be found from the following expression:

$$R = (Rpr + REE) / 2 \qquad (14)$$

which assigns the same relative importance to internal (with respect to the budgeted value) and external (compared to the other companies) profitability. This formula may, of course, be weighted differently.

To sum up, the degree of BUSINESS EFFICIENCY (EE) could be globally expressed as follows:

$$EE = (EO + \mathbf{R}) / 2 \qquad (15)$$

which would cover both its efficiency as a complex organization and its profitability as a company.

This final simplified version of the formula should not conceal its two chief characteristics: a) its complexity, as it usually comprises many indicators, and b) its necessity in determining the OVERALL EFFICIENCY of an organization or company.

To be applied to a company, then, Table 12.1 would be amended and completed as shown in Table 12.4 below:

Table 12-4. Requisites for business efficiency

1.	A company is Efficient if and only if it is EFFICACIOUS.
2.	A company is Efficient if and only if it is ECOLOGICAL
3.	A company is Efficient if and only if it is EFFECTIVE.
4.	A company is Efficient if and only if it is ADAPTED.
5.	A company is Efficient if and only if it is BALANCED.
6.	A company is Efficient if and only if it is PROFITABLE.

12.3 CONCLUSION

The present sketch of the operational concept of Organizational Efficiency begins and ends with the description of these six prerequisites, seen as a conceptual model to assist better decision making. Whether or not the definition so formulated represents progress towards such personal self-fulfilment is another question. Assailed by that doubt, perhaps the most that can realistically be hoped for is that others will be enticed to exploit this "axiological" concept of society more fruitfully.

REFERENCES

Baguley, P., 1994, *Improving Organizational Performance: A Handbook of Managers*, MacGraw Hill.
Bauner, R., et al., 1966, *Socil Indicators: A First Approximation*, MIT Press, Cambridge, Mass..
Buckley, W., 1999, "Preface" al libro de Parra-Luna, F. *The Performance of Social Systems: Perspective and Problems*, Plenum, New York.
Bunge, M., 1979, "Teatrise on Basic Philosophy", Vol. 4, *A World of Systems*, Redidell, Dordrecht.
Bunge, M., 1999, *Social Science under Debate*, UTP, Toronto.
Cameron, K.S., 1986, *Effectiveness as Paradox: Consensus and conflict in Conceptions of Organizational Effectiveness*, Managements Science, mayo.
Cameron, K.S. and Whetten, D.A., 1983, *Organizational Effectiveness: A Comparison of Multiple Models*, Academic Press.
Campbell, K.S., 1977, "On the natura of organizational effectiveness", in: *New Perspectives on Organizational Effectiveness*, P.S. Goodman & J.M. Pennings, eds., Jossey-Bass, San Francisco, CA.
Carr, E.H. et. al., 1973, *Los derechos del hombre*, Ed. Laia, Barcelona.
Chevalier, A. 1977, *El Balance Social de la Empresa*, Univ-Empresa, Madrid.
Churchman, W., 1950, "When do we start value research?", *Philosophy of Social Issues*, 6(4):61-63.
Churchman, W., 1959, "Organizations and goal revisions", Working paper CP-9, Center for Research in Management Science, University of California, Berkeley, Dec.
Churchman, W., 1974, *Qu'est-ce que l'analyse par les systemes?*, Dunod, Paris.
Churchman, W., ed., 1989, *The Well-Being of Organizations*, Intersystems Publications, Salinas, Calif.
Cortina, A., 1994, *La etica de la sociedad civil*, Anaya.
Dalton, R.D. and Kesner, I.F., 1985, "Organizacional performance as an antecedent of incide/outside chef executive sucession: An empirical assessment", *Academy of management Journal*, **28**:749-762.
Deustch, K.W. 1974, *Politics and Government. How People decide their Fate*, Houghton Mifflin Co., Boston.
Deustch, K.W., *Los nervios del gobierno: Modelos de comunicación y control político*, Buenos Aires.

Dithurbide, G. 1980, "Problemas en el análisis del conflicto laboral", in: *Economía, organización y trabajo: un enfoque sociológico*", Castillo Mendoza, C.A., Pirámide, Madrid, 1999. Paidos, 1980.

Dunderdale, P., 1994, "Análisis Effective Organizations", *Professional Manager*, Sep.

Edwards, R. et al., 1986, "The competing values approach to organizational effectiveness: A tool for agency administrators", *Administration in Social Work*, **10**, 4, 1.14.

Espejo, R. et al., 1996, *Organizational Transformation and Learning*, Wiley, Chichester.

Fernandez-Ríos, M. and Sanchez, J. 1997, *Eficacia Organizacional*, Díaz de Santos, Madrid.

Garmenida, J.A., 1987, "La empresa como organización e institución", in: *Socilogía industrial y de la empresa*, J.A. Garmendia y otros, Aguilar, Madrid.

Gigch van, J., 1978, *Applied General Systems Theory*, Harper & Row, New York.

Goodmand, P.S., Atkin, R.S. and Schoorman, F.D., 1983, "On the demise of organizational effectiveness studies", in: *Organizational Effectiveness: A Comparison or Multiples Models*, K.S. Cameron & D.A. Whetten, Academic Press, N.Y.

Handy, C.B., 1993. *Understanding Organizations*, Penguin.

Hannan, M.T. and Freemann, J.H., 1977, "The population ecology of organizations", *American Journal of Sociology*, 82.

Kaplan, R.S. and Norton, D.O., 1997, *Cuadro de mando integral*, ed. Gestion 2000, Barcelona.

Kreitner, R. and Kinicki, A., 1995, *Comportamiento de las organizaciones*, Irwin, Madrid.

Laswell, H., 1936, *Politics: Who get what, when, how*, MacGraw Hill, N.Y.

Lazarfeld, P., 1965, "Des concepts aux indices empiriques", in: *Le Vacabulaire des sciences sociales*, R. Boudon and P. Lazarsfeld, Mouton.

Mahoney, T.A., 1977, "Managerial perceptions of organizacional effectiveness", *Administrative Science Quarterly*, 14:357-365. (1967)

Maestre Alfonso, J., 1974, *Introducción a la antropología social*, Akal, Madrid.

Martín López, E., 1997, *Sociología Industrial Fund*, Form. A. Prof. Madrid.

Maslow, A., 1970, *Motivation and Personality*, Harper and Row, N.Y.

Miles, R.H., "Macreo organizational behaviour", Sta Monica, CA, Goodyer Pub. Co., 1980.

Mills, C., 1962, *Metodos estadísticos aplicados a la economía y los negocios*, Agular. Madrid.

Morgan, G., 1980, "Paradigms, metaphors, and puzzle solving in organizattional theory", *Administrative Science Quarterly*, **25**:605-622.

Mullins, J., 1996, *Management and Organizational Behaviour*, Pitman, London.

Olson, M., 1967, "Anuals of the American Academy of Political and Social Science", Social Goals and indicators for American Society, n.1, mayo.

Olson, M., 1969, "Social indicators and social accounts", *Socioeconomic Planning Scs*, **2**.

Parra Luna, F., 1993, *Elementos para una teoría formal del sistema social*, Univ, Complutense de Madrid. and *El Balance Integrado de la Gestión Estratégica*, Deusto.

Seashore, S.E. 1976, "Defining and measuring the quality of working life", in: *The quality of working life*, L.E. Davis and A.B. Cherns, eds., Free Press, N.Y.

Peters, T. and Watermann, R., *In Search of Excellence*, Harper & Row, N.Y.

Sudreau (Informe) Comité d'etudie pour la reforme de l'enterpirse, 1975, La Documentation francaise, Paris.

Terleckyi, N.E., 1970, "Meassuring progress toward social goals: Some possibilities at national and local levels", *Management Sciences*, **16**(12).

West Churchman, C. 1974, *Qu'est-ce que l'analyse par les systémes?*", Ed. Dunod, Paris.

Zammuto, R.F., 1982, *Assessing Organizational Effectiveness: Systems Change, Adaptation and Strategy*, Suny-Albany Press, Albany, N.Y.

SECTION E

NEW PARADIGMS: APPLICATIONS OF THE SCIENCE OF MANAGEMENT TO GOVERNANCE AND MANAGERIAL PRACTICE

Chapter 13

MOLAR AND MOLECULAR IDENTITY AND POLITICS
Working and re-working boundaries of identity: from citizenship to human rights and sentient beings

MCINTYRE-MILLS,J.

Abstract: The structures and processes of international relations and governance need to be re-considered to address diversity. The paper makes a plea for systemic governance. Policy makers and managers need to work *with* rather than *within* theoretical and methodological frameworks to achieve multidimensional and multilayered policy decisions. Conceptual tools can be used to enhance systemic governance. The closest we can get to truth is through compassionate dialogue that explores paradoxes and considers the rights and responsibilities of caretakers.

Key words: identity, capacity building, critical and systemic governance

13.1. INTRODUCTION: TWO-WAY LEARNING, SOCIO-ECONOMIC WELLBEING AND DIVERSITY

"Vision is always a question of the power to see- and perhaps of the violence implicit in our visualising practices" (Haraway, 1991: 192, in Fine et al 2000, 108).

Citizenship (as it is currently constructed) does not take into account the rights of young people, the marginalized within nation states, those who are stateless or the rights of those sentient beings who cannot speak for themselves, by virtue of age, mental or physical health or species limitations. Using conceptual tools could enhance our ability to design inquiring systems[1] that are more mindful and more conscious of both our rights and responsibilities not only to ourselves, but also to sentient beings. The structure of the chapter is to ask some questions and to address the questions by means of complex case studies and vignettes from the news media. The policy context is the legacy of development inequalities and the cold war, the 'fall out' of September 11th, the War on Terror, border control and Australia's role in 'the coalition of the willing'. It is also the epidemiological challenge of new epidemics, such as 'bird flu' and climate change (see volume 3). This has introduced the mantra of 'building the capacity' of 'weak states '(Fukuyama 2004). The questions are as follows:

- How do we define culture, identity, rights and responsibilities in a changing world?
- Can interconnected regionalism provide a more sustainable form of governance?
- What does this mean for understanding transnational governance and international relations and the caretaking role?
- What kind of capacity building is needed internationally and with the Asia–Pacific?
- Are we able to make policy or do we respond to policy decisions made on our behalf (Chomsky, 2003) in a bid to address 'failed states' (Fukuyama, 2004)
- How can so-called 'strong and weak states' build their governance capacity?

13.2. KNOWLEDGE, IDENTITY AND CULTURE IN A CHANGING WORLD

Capacity building for compassionate and sustainable governance needs to develop an ability to think about ontology (by asking what is the nature of reality?) and epistemology (by asking how do we know what we know?). Critical questioning could help improve our ability to move in this direction. Groups at the local level, networks, teams, and social movements could strive to model the sort of ongoing iterative dialogue that is vital for problem appreciation and contextual resolutions by and for the stakeholders who will need to live with the decisions.

Democracy needs to avoid becoming constrained and hidebound, because diversity is essential to the extent that it does not undermine the freedom of others.

---

[1] I work with and adapt Churchman, C. West, 1971, *The Design of Inquiring Systems: Basic Concepts of Systems and Organization*, Basic Books, New York.

13. MOLAR AND MOLECULAR IDENTITY AND POLITICS

I argue in this chapter that the concepts of consciousness and identity are responsive and contextual. The way forward to improve governance is to improve policy-making processes that appreciate complexity. This is only possible by means of a 'design of inquiring systems' (West Churchman, 1971) to support compassionate and sustainable governance. It is both idealistic and pragmatic to re-work governance and international relations.

Research by National Economics (National Economic/ALGA, 2002) shows that economic development has shifted from primary industry (fishing and forestry and mining) to secondary service industry and now knowledge creation, indicated by patent development, tolerance, technology and talent. All these factors go hand-in-hand. The reports provide qualitative and quantitative evidence for socio-economic and environmental well being as part of the same moment and context. This research benchmarks Australian regions against American and European regions for the pragmatic purpose of finding out the characteristics of successful regions, so that Australian local government and state government (to which local government is responsible in terms of the Australian constitution) and business can learn from the comparison. The gaps between the regions in Australia are much higher than in America[2] and in Europe (including United Kingdom). A number of reasons are given for this in these reports. One of the key suggestions being the way government operates.

Usually an argument begins with definitions. Definitions are based on drawing lines, based on decisions[3]. Culture, identity and politics are concepts that need to be defined contextually, so 'making a cut is a bad way to start'. Case studies and vignettes provide a way into understanding life chances associated with health, culture, race, education and gendered identity.

We can consider culture and identity as 'molar' or fixed or we can also consider culture and identity as 'molecular' and fluid, to quote Deleuze and Guattari (Bogue, 1989). An understanding of citizenship and human rights

---

[2] Sydney, the leading Australian region in terms of technology when benchmarked against the highest-ranking American region comes 24th, however in terms of composite diversity (defined in terms of cultural and social diversity as well as education, type of creative occupation and number of patents). This is problematic for access to knowledge that is commodified. But it is indicative of a realisation of the value of knowledge per se. Both Sydney and San Francisco are on a par.

[3] Decision is the Latin 'for cut', a point made by Churchman 1979a,b. West Churchman stressed methodologically that early decisions about frameworks and relevance could lead to designs that "cut of" options. He reminded us that "decision" is derived from the Latin root meaning "to cut off" and that making decisions sometimes leads to limiting our creative options and problem solving. Open communication systems are vital for managing the mess of complex human lives and tragedies, in so far as we need to work in a complementary way, not in isolation.

needs to encompass wider horizons than is currently recognized by democratic governments.

For some people locally and internationally a sense of who they are is based on *choice* and changes to their life styles and chances can and are made . For some mobility is impossible and their life chances are limited. For others, identity and culture is fluid and a sense of space and time is quite different. Today people live in many different worlds for some the worlds are connected.

Connections and barriers are social, political, economic and environmental. Lived experience and cultural knowledge have been given less power than professional knowledge in the past.[4] Gibbons and Limoges et al (1994) stress that the divide between the knowledges that they call mode 1 and mode 2 knowledge is unhelpful for development.

The Australian Research Council (2002, website cited the work of Gibbons et al, 1994, on the new nature of knowledge) and National Economics and ALGA (2002, 2003) have stressed the importance of enabling regions to draw on the tacit (non-codified)[5] knowledge of participants, by enabling networks of creative people to develop the region. This is a bottom up approach to development. The challenge in the future will be to base policymaking and governance on 'thick description'[6] based on the voices of many stakeholders and advocacy for the voiceless. This is the basis for revitalizing democracy.

The world comprises many different life chances and economies and as stressed in Volume 1 of the C.West Churchman series, conceptually information is not the same as knowledge. The challenge is to work with difference and different ways of knowing .The boundedness of knowledge is an issue for governance. It requires more than so-called knowledge management to address enhancing connections. In fact the term suggests that

---

[4] Gaventa, J. and Valderrama, C., 1999, "Participation, citizenship and local governance", Background note for workshop on *Strengthening Participation in Local Governance,* Institute of Development Studies, June.
Gaventa, J., 2001, "Towards participatory local governance: Six propositions for discussion", Paper presented to the *Ford Foundation*, LOGO Program with the Institute of Development Studies, June.
Gaventa, J. and Cornwell, A., 2001, "Power and Knowledge", in *Handbook of Action Research,* P. Reason and H. Bradbury (eds.), Sage, London.

[5] Gao, F. and Yoshiteru Nakamori, 2001, "Systems thinking on knowledge and its management", *45th International Conference for the Systems Sciences*, Asilomar, USA
Gao, F., Li, M. and Nakamori, Y., 2002, "Critical systems thinking as a way to organise knowledge", *Systems Research and Behavioural Science,* 3:19.
Mylonopolous, N. and Tsoukas, H., 2003, "Editorial: Technological and organizational issues in knowledge management", *Knowledge and Process Management*, 10(2):139-143.

[6] Geertz, C., 1973, "Thick description: towards an interpretive theory of culture", in *The Interpretation of Cultures. Selected Essays*, C. Geertz, Basic Books, New York, pp. 3-32.

knowledge and information are the same and that directing the flow of the right information for the task is the challenge. We need to acknowledge that many kinds of knowledge co-exist and finding ways to communicate across knowledge domains is the challenge.

Habermas[7] suggests respectful communication is the way forward. Whilst Derrida in the same publication suggests that not only do we need respectful communication (in order to co-create shared domains of understanding) we also need to accept that people will see things differently. Spaces for conceptual difference need to be respected to the extent that they do not undermine the rights of others.

Democracies are responsible for looking after their own people, their own citizens and for some categories of asylum seekers who are recognised to be refugees and not for others. But what about other people? Nation states draw boundaries around 'us' and 'them'. Australia has prided itself on "a fair go" democratic culture. We are a nation of immigrants. Ideally democracy requires 'working and re-working' the conceptual and geographical boundaries. This needs to be undertaken in a less cynical manner. Nevertheless the argument supports a form of revised democracy and enlightenment that respects the importance of freedom to the extent that it does not undermine the freedom of others.

The key concepts[8] are 'bonds' that draw us together, 'boundaries' drawn by individuals and groups and 'norms' that guide the behaviour of groups'[9] and 'transformation and emergence' to explore not merely 'culture in interaction'[10], but the processes for bringing about change. The concept of change is discussed[11] and are worked with, rather than applied, in order used to inspire creativity. Bonds are the connections we draw across self, other and the environment. The more inclusive and the wider we can draw the boundaries of protection of the other, the greater the potential for creating bonds or relationship and trust and the closer we can move towards Human and Environmental Justice that is supported by norms supported by international agencies such as the United Nations Aarhus Convention to

---

[7] See Habermas in Habermas, J., Derrida, J. and Giovanna, B., 2003.
[8] Drawing on the following papers by Brubaker, R., Loveman, M. and Stamatov, P., 2004, "Ethnicity as cognition", *Theory and Society*, **33**:31-64
Brubaker, R. and Cooper, F., 2000, "Beyond Identity", *Theory and Society*, **29**:1-47.
[9] Elias, N. and Lichterman, P., 2003, "Culture in interaction", *The American Journal of Sociology*, **108**(4):735-794
[10] Elias, N. and Lichterman, P., 2003.
[11] Such as: structural differentiation (Maturana), dissipative structures (Prigogine and Stengers), moving equilibrium (Parsons), autopoesis (Maturana and Valerela), eternal return (Deleuze and Guattari), boomerang effect (Ulrich Beck), dialectic (Habermas) and unfolding or sweeping in (West Churchman). They form the basis for what I call Gaian governance based on agape communication for emergence.

protect against environmental hazards across geographical boundaries (October 30, 2001)[12] and The International Criminal Court[13]. Australia signed up to the ICC after the deadline on 1 July 2002. The purpose of the ICC according to the Rome Statute is to support "peace and justice".

13.3. CASE STUDIES FOR EXPLORING BOUNDARIES

The following stories based on primary and or secondary research(see McIntyre 2005 a,b, McIntyre-Mills 2003). .

13.3.1 Boundaries, politics and identity

The Australian federal election held on Saturday the 9th of October 2004 was an election won by the Howard Government on the basis of Australian voters who considered that electing a tried, and trusted leader (in the sense of looking after their economic interests) was in their interests. Strong border protection and a strong economy based on fixed geographical and conceptual

---

[12] Svitlana Kravchenco, 2001, in a paper entitled: The doors to democracy are opened! Quotes Kofi Annan "The Aarhus Convention is the most ambitious venture in environmental democracy undertaken under the auspices of United Nations. Its adoption is a remarkable step forward in the development of international law as it relates to participatory democracy and citizens' environmental rights...Its entry into force today, little more than three years after it was adopted, is further evidence of the firm commitment to those principles of the Signatories- including States in Eastern Europe and Central Asia whose role in this process clearly demonstrates that environmental rights are not a luxury reserved for rich countries." He goes on to cite the UN High Commissioner of Human Rights Mary Robinson as follows" The convention is a remarkable achievement not only in terms of protection of the environment, but also in terms of the promotion and protection of human rights, which lie at the heart of the text. As such, Aarhus is a key step in the progress of integrating human rights and environmental issues...It entry into force is a key signpost for the future of human (sic) and the environment in all parts of the world...The great value of this Convention lies not only in the promise of protection it afford the people and the environment in Europe, but also in the model it provides for similar action in other nations and regions in the world"

[13] According to Human Rights Watch (http://hrw.org/campaigna/icc/ accessed 7/28/2004 "The statute outlining the creation of the court was adopted at an international conference in Rome on July 17, 1998. After 5 weeks of intense negotiations, 120 countries voted to adopt the treaty. Only seven countries voted against it (including China, Israel, Iraq, and the United States) and 21 abstained. 139 states signed the treaty by the 31st December 2000 deadline. 66 countries - 6 more than the threshold needed to establish the court - ratified the treaty on 11 April 2002. This meant that the ICC's jurisdiction commenced July 1 2002. From February 3-7, 2003, the court's Assembly of States Parties - the ICC's governing body-elected the court's first 18 judges. The resulting...Judicial bench (the judges include 7 women) were sworn into Office on March 11 in the Hague..."

boundaries won the day. A government that promised to maintain low interest rates and a 'firm hand' (based on non- negotiation with terrorists was considered the best choice to secure their futures) in the wake of the Bali Bombing and the bombing of the Australian Embassy in Jakarta. The only Australian citizen injured in the latter tragedy was a small girl born to an Indonesian woman, who had recently married an Australian. They were en route to collect documentation from the embassy.

Currently governance supports citizenship and nationalism that is based on thinking and practice that encourages looking after ourselves and not others. Selfishness may seem to make sense in the short term, until we realize that our social, economic and environmental decisions could lead to a 'boomerang affect' (Beck 1992). Our mutual survival depends on co-operative and compassionate governance. Compassion for self, other (including all sentient beings) and the environment is both pragmatic and idealistic.

Critical and systemic thinking is helpful to enable us to think ethically (McIntyre-Mills, 2003), not only because it exposes the contradictions and helps us to think about their implications for Australia and Australians, but also the contradictions in other Western democracies.

Internationally democracy as it is currently constructed in the powerful Western nations (comprising the so-called coalition of the willing lead by the United States) does not support a sustainable future. A revised enlightenment approach is needed as argued in volume 1. The ideal version of open and participatory democracy and enlightenment go hand-in-hand. Closure leads to entropy (Flood, R. and Carson, E., 1998).

When are Australians, Australians? We can look at any topic in terms of what Bateson (1972) calls Level 1, 2 or 3 thinking in the 'Ecology of Mind'. Level 1 thinking means thinking in terms of one existing framework. It repeats what is taken for granted. It is rote learning. So citizenship could be discussed in terms of the legal status of Australian citizenship that came into being on 26th of January 1949 under the Nationality and Citizenship Act of 1948 (the Act). It was renamed on the 17th of September 1973 as Australian Citizenship Act the Australian Constitution. Full commonwealth voting rights were 'conferred 'in 1962 – Commonwealth Electoral Act – sixty years after the initial act of parliament had removed these rights. Only in 1967, however, were Aboriginal citizens able to vote.[14]

---

[14] The Commonwealth Referendum of 1967 and Indigenous Australian Citizenship: an interpretation of historical events, 1-19. This paper was presented to the Aboriginal Nations and the Australian Constitution Conference, Old Parliament House, Canberra, 23-24 May 1997. It was amended by penny Tripcony, 2001. Downloaded on 10/07/2004 http://www.qiecb.qld.edu.au/html/PPL1.htm.

Level 2 learning could compare what it meant to be an Aboriginal in 1894 in South Australia- where the vote was given to all women, including Aboriginal woman (in theory anyway). Also Pre Federation Aboriginal people could vote as British citizens in South Australia, Victoria, Tasmania and New South Wales, but it was only exercised by a few.[15] It was taken away by the Commonwealth of Australia Constitution Act of 1900, which stressed that any Aborigines who had voted prior to federation could continue to vote, Aborigines as a group could no longer vote. Only in 1962 the clause that excluded them was repealed after debates and deferments. A referendum was held in 1967 that enabled them to be voting citizens.

What does it mean currently to be an Aboriginal citizen without a representative Council such as the Australian and Torres Strait Islander Council? Does the practical reconciliation of the Howard Government meet the international human rights concerns outlined in the Universal Declaration of Human Rights and the two covenants on civil and political rights and economic Social and cultural rights?[16] What does comparing these different experiences tell us about democracy? Level 3 remakes definitions and leaps out of existing frameworks.

This section reflects on being a citizen with minimal rights and reflects on current examples of Australian citizens who have been incarcerated or deported, because as a result of their mental illnesses they have been unable to speak out on their own behalf and they have been without advocates.

Two cases illustrate the powerlessness of the mentally ill and the powerlessness of those who are without citizenship rights in a context of suspicion and the will to control the boundaries. Vivienne Young and Cornelia Rau were incarcerated because they were unable to argue that they had rights. The latter is;

> "the mentally ill Australian resident accused of being an illegal immigrant and held in detention for nine months, the former is a woman who had suffered a car accident after dropping off her child at child care. She was suffering a mental illness and she was 'deported in 2001 after NSW and Queensland police failed to realize that she was listed as missing'. The woman has not been found and her children remain in Australia.

Five months after Mrs. Young's son was abandoned, immigration officials were alerted to the case of a Filipina woman being treated at a Lismore Hospital...She told hospital staff she had lodged a citizenship application, but otherwise gave little detail of her identity. Believing the

---

[15] http://www.qiecb.qld.edu.au/html/PPL1.htm
[16] Ibid.

woman was in Australia illegally, immigration officials sent her to a holding facility at Coolangata, before flying her to the Philippines..."[17]

Rau was eventually identified in part because of the concerns raised for her health by her fellow inmates. In Central Australia, Alice Springs citizens – including Aboriginal citizens have varying life chances. The second case study is based

> "on an experience of undertaking a study of the life chances of citizens in a remote region of Australia. Indigenous people make up 16-20% of the 27,000 population on any one day in Alice Springs. ...Research data from a range of sources underlines that Indigenous people score lowest in terms of employment, health and education outcomes (the pillars of citizenship rights, that are now sidelined in a 'post welfare state',[18] and the highest in terms of incarceration rates. Self-determination is still a goal for this nation within a nation, which is hardly surprising as a result of their feeling of marginalisation. Isolation for some from the rest of the world (as a result of illiteracy and innumeracy) and connectedness for others (through being mobile knowledge workers is a reality). Added to this is the marginalisation Indigenous people feel from local public spaces and a lack of respect for some Indigenous people in banks, on pavements and in shops. In places where their separate, parallel lives are visible...Politically they have a history of colonization, dispossession, missionary settlements, control of movement, land rights, mandatory sentencing, and repealed in the NT in 2001 by the incoming labour government in NT. Many lobbyists were aware that the labour party in WA had helped to introduce mandatory sentencing, however! Aboriginal Australians have the lowest standard of living in Central Australia. An indicator of their quality of life can be given by looking at housing and health and the rate at which they are incarcerated. (McIntyre-Mills, 2003: 20-21)

Any plans need to take into account the historical and global context and the regional service nature of Alice Springs and the way in which the health services are accessed. Overall in Alice Springs Indigenous people cover the full spectrum from owners of property to relying on the social wage, but the majority is unemployed or on CDEP programs. Culture is defined as being the basis for the 'nation within a nation' concept outlined in the Kalkaringi

---

[17] Parnell, S. 2005 Wrongly deported mum left toddler behind . *Weekend Australian,* May 7-8
[18] Jamrozik, A., 2001, *Social Policy in the Post Welfare State: Australians on the Threshold of the 21st Century*, Longman, NSW.

Statement. 'Land and life' are defined in existential terms. Culture and class can be seen as proxies.[19]

Geographically, when is Australia Australia? This is no longer a nonsensical question. Australian rights as a nation can be expanded to dictate who can be allowed into our waters (and perhaps nearby international waters) as in the case of the Danish Ship, the Tampa carrying refugees who were rescued at sea. Australia has tried to suggest that the taken for granted or categorical moral law that those "in peril on the sea"[20] ought to be rescued immediately, should now defer to Australian law. Where do the boundaries of Australia begin and end? The exclusion of immigration zones could be raised as a question of when is an island a part of Australia and when isn't it? The Pacific Solution is based on using islands such as Nauru and Manus Island as places to incarcerate unwanted refugees whilst they are being processed in terms of international human rights regulations.

Nauru is an island that has become dependent on Australia, because it is financially bankrupt, according to the Head of State (ABC radio news June 23rd). The so-called "New Border Regime" (Executive Summary of the Select Committee for an Inquiry into a Certain Maritime Incident) is discussed in terms of four issues:

> "1. The so-called 'children overboard' incident involving the HMAS Adelaide and the vessel known as SIEV 4 and the management of information concerning that incident by the Federal Government and Commonwealth agencies; 2. Accountability issues arising from the children overboard incident, including the adequacy of administrative practices in certain Commonwealth agencies, and the accountability framework for Ministers and their staff; 3. Other matters arising out of the Australian Defence Force operation 'to deter and deny' asylum seekers from arriving in the Australian migration zone in an unauthorised

---

[19] In Central Australia cross cutting language groups and competition for resources through a range of organizations such as Arranta, Council Central Land Council and Tangentyere Council is not unusual, but they were united against the problem of mandatory sentencing and all worked together as Aboriginal Australians. Life chances of Aboriginal Australian Citizens[19] can only be understood by considering social, historical, cultural, political, economic and environmental concerns. Only 30% of Indigenous housing in Central Australia meet the most basic functional pre-requisites acc to Menzies School of Health (i.e. cooking, washing, storage, rubbish removal, drainage. Safe spaces for sleeping, studying not included in the study). The characteristics of housing have been directly impacted by both the 2 km law that makes it illegal to drink alcohol with in 2 kilometers of town in a public place. Direct impact on the quality of life and life chances, because Aboriginal people cannot afford to go to restaurants and they drink in public. In order to avoid arrest they congregate in town camps to avoid prosecution for public drinking within 2km of the town centre. This impacts on safety, local and regional governance

[20] From Anglican hymnbook.

13. MOLAR AND MOLECULAR IDENTITY AND POLITICS

manner by boat, with particular reference to the vessel now known as SIEV X; and 4. The nature of the agreements reached, the operation and cost of detaining persons in Nauru and Papua New Guinea as part of the so-called Pacific Solution'.

Our borders are being protected in terms of the Border Protection (Validation and Enforcement Powers) Act 2001[21] by Commonwealth officers.[22] According to the Australian Migration Act of 1958 all non-Australians who are unlawfully on the mainland must be detained. According to the immigration minister:

"People being held in immigration detention have broken Australian law, either by seeking to enter Australia without authority, or having entered illegally, failing to comply with their visa conditions."[23]

A non-citizen is classified in terms of the Migration Act 1958 Section 13. Between 1999 and 2003 nine thousand, five hundred people were classified as asylum seekers, according to Senator Amanda Vanstone:[24]

"People who apply to the Australian Government for recognition as refugees are called 'asylum seekers'. They are not 'refugees' until their claims for protection are assessed against the UN Convention and Protocol relating to the status of refugees, and they are accorded such status. Although some asylum seekers are currently in immigration detention because they arrived in Australia illegally, the majority of asylum seekers are free in the community while they pursue their claims because they arrived lawfully with a valid visa."

A critic of the current policy argues as follows:

"Under the 'Pacific Solution Australia's navy intercepts asylum seekers at sea, and forcibly moves them to detention centres on Nauru and Manus Island, Papua New Guinea. In exchange for a 20 million assistance package, which not only included payment for providing the detention

---

[21] No 126, 2001. An act to validate the actions of the Commonwealth and others in relation to the MV Tampa and other vessels, and to provide increased powers to protect Australia's borders, and for related purposes.
[22] That includes a person who: "is in the service or employment of the Commonwealth or an authority of the Commonwealth; or holds or performs the duties of any office or position under a law of the Commonwealth; or is a member of the Australian Defence Force".
[23] http: www.minister.immi.gov.au/borders/ accessed 6/25/2004
[24] http:www.minister.immi.gov.au/borders/ accessed 6/25/2004

services, but also measures to improve the living conditions for the local population of the cash strapped nations." (Rogalla 2004: 2)[25]

Jamrozik (2004) talks of the colonial heritage and "the chains" that continue to bind us to our British links and to the ANZUS alliance with America that has led to our becoming part of the British, American and Australian force that invaded Iraq. We have been called, as a result of our involvement post September 11th in the Afghanistan war and the war in Iraq as the "deputy sheriff" of South East Asia and the South Pacific.[26] To what extent have we (and other Western democracies) become less compassionate in recent years as post September 11? To what extent is Australia's current position a continuation of our colonial heritage? According to Jamrozik the Australian government has defended its decision to excise Melville Island from its migration zone after the arrival of a boat carrying 14 asylum seekers:

> "In 2001, under the pressure of world events and using those events as reasons for action, some of the measures taken have affected the attitudes and values that people believed were fundamental to the Australian political and social system. The use of SAS forces to board a ship of another country, in international waters, for the purpose of preventing asylum seekers from entering Australian shores, and establishing prison-like detention centres on faraway Nauru and Manus Islands, were measures that a few years ago Australians and Australian governments would have protested against if they had been taken by another country. Then defining the neighbouring islands that Australia claims to be its integral part as 'not Australia for the purpose of immigration creates an interesting claim for Australia's legal claims...what is their status in international law?" (Jamrozik 2004: 177).

Immigration Minister Amanda Vanstone told CNN:

> "...the government's move was designed to make it harder for people – smugglers to operate. She said Australia had a dedicated humanitarian program for refugees, taking 12000 a year, but wished to control that intake in an orderly and sensible fashion and ensure those most in need were accommodated first.... The policy has been a success with just two boats arriving since the crackdown began in August 2001. Before that at least one boatload of asylum seekers was arriving in Australia each month...Recent data released by Australia's Department of Immigration

---

[25] Modern–day torture: government–sponsored neglect of asylum seeker children under the Australian mandatory immigration detention regime. Online version accessed 6/24/2004: http://members.westnet.com.au
[26] Jamrozik 2004: 179-183.

showed around 90 percent of asylum seekers who make it to Australian soil are eventually granted temporary protection status and allowed to remain in the country" (CNN, November 6 2003)[27]

According to Rogella (op cit):

"Therefore the prevention of their arrival is merely stalling at best and at worst an attempt to send out a message that Australians are serious about Border protection and is not hospitable to those labelled 'queue jumpers'. This is of course the whole point of the political exercise. As a result of the reduction in the number of asylum seekers policy has shifted and led to awarding temporary protection visas to asylum seekers.

When Afghans, Iraqis and Iranians arrived on boats seeking Australia's protection, the immigration minister at the time, Philip Ruddock, denigrated them as 'those who are prepared to break our law, those who are prepared to deal with people smugglers and criminals'. His successor in the portfolio, Amanda Vanstone, says those law-breakers have now blended well into the community and are 'contributing to the economies of regional Australia'... The changes in rhetoric and policy may partly reflect the change in ministerial personalities; more important is the fundamental shift in the politics of the refugee–asylum seeker issue since the boats stopped coming in late 2001.

...The ... policy change is welcome to the extent that it may allow some refugees to stabilise their damaged lives. But the devil may be in the detail of the regulations, which are yet to be released. Under current migration categories meat workers, farm hands and fruit pickers would have no chance of securing a visa. In policy terms Vanstone's TPV changes is a muddle, reminiscent of the cumbersome 'work around' solution devised for dealing with asylum seekers from East Timor. The East Timorese arrived in the mid-1990s but had their cases frozen pending the outcome of protracted legal proceedings. By the time their cases were considered, Indonesia was no longer occupying East Timor and their chances of being recognised as refugees were slim. When Philip Ruddock failed to secure Cabinet approval for a special visa for the East Timorese, a messy compromise emerged. Each asylum seeker had to go through the motions of applying for refugee status, then appealing to the tribunal, even though there was almost no chance of success. After jumping through these unnecessary hoops, East Timorese applicants could then ask the minister to grant a visa on humanitarian grounds. Although successive ministers have generally taken a sympathetic view

---

[27] World: Australia Defends Border Laws.

of the East Timorese cases on humanitarian grounds, this protracted procedure is hardly good public policy."[28]

The 'molar' fixed identity of asylum seekers needs to be redefined before they have a chance to access what some would argue are human rights, let alone citizenship rights. Changes in governance have occurred as a result of lobbying, despite the initial official responses from government to the outcry from citizens and professionals concerned about human rights. Rogalla(2004), in the same paper stresses that an investigation by the Human Rights Commissioner did not cover Nauru and Manus Island. Mandatory sentencing of refugee minors has been the subject of criticism by the Sev Ozdowski, the Human Rights Commissioner.[29] The report tabled in parliament stressed those children who arrived by boat were accorded treatment that was unacceptable:

> "The commonwealth's failure to implement the repeated recommendations by mental health professionals that certain children be removed from the detention environment with their parents amounted to cruel, inhumane and degrading treatment of those children in detention."

The report was rejected by parliament. In some states in Australia there has been more concern for mandatory sentencing and children's rights than in other states. Labour and Liberal politicians have used mandatory detention at a state/territory and federal level as a vote winner. The mandatory incarceration of Aboriginal children in Northern Territory was the subject of protest leading to the repeal of the law by the incoming Territory level Labour Government on 23rd August 2001. Ironically the labour government had helped to introduce mandatory sentencing in Western Australia.(Harding 1993) The ex Liberal Prime Minister Malcolm Fraser helped to lobby against mandatory sentencing that was outlawed in the NT and John Howard, the incumbent Liberal prime minister, stressed that it was a matter for the states and not a federal matter.

At a Federal level being tough on "border protection" against illegal immigrants, called "queue jumpers" has been important as a vote winner in the last Federal Election post September 11th and post the so-called Tampa incident and the "children overboard affair", called "A Certain Maritime Incident" by the Select Senate Committee (ended 30th July 2002). Children

---

[28] "Refugees: Quick fix better than nothing". Mares, P., Australian Policy on Line. 7/26/2004. Peter Mares is a senior researcher at the Institute for Social Research at Swinburne University and author of Borderline: Australia's response to refugees and asylum seekers in the wake of the Tampa (UNSW Press, 2002). A version of this article appears in the 15 July edition of the *Australian Financial Review*.

[29] Bob Burton 18th June 2004. Refugee Day – Australia Global Information network http://global.factiva.com/en/arch/display.asp 6/24/2004

were not in the water as the media photographs indicated, because their parents had thrown them overboard. The executive summary[30] stresses that the facts of the matter were that no children were thrown overboard and that a conversation between a commander and a brigadier on the 7th of October 2001was the basis for the misinformation. By the 11th of October the Chief of the Defence Force was informed that the information was doubtful, but by then photographs of the children being rescued from the sinking SIEV 4 had been splashed in the media. Minister Reith was briefed, but the chain of command did not reach the Prime Minister and Cabinet. The conclusion of the Senate investigation was that the problem was with the lack of information management and poor accountability in the case of the first two matters. In the case of the second matter, the sinking of SIEV X, insufficient evidence was cited as a reason for not taking this further and in the last instance it was stated that if the refugee status could be proven, then those asylum seekers off the mainland in detention camps, should be afforded all the rights of refugees. Poor communication and the lack of compassion are self evident in this series of events, but the issue of border protection helped to win the two last elections in Australia.

To what extent does the Border Protection Act of 2001 undermine breaches of human rights, criminal and civil law in Australia and in the Pacific on the islands that form part of the so-called solution? Following this lack of logic we could ask whether Nauru – bankrupt and dependent on Australia is part of Australia – after all it is part of our Pacific Solution. The tough stand during the last Federal Election was bipartisan during the election. The stand by Amanda Vanstone was later criticised by the Labour Left, who thought after their election defeat that the Labour Party had alienating voters concerned about human rights. The current election shows that the electorate was more concerned about the economy and border protection than old growth forests in Tasmania. Democratic majority does not necessarily co-incide with socially and environmentally just concerns.

Narrow pragmatism[31], political votes, lobbying and convenience drive policy. Australia seems to be able to expand her borders in preventing entry, by incarcerating even those at sea and to contract her borders by removing islands from the immigration zone. Those who are part of the Pacific Solution are not in Australia and are said to be outside Australian Law for some purposes and under our immigration act for other purposes.

---

[30] Page xxiii cites Findings of Fact. The summary is drawn from pages xix- xlvii

[31] In vol 3 McIntyre-Mills et al develop the argument for an expanded form of idealism and pragmatism, based on considering the contextual consequences for all the stakeholders as caretakers.

13.3.2 Consideration of the European Union model of federalism based on subsidiarity

"Today, Australia is erecting a transnational border in cooperation with European Union states that have collectively defined international refugee flows as a crisis of sovereignty. The erection of the new 'transnational border' is part of a wider process of enlargement and containment through' hierarchical integration or asymmetric incorporation' (Pieterse, 2001: 11) According to the UN Commissioner for refugees and human rights organizations, the international refugee protection system is in crisis, because signatory states are not living up to their obligations, under the 1951 Refugee Convention. Restrictive policies on asylum and migration in the West are narrowing the definition of who is a refugee, and strategies to prevent asylum seekers getting the chance to make claims are adopted through detention policies, border protection and containment at a distance..." (Humphrey, 2003, 33.)

No model has all the answers and European Union has an immigration policy that leaves much to be desired (op cit). Integration of policy and law can be either liberative or so-called conservative. I argue that conservatism is a policy approach that is ironically named, as it is likely to lead to undermining sustainable options in the future, by driving wedges between the privileged of the world and the disadvantaged who will be mobilized by means of religious politic [32].

13.4 TRANSBOUNDARY IMPLICATIONS FOR UNDERSTANDING NATIONALITY, CITIZENSHIP HUMAN RIGHTS AND ANIMAL RIGHTS?

Geoff Shaw (1998)[33] talks about the need for co-intelligence as the starting point "When you watch the spider build its web, the whole structure interacts and binds the web together with diagonal, horizontal and vertical

---

[32] See Hindess, B., 2003, "Responsibility for others in the modern system of states", *Journal of Sociology,* **39**(1):23-30.
Humphrey, M., 2003, "Refugees an endangered species", *Journal of Sociology,* 39(1):23-30.
[33] Tangentyere Council Annual Report.

ties to give it strength and stability...we must continue to be like the spider's web...linked to each other...giving strength to each other".

"At a public forum in London he [Jack Straw, Britain's Foreign Secretary] put his finger on the fundamental contradiction at the heart of the 1951 Convention on Refugees: it gives people facing persecution the right to claim asylum, but it does not oblige any nation to admit them so that they can make that claim. The consequence of this contradiction, Jack Straw admitted, is that refugees are forced to break the law to escape the threat of persecution in their home country and to seek safety elsewhere."[34]

Yeatman (2003) in her paper 'Global Ethics, Australian citizenship and the boat people' - a symposium argues that the Treaty of Westphalia is no longer a suitable basis for international law. The Thirty Years War in 1648 was ended by the Westphalian treaty to limit interference from other states. If people are only treated as citizens with rights, by virtue of their membership of a nation, where does this leave asylum seekers. Singer (2002), Beck(1992,1998) and Chomsky(2003)[35] would argue that hegemonic decisions or transgovernance decisions in a globalised world have an impact that goes beyond borders of any one nation. Pollution and war, corruption and greed, have no boundaries and are likely to be reciprocated by poor quality of life for the perpetrators in the long term if not in the short term.

The systemic connections across animal husbandry for protein that is costly to produce, slaughter and transport, the pressure on the environment from introduced species, the value of moving towards vegetarianism in terms of limiting production costs, by using less water and other grazing resource and addressing the opportunity costs of using resources more sustainably have already been drawn by Singer and these ideas form the basis of a policy paper for 'Animals Australia'[36] on the ethics of transporting live sheep from Australia to the Middle East. The paper is a measured argument based on an acceptance that animals are conscious, because consciousness is a continuum across all organic life (this is supported by the neuroscientist Baroness Professor Green field (2000) and in a roundtable discussion on – different

---

[34] Mares, P., 2002, *Borderlines*, 2nd ed., UNSW Press Books.
[35] Beck, U., 1992, *Risk Society Towards a New Modernity*, Sage, London.
Beck, U., 1998, *Democracy Without Enemies*, Polity, Cambridge.
Chomsky, N., 2003, *Hegemony or Survival: America's Quest for Global Dominance*, Allen and Unwin, Crows Nest, NSW.
Singer, P., 2002, *One World: The Ethics of Globalisation*, The Text Publishers, Melbourne.
[36] Levy, N. 2005. *The Ethics of Live Sheep Export Trade*. Animals Australia http://animalsaustralia.org/

ways of knowing[37] Animals Australia (op cit) argues for a policy of banning live meat trade for a number of reasons:

> "It is not economically necessary as humane slaughter yards in Australia provide jobs for local people; It causes inhumane pain and suffering to animals on the voyage; Methods of slaughter in the Middle East are not as humane as the methods of slaughter used in Australia."

Having higher intelligence (in some areas) and power is also no excuse for cruelty. In fact it has been show that cruelty to animals is indicative of an ability to be cruel to other forms of life and mandatory reporting of cruelty to animals in the United Kingdom and New Zealand has revealed that abused people are more likely to abuse animals as a precursor to acts of violence on human beings (ABC news Sunday 18th of July). Mental health is linked with respect for all forms of life. This is the basis for ethical behaviours that is concerned about others. Ethical thinking is mindful that we are part of a wider system and that each part is systemically linked with others.

The following diverse factors played out systemically in the international context of the so-called 'war on terror', suspicion, and the poor relations with the Middle East exacerbated by the apparent attempts to initially cover up the rejection from the first port of call (Longo 2004) added to the problem of the commodification of animals. From September to October[38] the news

---

[37] Adelaide, July 2004, Adelaide University

[38] "Sheep trade: no room for woolly thinking", *Financial Mail*, 1-2 November.
"Australians agricultural heritage imposed itself on the international stage. Few noticed when a livestock transport vessel I, the MV Cormo Express, sailed from Fremantle, Western Australia on 6 August 2003. It made a far more dramatic public impression later that months once it arrived at its destination in Jeddah. Authorities in the Saudi Arabian port city rejected the ships cargo of 57000 sheep, citing unacceptable levels of scabby mouth, a viral infection common in Australian flocks. Australian on board vetinarians denied the charge, but the Saudis were uncompromising and refused the shipment. Mostly older wethers, the sheep were intended for slaughter and had already been through an arduous journey. The livestock trade …represents around 1 billion in exports annually. Seeking to protect this trading reputation, Australian officials set off a frantic diplomatic maelstrom as they scoured the planet in search of a new port that would take the stranded animals. In all Australia approached fifty-seven countries. There were fleeting rumours that Kenya, Kuwait, Oman, Bangladesh, or even Iraq might take the sheep as a form of live food aid…In Iraq, despite initial confidence an option had been found, the prospect collapsed when the British forces in control of the Southern region realised the daunting logistical task of such a mass slaughter, coupled with the prospect of civil unrest. Reports of diseased sheep arriving provoked local fears of a plot to spread pox among Iraqis. Domestic pressure mounted in Australia to expedite a humane outcome, with some suggesting that the shipment be returned home and others urging a butchering at sea. The farm lobby vehemently opposed repatriation, fearing the introduction of foreign pests, which the animals might have picked up during their unhappy voyage. The option of slaughter on the open sea was also fraught with problems. In the end Eritrea accepted the

saga of sheep gained attention. But the issue is not the rejection by the Saudi Government, because of scabby mouth disease; it is the commodification of animals, treated as means to an end. For two months they wandered from port to port. Sheep died from heat and thirst no one would take them and the Australian government did not want them back for fear of jeopardising their meat market and so they were given as "aid" to Eritrea in Africa.

13.5 SUBSIDIARITY AND FEDERAL GOVERNANCE IN THE EUROPEAN UNION IS UNDERMINED BY BOUNDED THINKING

The argument shifts in this section from discussing the potential of the EU federation that tries in principle to address human rights (and fails in some instances where it tries to balance collectivism and individualism) to reflecting on the experience of working in the most extreme context of the abuse of human rights. Wider regulation by the EU can however have advantages according to Long (2002: 9) who cites "the European Union's recent pressure on Turkey resulting in the grudging establishment of a Kurdish-language broadcasting service, in accordance with the union's policies on the cultural rights of minorities."

Nevertheless despite the positive aspects the EU needs to consider the implications of local democracy where the most powerless are the subject of decision making over which they have no control as they do not have the vote.

Singer emphasizes the potential of the European model, showing how it has been used to protect animal rights, nevertheless the model has limitations as the following case study demonstrates, when citizenship rights are allowed to override human rights. Fortunately international interests enabled citizens to think in terms of international supranational concerns as a result of the capture of French journalists, as I will explain. This example demonstrates the importance of being able to work beyond bounded categories in order to make ethical decisions in context. A design for inquiring systems to support sustainable governance needs: both qualitative and quantitative data from many stakeholders (empiricism) an appreciation of the logic of many stakeholders, idealism adapted from Kantian ethics to

---

shipment nearly two months later. In mid-October Australia donated the sheep to this small, chronically poor country on the Horn of Africa, along with 3000 tonnes of feed and $ 1 million to meet transportation and abattoir costs. More than 5500 sheep, or 9.6 percent of the original cargo, had died from stress, heat and fatigue. The Australian taxpayer copped a bill estimated well in advance of $10 dollars. ...."(Long, 2004: 239)

contextual situations through dialectical engagement and a pragmatic synthesis.[39]

> "For several months now France has been obsessed with an item of woman's clothing. The garment in question is not the skimpy lingerie modelled in a Paris Metro ad ... but the Islamic hijab, increasingly in vogue among French Muslim woman." [40]

The tone of this quotation makes it clear that "the French parliament's recent vote (494 to 36) to ban the wearing of 'conspicuous' religious symbols in public schools" (ibid) is regarded as unfair and immoral.

> "...The French Law is meant to protect the republican principle of laicite, a strict form of secularism established after bitter struggles at the beginning of the last century to keep the Catholic Church out of politics. Nearly everyone agrees that laicite must be preserved- including most of the far right, who take the Catholic jihardi Joan of Arc as figurehead for their anti-immigrant campaign. At a moment of perceived crisis it is a powerful rallying cry. Exactly what the crisis is depends on who you ask...France's Muslims, most of them children (sic) of its colonial adventures in North Africa, make up about 7 percent of the population, no government has challenged the racism that keeps so many of them in the windswept, high rise suburbs (the banlieues) on the margins of the cities. A second-generation Algerian is three to four times more likely to be unemployed than a 'native' French person; schools in the banlieues are bleak and badly funded".

In France about 5 million are Muslim[41] and 15 million live in Europe[42] and 1 in 15 British residents[43] are Muslim. The hijab, a head cloth worn by Muslim woman has become a site for Western Europe and United Kingdom's concern for preserving western culture and political interests. It is poignant that the site for the contest over democracy and the enlightenment[44] should be played out through attempts to control young (relatively poor) Muslim women of school going age who need to attend public schools and to obtain an education if they are to improve their own and their children's life chances. There is no formal representation of North

---

[39] Derived from the work of Churchman 1971 and Van Gigch 2003
[40] Margaronis, M., 2004, *The Nation*, 15th March.
[41] Margaronis, M., 2004, *The Nation*, 15th March.
[42] Alaa Bayoumi, 2004, *Europe taking wrong route to integrate Muslim Population.* 10 March.
[43] Bertram, T. and Pascal, C., The OECD Thematic Review of Early Childhood education and care: background report for the United Kingdom. http://www.oecd.org/copyr.htm/
[44] Margaronis, M., 2004, *The Nation*, 15th March.

African interests in the French National Assembly "only seven on local and regional councils" (ibid).

Young Muslim girls who are not of voting age are subject to the rules of adults (often males) and the fundamentalist, patriarchal decisions to wear the hijab (headscarf). The hijab has become shorthand or a potent symbol of enlightenment thinking and democracy. The day after 12 Nepalese civilians were killed by Iraqi hostage takers, two French journalists were taken as hostages (September 1 2004).

> "The hostages demanded that the policy on the banning of the hijab ... be revoked, but the ban was implemented. French Foreign Minister Michel Barneier said he understood that Mal Bruner and Chesnot were ' alive and getting good treatment...the hostage takers were split between the radical foreign fighters who wanted to keep the journalists captive and Iraqi elements opposed to the Baghdad authorities who supported their release."[45]

French Muslims protested against the capture of the journalists and called their captors unIslamic. Paradoxically this has lead to greater solidarity with France, because the French were against the war with Iraq.[46] It stood the risk of becoming a site for playing out the clash between cultures expressed in religious terms. It is also a clash between so-called democracy and enlightenment versus fundamentalism. It reveals the shortcomings of the notion of subsidiarity - or at least in the way it is interpreted.

Governance based on subsidiarity is supposed to be taken at the lowest level possible, unless the whole would be better served by a group response. In this case human rights could be undermined when young girls are deprived of an education. The fears post September 11th aftermath and the aftermath has lead to a strong sense of wishing to "protect democracy and the enlightenment" against all others. It was possible that the unintended affect could have entrenched fundamentalist identity. According to an editorial[47] in three German states (lower Saxony, Bavaria and Baden Wuerttemburg) they were considering banning teachers from wearing the headscarf at public schools. Other religious symbols are permitted. "a German Catholic organization is urging that the scarf be viewed as a political symbol, insisting that any law banning it shouldn't be applied to symbols that are part of the country's Christian tradition" (ibid).

---

[45] "Release imminent for French Hostages", *Weekend Australian,* Sept 4-5 2004, Reuters AFP-AP.

[46] 9 January 2004, Unveiling discrimination Times-*Picayune*

In another editorial[48] it is argued that Muslim asylum seekers in Netherlands had not been accepted and 26000 would be returned over three years. The same writer argues that Denmark has proposed legislation to limit the number of Muslim religious leaders as part of a deal between Denmark's Liberal–Conservative government an d its far right ally, the Danish People's Party. The law applies to all religions.[49]

In the United Kingdom a scholar at Denbigh High School where 80% of the students are Muslim, requested to wear a long loose flowing garment instead of the culottes and tunic worn by the other Muslim students. She lost a court case to uphold her right to wear this garment.

> "Fifteen-year-old Shabina Begum is like any normal girl her age, and given a chance she would, presumably, be doing what teenagers do - go to school, have fun in the playground...."[50]

Hasan Suroor argues that she has been used as a pawn. Her family (in particular her brother, is a member of Hizb-ur-Tahrir) insist on her remaining at home if she is unable to wear the long robe, as it was not part of the dress code. This was upheld by the High Court in London. The hijab issue in the E Union brings into focus the rights of the least powerful young women in the community in need of a public education. It has become a site for playing out the clash between cultures expressed in religious terms. It is also a clash between so-called democracy and enlightenment versus fundamentalism. It reveals the shortcomings of the notion of subsidiarity or at least in the way it is interpreted. Governance based on subsidiarity is supposed to be taken at the lowest level possible, unless the whole would be better served by a group response.

13.6 CASE STUDIES PROVIDE INSIGHTS INTO COMPLEXITY, CARETAKING AND COMPASSION

The argument developed from case studies addresses conceptual and contextual slippage and readings across embodied social actors and sentient beings, life chances, communication, social and environmental justice and governance. More specifically it addresses the central paradox, namely that trust is a risk for people who make themselves vulnerable to others (Warren,

---

[48] Alaa Bayoumi, 2004, "Europe taking wrong route to integrate Muslim Population", *Seattle Post-Intelligencer,* 10 March.

[49] Schofield, M., 2004, "Headscarf furore covers a deeper issue; Support for anti -Muslim legislation is growing in Europe", *The Seattle* Times, 7 March.

[50] Suroor, H., 2004, "From hijab to jilbab", *The Hindu,* 24th June.

1999).But without trust that is developed through respectful communication (governance is unsustainable.

Any one model may make a contribution, but it cannot provide all the answers. Critical systemic thinking is needed to 'unfold' the values of the stakeholders and to 'sweep in' the social, political, economic and environmental aspects.[51] Open, not closed communication is needed and appropriate communication techniques for participatory design are needed. This has implications for the way we govern and design options for the future. One- way communication undermines the potential of education and of democracy (albeit always a compromise at best) to pool ideas and to be accountable to citizens on the basis of dialogue, which is by definition at least two- way communication.

The closest we can get to truth is through dialogue that explores paradoxes and compassionate advocacy for the voiceless. Truth is a process, just as an ideal form of democracy is a process supported by social structures. 'Frank and fearless' participation by members of the public and public sector bureaucracy needs to be ensured for participatory democracy to exist. This is the so-called mantra of the public service- that is currently being ignored in many Western democracies today, such as Australia. According to Mike Scrafton, a member of Australian civil service, who argued prior to the 2004 federal election, that the "Children overboard" incident was not rectified although the government was aware that asylum seekers had not thrown their children into the water, because there was bipartisan support at that time for border protection. Members of government bureaucracy and civil society need to feel that participation and debate is acceptable and not an indication of "being ratbags", "members of the chattering classes, such as doctors wives", in the minority or spoilers on the side of 'illegals'. Dialogue helps to identify the paradoxes, which in turn provide portals for transformation. The enlightenment and democracy need to be seen not so much as a static universal law, but as a dynamic structure and process for balancing the eternal paradox that:

- On the one hand: openness to debate and to other ideas and possibilities is the basis for both the enlightenment process of testing and for democracy and
- On the other hand, for openness to occur there has to be some trust that voicing new ideas will not lead to subtle or overt marginalisation of oneself or one's associates. The West faces the challenge of preserving this openness and trust and redressing the imbalances in wealth and power caused by centuries of colonization, modernization and globalization that is based on the single bottom line[52] of profits for

---

[51] To draw from West Churchman's work 1971, 1979a, b, 1982).
[52] Elkington, J., 1997, *Cannibals with forks,* Capstone, Oxford, for a critique.

competitiveness in markets, that support hegemony, rather than on a multiple dynamic awareness of socio-cultural, political, economic and environmental factors that when considered together support a sustainable future. Knowing is based on a range of experiences, senses and on recursive communication, so knowing is a process. (Greenfield 2002) To know has transformative and recursive relationship and is not merely about representation, but about change. Governance needs to be constructivist, not based exclusively on old notions of science[53] based on fixed representation of rationality, but instead an appreciation of the many domains of knowledge that can be addressed through an inquiring system that takes the objective, the subjective and the intersubjective into account when attempting to understand the nature of governance and international relations challenges. To know is a process based on the senses, emotions and the contextual experience. It is not merely about representing reality "that is out there".

Knowing is a potentially transformative experience. Democracy only works when those affected by decisions can be party to them. Openness to ideas is vital for re-working democracy to take into account the balance between diversity and collectivism and to avoid entropy. Currently international relations face entropy, because of closed conceptual and geographical boundaries. Two-way communication that is respectful of diverse ideas and helps to build relationships is vital for managing risk. This is the key point made by Habermas and Derrida (2003) in their conversation about thinking and its relevance to preventing terrorism. Trust develops further networks of co-operation.[54] But as Edgar (2001) stresses in his book on governance in Australia, we need space for difference and space for cooperation.

'Molar' or fixed identity politics and molecular fluid politics and identity can be useful in different political contexts.[55] Conceptual diversity is vital. This means that it could be important to have different social spaces and different networks. As Bourdieu (1986) stressed in his work on cultural and

---

Gallhofer, S. and Chew, A., 2000, "Introduction: accounting and indigenous peoples", *Accounting, Auditing and Accountability Journal,* **13**(3):256-267.

[53] See Bausch 2001, Banathy 2000, Van Gigch 2002.
Van Gigch, J.P., 2002,. "Comparing the epistemologies of scientific disciplines in two distinct domains: modern physics versus social sciences", Part 1: the Epistemology and knowledge characteristics of the physical sciences. *Systems Research and Behavioural Science,* 19(3):199-210.
Van Gigch, J., 2003, *Metadecisions: rehabilitating epistemology,* Kluwer Plenum.

[54] See Putnam, R., 1995, "Bowling alone", *Journal of Democracy,* **6**(1):65-78.
ABS, 2002, "Social capital and social wellbeing: discussion paper", Commonwealth of Australia, Australian Bureau of Statistics Publication.

[55] See Deleuze and Guattari, 1989, in Bogue, Buchanan and Colebrook 2000.

13. MOLAR AND MOLECULAR IDENTITY AND POLITICS

social capital some networks are in the interest of some rather than others. Networks per se do not build social capital for everyone. They can establish 'in groups' and 'out groups'. It is the role of governance, particularly local and regional governance to address representation across self-other and the environment. Better opportunities for communication are conducive to sustainable democracy and to economic development. This can require separate as well as joint meetings hosted in the interests of achieving participation by those who are to be at the receiving end of decisions.

13.6.1. ANALYSIS USING A CRITICAL AND SYSTEMIC APPROACH

Critical and systemic thinking and practice can assist in building governance capacity. The area of concern is enhancing social and environmental justice through sustainable and compassionate governance, not only because it is an idealistic concern, but also because it is pragmatic.

Some of the essential characteristics of systemic governance are:
- Openness to many people and many ideas,
- Representation of all the sentient stakeholders,
- Respectful communication with those who have a voice and mindfulness of those who do not,
- Working across organisations and across sectors (health, education, employment, for example) and addressing the challenges of managing networks[56] Forming responsive transdisciplinary approaches to areas of concern and working across : a)conceptual boundaries of professional and lived experience b)elected representatives, civil society and corporate public and private sectors in a range of socio-geographic arenas locally, nationally and internationally, in order to achieve sustainable social, economic and environmental futures.

Systemic governance recognizes that reality is not fixed and cannot be represented by essentialist, unresponsive categories. Testing for truth needs to be based on testing or falsification of hypothesizes generated and tested not only by experts, but also by all those contextually who are to be at the receiving end of a decision[57] as well as by advocates for voiceless, sentient beings. The role of the expert changes to the role of a caretaker and

---

[56] Kavanagh, D, and Richards, D., 2000, "Can Joined-Up Government be a Reality?" Paper presented at the *Australian Political Studies Association 2000 Conference*, Australian National University, Canberra, 4-6 October, 2000.

[57] Wadsworth, Y., 2001, "The mirror, the magnifying glass, the compass and the map: Facilitating participatory action research", in *Handbook of Action Research*, P. Reason and H. Bradbury, Sage, London.

facilitator, because this is appropriate for social and environmental risk management.

One of the techniques for achieving sustainable governance at the local, national and international level is expanding the concepts of accounting and accountability to incorporate: social, cultural, political, economic and environmental indicators.[58] The process for governance involves working across or communication across civil representatives, elected representatives and corporate structures to link public, private and non-government organizations. In Australia and internationally the categories of culture and identity and access to citizenship remains important. Today the story of culture and identity is played out in terms of self-determination, reconciliation and treatment of asylum seekers. The wider supra national identity of human rights remains a challenge.[59]

National Economics (2002, 2003) found that mobile knowledge workers settle in desirable areas with environments that support a good quality of life, but conceded that professional, creative people drive up property prices and create wealthy areas and regions. The message for governments, non-government sector (both business and volunteer organizations) is clear. Conceptual diversity thrives in environments that are open to new ideas and that support open communication. Local and regional governance can provide a context for fostering a sense of geographical, face-to-face community, by providing opportunities for people to become "makers and shapers of their community, rather than being just users and choosers"[60] through participating in the design of space and place, rather than merely voting or using services.

This approach to engaging people in the process of creating their own communities using participatory action research and action learning shifts the power, knowledge and control from the expert to those with lived experience[61] . This can be vital for building capacity in these environments. Democracy can be debased. It can be merely about winning the votes of the majority. The concerns of the majority may or may not be about social and

---

[58] Gallhofer, S. and Chew, A., 2000, "Introduction: Accounting and indigenous peoples", *Accounting, Auditing and Accountability Journal*, 13(3):256-267.

[59] Gay rights have only recently been achieved in Tasmania, whilst young people locally and internationally do not have a voice in shaping their life chances. We are still debating whether gay people should have the right to marry.

[60] Gaventa, J., 2001 "Towards participatory local governance: Six propositions for discussion", Paper presented to the *Ford Foundation*, LOGO Program with the Institute of Development Studies, June, p4.. Gaventa, J. and Valderrama, C., 1999, "Participation, citizenship and local governance", Background note for workshop on *Strengthening participation in local governance,* Institute of Development Studies, June.

[61] McIntyre, J., 2003, "Participatory democracy: Drawing on C. West Churchman's thinking when making public policy", *Systems Research and Behavioural Science*, 20:489-498.

environmental justice. Sometimes elections can be about protecting short-term economic interests, such as maintaining low interest rates to support home ownership. Mobile professionals who move regularly for work reasons are often in relationships where both partners work and thus have less time to be actively involved in their communities Putnam, 1995, McIntyre, 2003.People engage less in face-to-face interactions as they become more mobile as workers and recreational travelers.[62] Local social capital can be threatened. Putnam (1995) argues that people spend less time volunteering service in their community and more time within their homes using communication technology or on personal interests. In this context the Australian electorate voted for interest rates (for home mortgages) to be kept low.

Castells (1996-8) argues in his work on access to 'networked society' that the divides between the networked and non-networked will lead to different societies. Knowledge and access to technology are the basis for wealth and wellbeing in networked society. For example, the Australian Bureau of Statistics (2004)[63] argues that the gaps between rich and poor in Australia continue to grow. This sort of diversity is not socially just or sustainable and needs to be addressed through development at the regional

---

[62] Challenges for democracy need to be addressed. Formal participation in democracy (if measured by voting) has fallen in Western democracies. Putnam (1995) argues that this is because people are more mobile and less engaged in service in their communities in a face-to-face (community-based) manner and more involved in more abstract social issues, through their membership in electronic networks. This has the potential to enable people to think beyond the local context, provided they feel sufficiently connected to the places they live in for a few years, as opposed to a life time's sense of place that is traced back generations and which can be considered for future generations. Mobility has advantages and disadvantages that need to be addressed through local governance that encourages participation of all who are living within an area. Bentley (2003), Skidmore and Hakim (2003) argue that some people are disenchanted with the political categories or parties that are available. Disengagement is prevalent in politics today. According to Bentley (2003). "The clearest illustration of the problem is the steady decline, across the industrialised world, of people's engagement with formal politics. In eighteen of the world's twenty most industrialised countries election turnout has declined since the 1950s, on average by 10 percent. At the same time and with the same consistency, people have become far less likely to identify strongly with a political party."

This is not necessarily a problem, because people are engaged in political activity in many fluid and non-formal contexts. Social movements are not fixed to specific contexts or organisations or groups and can provide useful ways to bring about change to political institutions that are no longer in touch with people's values. Participatory democracy needs to go beyond voting and good governance needs to go beyond the organisational context. Voting for a candidate or following sound guidelines for good corporate governance, although important is insufficient to ensure social and environmental justice Elkington, 1997, Beck 1992, Chomsky 2003, Pilger 2002, McIntyre-Mills 2000, Gaventa and Valderrama 1999).

[63] Financial Review, 2004, 28˜29 February.

level. Edgar (2001) stresses the risk of divisions in society and the way that divisions can lead to 'tribalism'. He also discusses how development at the local level needs to be managed so that the negative aspect of decentralization can be redressed.

Local development needs to provide a) space to be different and b) for co-operation with a wider group to avoid the problems of stereotyping that lead to racism, ethocentricism locally and internationally to versions of what Baruma and Margalit (2005)call: Orientalism (stereotyping the Eastern culture) and Occidentalism (stereotyping Western culture).

According to the State of the Regions Reports (2002, 2003) the higher the level of diversity, education and the greater the 'tolerance' for difference- the better the socio-economic outcomes.

We need to see the patterns of what constitutes good and bad governance locally and to apply these insights more widely. If people are excluded and marginalized it leads to low socio-economic outcomes and they are unable to move easily from areas of low development within a nation state. Governance implications at the local level should 'sweep in' (Churchman, 1979; McIntyre, 2003) the consideration of wider issues. According to these reports the gaps between regions in Australia are not caused by differences in the number of employed, but in the size of the salaries paid and the differences in the cost of living. Younger people in remote areas who wish to move to more urban regional centres are affected by property prices. The so-called intergenerational disparities have been highlighted as an area of concern that could best be addressed through developing the lagging regions through capacity building. The notion that market forces will solve all the problems has been criticized and participatory democracy has been given the big tick, alongside the importance of sustainable development.

13.6.2. TOWARDS A PROCESS FOR SYSTEMIC GOVERNANCE

In Australia formal governance is through institutions and the amount of networking and formation of partnerships across the public, private and NGO sectors needs to improve for the purpose of lifting the quality of life in regions in social, economic and environmental terms. Governance in Australia tends to work on partisan or party lines at state and commonwealth level. This means that the party line is followed (Dean Jaensch Dec 4th 2003 at a public lecture). Other options are for elected representatives to act as trustees who decide what will be done once elected, or to decide as delegates or conduits for public participation. Local Governance can choose amongst these options and consider the situational context. It allows considerably more flexibility. This is what is needed in a fast changing globalised world,

13. MOLAR AND MOLECULAR IDENTITY AND POLITICS

where not only international competition for resources prevails, but also regional competition for resources.

We need to be able to see the world through multiple sets of lenses and understand the implications for the way people think and act. This 'appreciation' can help us to avoid what Vickers calls 'mind traps'[64] of just seeing the world in terms of one set of values.[65] Although we may need to make a decision one way or another, it is vital to be able to think/appreciate multiple viewpoints and to "hold more than one idea in mind or more than one big idea simultaneously".[66] This is vital for governance that manages socio-economic and environmental risks.

Policy makers informed by old style enlightenment thinking and rational knowledge control and engineers outcomes through linear logic of cause and effect. Crude versions of enlightenment thinking (based on misinterpretations of Cartesian[67] thinking) tend to compartmentalize, specialize and professionalise knowledge domains.[68] To be objective the policy researcher distances herself from the researched. She does not see herself as part of her subject matter so that she can shape and reshape the research context. Policy researchers need to be informed by new social science that addresses the anomalies of recursiveness, 'strange loops'[69] and chaos theory.[70]

---

[64] Vickers, 1970, "Freedom in a rocking boat" in Flood and Romm, 1996, pp. 128-129, for a discussion of this concept of mind traps being similar to being trapped in a lobster pot!

[65] McIntyre-Mills, J., 2000, *Global Citizenship and Social Movements: Creating Transcultural Webs of Meaning for the New Millennium*, Harwood, Netherlands.
McIntyre, J., 2002, "A community of practice approach to knowledge management and evaluation re-conceptualized and owned by Indigenous stakeholders", *Evaluation Journal of Australasia*, 2(2): 57-60.
McIntyre, J., 2003, "Yeperenye dreaming in conceptual, geographical and cyberspace: a participatory action research approach to address local governance within an Australian Indigenous Housing Association", *Systemic Practice and Action Research*, 16(5):309-338.

[66] Jones. B., 1990, *Sleepers Wake! Technology and the Future of Work*, 3rd.ed., Oxford University Press, Melbourne.

[67] Veitch, J. 1977, *Descartes: A Discourse on Method: Meditations and Principles*, Everyman, London.

[68] Bausch, K., 2001, *The Emerging Consensus in Social Systems Theory*, Kluwer/Plenum.

[69] Hofstadter, D., 1979, *Godel, Escher, Bach an Eternal Golden Braid*, Basic Books, New York.

[70] Bausch, K., 2001, *The Emerging Consensus in Social Systems Theory*, Kluwer/Plenum.

Table 13-1. Towards a model of systemic governance

From frameworks to dimensions	Methodology based on participatory action research[71]	Areas of concern
Social, economic and environmental frameworks or triple bottom line Elkington, 1997. become dimensions of a dynamic whole	Sustainable governance based on participatory action research and a design of inquiring systems West Churchman(1979)	Enlightenment and democracy is in the interest of some stakeholders and not all. How can social and environmental justice be pursued?
Working the hyphens of self-other[72] – environment[73]	Compassionate consideration of the powerful and powerless – including sentient beings that cannot speak for themselves[74]. A measure of sustainable culture and civilisation is the way that the powerless are treated in society	Consciousness is a continuum across all organic life(Greenfield 2002)
Logic, empiricism, idealism, the dialectic and pragmatism (West Churchman 1979)	Addressing projections that lead to Occidentalism and Orientalism [75]in the interests of international relations that sustain peace.	Identity can be fixed or fluid[76]- conceptual and geographical boundaries can be made and remade contextually to address the concerns of all the stakeholders. Decisions need to be made on the basis of compassion and sustainability

Research in the area of social cybernetics has lead to the formulation of Ashby's Law (1956) and we can infer that good decision making and good governance can only be achieved if the sort of diversity amongst the decision makers matches the diversity of the population. Nevertheless decisions have

---

[71] Checkland, P. and Holwell, S. 1998, "Action research: Its nature and validity", *Systemic Practice and Action Research*, **11**(1):9-21.

[72] Haraway, D., 1991, *Simians, Cyborgs, and Women: The Reinvention of Nature*, Free Association Books, London.

[73] Adapted from Fine, in Denzin and Lincoln, 1994.

[74] Singer, P., 2002, *One World: The Ethics of Globalisation*, The Text Publishers, Melbourne.

[75] Baruma, I. and Margalit, A., 2004.*Occidentalism: The West in the Eyes of its Enemies*, Penguin, New York.

[76] Bogue, R., 1989, *Deleuze and Guattari*, Routledge, London.
Buchanan, I. and Colebrook, C., 2000, *Deleuze and Feminist Theory*, Edinburgh University Press.

to be made by governments. Sometimes extreme diversity is positive, sometimes it can have negative implications and these have to be weighed up by governments. What is clear, however is that the greater the level of participation in lobbying and the greater the level of representation, the better the quality of decisions and the quality of life. Hugh Stretton (2001) for example makes a comparison between Green Valley in Sydney, Australia and Elizabeth in Adelaide, Australia. Both are planned cities, but Elizabeth is more diverse in that it has both public and private housing and a high level of public participation in local government. Also and most importantly the council members live in Elizabeth. 'Tolerance' is a word used in the State of the Regions Report without any irony. It is also a word used by Habermas and criticized by Derrida for being paternalistic, at best and at worst rooted in the arrogance of the powerful who can and do decide what and who should be tolerated and under what circumstances. Derrida and Habermas discuss the importance of communication and critical thinking for democracy. They concentrate on discussing two concepts, namely: tolerance and hospitality and their relevance for democracy with Borradori[77] in the wake of September the 11th. Derrida avoids accepting the issue as a single event in time - namely 9/11 and stresses that drawing boundaries in this discussion can lead to problematic conclusions. Instead the globalised world is still experiencing the aftermath of the cold war and the ramifications of decisions made by superpowers.[78]

The grass roots arena remains important. People operate conceptually in a range of contexts and travel widely geographically, but the majority still live in one place. The mobile knowledge workers[79] are those with wider options.

We need to rescue the enlightenment from itself though re-working the conceptual boundaries (of democracy and the enlightenment) and the

---

[77] See Habermas, J., Derrida, J, and Borradori, G. 2003, *Philosophy in a Time of Terror Dialogues* with Jurgen Habermas and Jacques Derrida, interviewed by Borradori, Giovanna, University of Chicago Press, Chicago, pp, 16-18, 72-74.

[78] Pilger, J., 2002, *The New Rulers of the World*, Verso, London.

[79] The mobile groups can live in higher density places and holiday elsewhere (Stretton 2001). It is important to have different density options for different age groups and those with limited income. He argues that young children and young families need space, but cost saving, in terms of saving for infrastructure costs can be enhanced by high density living needs to be approached carefully. The long-term implications for quality of life need to be considered in terms of triple bottom line accounting. It is important to note that the kinds of environment s that the highest paid knowledge workers and the super rich capitalists choose are unspoiled and unpolluted, green and leafy – or if the inner city life is chosen – then regular breaks away are affordable (Stretton, 2001). He is scathing about systems modellers who think that they can solve all the problems. Insider knowledge is as important as outsider or expert knowledge and openness remains essential at all times as a means to find out whether ideas can stand up to testing.

geographical boundaries (of governance). The challenge is to balance individualism and collectivism by means of participation, advocacy and mindfulness;because no single model or approach can provide all the answers.

13.6.3. DESIGN OF INQUIRING SYSTEM FOR COMPASSIONATE, SYSTEMIC GOVERNANCE

Capacity building is needed in many arenas to enhance the ability of people to think holistically. A design of inquiring systems could assist in this regard. The inquiring system[80] addresses domains of knowledge, as follows:
- Logic or patterns of meaning of many stakeholders, qualitative and quantitative empirical data,
- Idealism of ethical praxis within context[81],
- Dialectic of working on thesis, antithesis and synthesis can help to achieve reasonable[82] ethical outcomes that stand up to testing by those who are to be at the receiving end of the decision and by those who are advocates for the voiceless. This approach is based on the assumption that ethical idealism and sustainable pragmatism are one. The 'boomerang affect' (Beck 1992) ensures that poor policy decisions will impact both the powerful and the powerless, because social and environmental pollution cannot be contained by means of artificial governance boundaries. West Churchman[83] stressed the importance of considering the social, political, economic and environmental factors when undertaking an analysis. Questioning within context helps to move closer to the ideal, because testing is conducted within context by those who are to be at the receiving end of a decision. Thus the pursuit of truth is not abandoned. 'Sweeping in'[84] a range of private, personal and public, political factors can help us to ensure that we do not leave out areas that are relevant in an analysis. He stresses that the categories of 'religion, politics, ethics and aesthetics', with a strong value basis are the challenges that we face and need to be 'unfolded'. They are to use his words our 'enemies', but they are within every one of us, by virtue of our

---

[80] Adapted from West Churchman, 1971; Habermas, 1984; Habermas and Derrida, 2003.
[81] Adapted from Kant Paton, H.J., 1976, *The Moral Law: Kant's Groundwork of the Metaphysic of Morals*, Hutchinson, London.
[82] Habermas, J., 1995, "Reconciliation through the public use of reason: Remarks on John Rawls's Political Liberalism", *The Journal of Philosophy*, **XC11**(3):109-131.
[83] Churchman, C. West, 1979a, *The Systems Approach*, Delta, New York.
Churchman, C. West, 1979b, *The Systems Approach and its Enemies*, Basic Books, New York.
Churchman, C. West, 1982, *Thought and Wisdom*, Intersystems Publications, California.
[84] Churchman drew on Edgar Singer.

humanity. We need to be aware of the implications of assumptions on our ability to think critically and rationally and avoid projecting these perceptions onto others. Any framework of thinking or methodology we select should be in relation to an area of concern and iterative consideration of the implications for selection and application should involve 'unfolding', to use West Churchman's concepts(1979a,b,1982) 'the enemies within' and 'sweeping in' the social, political, economic and environmental implications of our choice. This is a participatory process. We can accept that diversity in thinking, conceptualization and methodology are as important to human understanding as biodiversity is to evolutionary potential. The closest we can get to truth is through the dialectic of thesis- antithesis and synthesis or self-other and the environment[85].

When we are able to think about our thinking and its implications we can do better governance and decision-making. Consciousness has implications for governance. The brain is not divided into compartments that can be allocated specific functions, one of which can be called consciousness. Brain, mind and matter are interconnected. Mindfulness or consciousness is the connection across self, other and the environment. If consciousness is the result of seeing connections, then governance also needs to make connections not divisions. As Greenfield, thinker in residence in Adelaide stressed:

> "My particular definition of mind will be that it is the searing morass of cell circuitry that has been configured by personal experiences and is constantly being updated as we live out each moment" (Greenfield, 2002: 13).

Mindfulness however can be understood in terms of Bateson's level 1, 2, 3 learning (Bateson 1972 in McIntyre, 2004). Although context is all-important to understanding, it needs to go beyond frameworks to 'unfold' and 'sweep in' a range of considerations. What motivates or energizes us to act? Information that drives and restrains us comes both from our emotions and our rational thinking and involves us as embodied social actors – a series of systemic webs or feedbacks can be moderated by mindfulness or inflamed by passion. Passion may be generated by fear or desire or equally by rational thought about the implications of greed and pollution. Compassion may also be generated by emotion and rational thought. The notion that rational

---

[85] In this sense I revise the work of Michelle Fine (see Fine, Michelle (1994) "Working the hyphens: Reinventing self and other in qualitative research" in *Handbook of Qualitative Research*, Denzin and Lincoln, Sage) and the work of Elkingon by drawing on Singer, Peter and Haraway, D., 1991, *Simians, Cyborgs, and Women: The Reinvention of Nature*, Free Association Books, London.

thought or emotions are primary is a false dichotomy based on binary thinking or Cartesian thinking.

Greenfield (2002) argues that consciousness is a continuum and that some animals do have personality and as a result of thinking and reacting they have individualized their brain circuitry and made meanings of their world in a particular way. At a certain point in organic life a degree of mindfulness or self-awareness is created. This extends Peter Singer's (2002) argument about human rights and animal rights and underlines the importance of the need for human beings to see themselves as carers or sustainers of social and environmental justice. It also reminds us that the human animal could evolve in ways that the powerful can choose.[86] Systemic governance needs to be aware of the potential and pitfalls of the way we 'work the hyphens' (Fine, 1994) of self-other and the environment.

The context of community, region, nations and international arenas has been considered. Supranational identity relates to human rights but transboundary identity[87] relates to self-other and the environment'[88]

A shift from 'either or' thinking that is bounded to 'both and' thinking is needed. This requires an ability to think critically and systemically about the future.[89]

Human consciousness is a continuum from compassionate caretakers who are mindful or conscious of the many factors that are required for sustainable governance to passionate fundamentalists in West and East who are ruled by either religion or the market. Zealotry is a symptomatic of emotive decision-making that takes into account only some connections and not others. Critical and systemic thinking and practice are required for sustainable international relations and governance. Thinking can change practice and practice can and does change thinking. 'How can we frame areas of policy concern?(McIntyre-Mills 2000, 2003, McIntyre 2003a,b,c) This is the challenge raised by West Churchman.(1971,1979a,b, 1982).

---

[86] Haraway, D., 1991, *Simians, Cyborgs, and Women. The Reinvention of Nature*, Free Association Books, London.

[87] Otto, D., 1999, "Everything is dangerous: some poststructuralist tools for rethinking the universal knowledge claims of Human rights law", *Australian Journal of Human Rights*, accessed 9/22/2004

[88] Fine, Michelle, 1994, "Working the hyphens: Reinventing Self and Other in Qualitative Research" in: *Handbook of Qualitative Research*, Denzin & Lincoln, Sage.

[89] Banathy, B., 2003, "Self-Guiding Evolution of civilisation", *Systems Research and Behavioural Science*, 20:4309-323.

Banathy, B., 2001b. "The Agora project: Self guided social and societal evolution: the new agoras of the 21st Century", 45th International Conference International Society for the Systems Sciences, July 8th-13th 2001.

Bateson, G., 1972, *Steps to an Ecology of Mind: A Revolutionary Approach to Man's Understanding of Himself*, Ballantine, New York

Bausch, K., 2001, *The Emerging Consensus in Social Systems Theory*, Kluwer/Plenum.

13. MOLAR AND MOLECULAR IDENTITY AND POLITICS

What are the right questions to ask? Midgley[90] writes of the way thinking, methodology and practice are bounded and the implications for those who fall outside the boundaries of what is acceptable and what is not. The challenge is for policy makers to be better boundary workers if we are to enhance ethical governance or at the very least to avoid the risk of the 'boomerang affect' (1992) of shifting the burden of poverty, pollution, 'fall out' of war onto others. A conscious governance decision needs to be made to address social and environmental justice.[91]

The way forward to improve capacity building is to improve policy-making decisions through appreciating, not denying complexity.[92] This is only possible by improving our ability to hold in mind many ideas of many people at any one moment[93] and to explore innovative ways to enhance sustainable life chances of human beings. This requires critical and systemic thinking beyond conceptual and geographical boundaries. It is also based on an understanding that knowledge is made and remade contextually. It is fluid and changing and this has implications for governance.

---

[90] Midgley, G., 1992, "Pluralism and the legitimation of systems science", *Systems Practice*, 5(2):147-172.
Midgley, G., 1997, "Dealing with coercion: critical systems heuristics and beyond", *Systems Practice,* 10:(1):37-57.
Midgley, G., 2000, "Systemic intervention: philosophy, methodology, and practice", *Contemporary Systems Thinking*, Kluwer, New York.
Midgley, G. and Ochoa-Arias, A., 2001, "Unfolding a theory of systemic intervention", *Systemic Praxis and Action Research*, 14(5): 615-649.
Midgley, G., 2003, *Second Order Cybernetics, systemic therapy and soft systems thinking*, Sage, London.

[91] Laslow, K. and Laslow, A. 2004, "The role of evolutionary learning community in evolutionary development: the unfolding of a line of inquiry", *Systems Research and Behavioural Science*, 21:269-280.
Laszlo, E., 1991, *The Age of Bifurcation: Understanding the Changing World*, Gordon and Breach. Philadelphia.
Laszlow, A., 2001, "The epistemological foundations of evolutionary systems design", *Systems Research and Behavioural Sciences*, 18:307-321
McIntyre Mills, J., 2003, *Critical Systemic Praxis for Social and Environmental Justice: Participatory Policy Design and Governance for a Global Age*, Kluwer, London.
McIntyre, J., 2004, "Facilitating critical systemic praxis by means of experiential learning and conceptual tools to enhance our capacity to design systemic policy and praxis that promotes understanding through better governance", *Systems Research and Behavioural Science*, 21:37-61.

[92] White, L., 2001, "Effective Governance" through complexity thinking and management science", *Systems Research and Behavioural Science*, 18(3):241-257, Wiley.
McIntyre, J., 2004, "Facilitating critical systemic praxis by means of experiential learning and conceptual tools to enhance our capacity to design systemic policy and praxis that promotes understanding through better governance", *Systems Research and Behavioural Science*, 21:37-61.

[93] Edwards and Gaventa, J., 2001, *Global Citizen Action,* Colorado, Boulder.

Using conceptual tools could enhance our ability to design inquiring systems that are more mindful and more conscious of both our rights and our responsibilities to not only ourselves, but also to all sentient beings. Consciousness is a continuum from inorganic to organic life (Greenfield 2002) and at a certain point life becomes self-aware. As conscious caretakers we have a role of achieving balance between the freedoms of individuals and the need for collective responsibility for social and environmental justice.

Looking after ourselves, our own gene pool, whether we are human or animal(Singer, 2002, pp.119-164.) and our immediate community first may seem to make sense in the short term, until we realise that Gaia – the earth is interconnected(Elkington 1992, Singer 2002) and our selfishness could lead to a 'boomerang affect'(Beck 1992) Our mutual survival depends on co-operative governance of our environment. Critical and systemic thinking is useful in this regard, not only because it exposes the contradictions and helps us to think about their implications for Australia and Australians, but also the contradictions in other Western democracies. Hindess(2003) discusses "responsibility for others in the modern system of states". He argues that democracies are responsible for looking after their own people, their own citizens and for some categories of asylum seekers who are recognised to be refugees and not for others. But what about other people? Nation states draw boundaries around 'us' and 'them'. Australia has prided itself on "a fair go" democratic culture. We are a nation of immigrants. Ideally democracy requires working and re-working the conceptual and geographical boundaries. This needs to be undertaken in a less cynical manner. This argument supports a form of revised democracy and enlightenment that respects the importance of freedoms to the extent that it does not undermine the freedoms of others. The challenge is to find ways to work with the subjective, technical and communicative domains of knowledge.[94]

---

[94] Habermas, J., Derrida, J. and Giovanna, B., 2003, *Philosophy in a time of terror dialogues with Jurgen Habermas and Jacques Derrida interviewed by Borradori, Giovanna.* University of Chicago Press, Chicago.
McIntyre, J., 1998, "Consideration of categories and tools for holistic thinking", *Systemic Practice and Action Research*, **11**(2):105-126.
McIntyre-Mills, J., 2000, *Global Citizenship and Social Movements: Creating Transcultural Webs of Meaning for the New Millennium*, Harwood, Netherlands.
McIntyre, J., 2003, *Critical Systemic Practice for Social and Environmental Justice*, Kluwer, New York.
McIntyre, J., 2004, "Facilitating critical systemic praxis by means of experiential learning and conceptual tools to enhance our capacity to design systemic policy and praxis that promotes understanding through better governance", *Systems Research and Behavioural Science*, **21**:37-61.

13.7. CAPACITY BUILDING FOR A SUSTAINABLE FUTURE

Governments need to help to make places and regions competitive and cannot rely on the market alone. In Australia, like elsewhere the challenge for governance is to build civil society. The market has in fact created enormous divides between high priced areas with expensive properties and so-called 'regional gulags' where technology, talent and tolerance are more limited. Supporting local communities to learn from one another and to support one another in practical projects that enable people to draw on their experiences whilst trying out new projects, in other words to learn by doing and to be supported by means of a range of networks locally and regionally. Action learning as individuals, groups and communities is important.[95] In terms of this argument the idealistic notion of social, economic and environmental justice makes sense. Thus idealism and pragmatism form part of one sustainable cycle. Managed diversity has been argued as being good for development and for governance. Good governance from the point of elected representatives, corporate governance (institutional governance) and citizen representation needs to address the point of view of multiple stakeholders. The values of all the stakeholders need to be 'unfolded'[96] in the process of making policy. This requires an ability to communicate appropriately in a range of arenas. Nevertheless, the grass roots arena remains important. People operate conceptually in a range of contexts and travel widely geographically, but the majority still live in one place. The more mobile knowledge workers[97] are those with wider options and more rights than those who are less mobile.

---

[95] See Reason, P. and Bradbury, H., 2001, *Handbook of Action Research: Participative Inquiry and Practice*, Sage, London.
Reason, Peter, 1994, "Three approaches to participative inquiry", in *Handbook of Qualitative Research*, Denzin & Lincoln, Sage.
Reason, P. and Bradbury, H., 2001, *Handbook of Action Research: Participative Inquiry and Practice*, Sage, London.

[96] McIntyre, J., 2003, "Participatory democracy: Drawing on C. West Churchman's thinking when making public policy", *Systems Research and Behavioural Science*, 20:489-498.

[97] The mobile groups can live in higher density places and holiday elsewhere (Stretton 2001). It is important to have different density options for different age groups and those with limited income. For example young children and young families need space (Stretton 2001) and the argument that cost saving, in terms of saving for infrastructure costs can be enhanced by high density living needs to be approached carefully. The long-term implications for quality of life need to be considered in terms of triple bottom line accounting. It is important to note that the kinds of environments that the highest paid knowledge workers and the super rich capitalists choose are unspoiled and unpolluted, green and leafy – or if the inner city life is chosen – then regular breaks away are affordable (Stretton 2001). He is scathing about systems modellers who think that they can

13.8. CONCLUSION: CAPACITY BUILDING IS NEEDED IN BOTH STRONG AND WEAK STATES

Capacity building for better governance is needed internationally, not merely in the so-called weak states as Fukuyama(2004: 11-112) stresses. His work assumes the superiority of the United States as a strong state that has to intervene, because Europe chooses not to do so and instead chooses to rely on international law, not state force. According to Fukuyama:

> "It is of course undeniable that small, weak countries that are acted on rather than influencing others naturally prefer to live in a world of norms, laws, and institutions, in which more powerful nations are constrained. Conversely, a sole superpower like the United States would obviously like to see its freedom of action is as unencumbered as possible. But to point to differences in power is to best the question of why these differentials exist. The EU collectively encompasses a population of 375 million people and has a GDP of $9.7 trillion, compared to a US population of 280 million and GDP of $10.1 trillion. Europe could certainly spend money on defense at a level that would put it on a par with the United States, but it chooses not to. Europe spends barely 130 billion, which is due to rise sharply. Despite Europe's turn in a more conservative direction in 2002, not one rightist or centre-right candidate is campaigning on a platform of significantly raising defense spending. Europe's ability to deploy its power is of course greatly weakened by the collective action problems posed by the current system of EU decision making, but the failure to create more useable military power is clearly a political and normative issues ... the continent that invented the very idea of the modern state built around centralized power and the ability to deploy military force has eliminated the very core of stateness from its identity. This was the case above all in Germany where, as Peter Katzenstein (1997) has shown, post war identity was constructed around an anti sovereignty project. German freedom of action would henceforth be constrained by multiple layers on international constraints, above all the EU but including the United Nations. Germans for many years after World War 11 taught their children not to display the German flag or cheer too loudly for German teams at football matches. The kind of patriotism Americans displayed in the aftermath of September 11 is thus

---

solve all the problems. Insider knowledge is as important as outsider or expert knowledge and openness remains essential at all times as a means to find out whether ideas can stand up to testing.

quite foreign and, indeed, distasteful to them- and would, if displayed by the Germans themselves, be distasteful to everyone else."

Fukuyama (2004: 113) does not begin to engage in a critique of America's economic hegemony or fundamentalism. Instead he argues that the European idea of legitimacy and democracy is based on abstract concepts of human rights and goes beyond the nation state, but then he lapses into an argument that it is America, the strong state that supports world order. He does not consider that perhaps the American nation state can be seen quite differently. Fukuyama argues that Western development has been used to achieve particular outcomes and it is measured in terms of performance measures that do not take into account the need to build people skills in the public and private sector. He could argue that these people skills are needed in the West as much as they are in the East, but he is too busy projecting problems on the other to see the lack of capacity in the West! Capacity building has become a catch cry, but not the understanding that we need it as much in the West as in the East!

Fukuyama (2004) argues, that the European approach to governance is based on human rights, which is a transboundary concern, and the American is based on a bounded national view of citizenship rights. According to him, this is because the US has the power (at the moment) to act as a powerful state. Perhaps if nation states were based on social and environmental justice, rather than power and might and if there was a realization that justice is preferable to power (and not merely a sign of weakness) this could be a step forward. A world where justice rules is preferable to a world where might rules. Unfortunately implementing justice requires some power, as Fukuyama is keen to point out. He agrees that legislative soft power based on trust would be preferable. For this to occur dialogue across interest groups is essential. This point is vital and needs to be increasingly recognized and implemented at a local, national and international level. Public administration and leadership in building capacity to work across multiple maps of reality is also vital for caring governance and discursive international relations with diverse stakeholders.

I argue that the capacity to think critically and systemically could help policy makers think in terms of the 'boomerang affect' (Beck, 1992) of selfish, short term policy solutions.

Expanded Pragmatism is about being mindful or conscious of the implications -not just for **some** stakeholders-but for **all life** in the short and long term (See McIntyre-Mills 2006, vol.3). The chapter argues that the economic rationalist agenda needs to be balanced by a capacity to think about a sustainable future. This requires asking questions. Could *pain, poverty and pollution* have feedback loops for this generation and the next? How will cruelty and commodification of people and animals impact on the web of life?. Are trans species diseases the result of weakened farm animals (Singer 2002) whose immune systems are affected by incarceration and poor food? The latter question highlights how much more work is needed to achieve human rights, let alone ecological mindfulness.

REFERENCES

Ashby, W. Ross. 1956, *An Introduction to Cybernetics*, Chapma and Hall, London.
Banathy, B., 2001b. "The Agora project: Self guided social and societal evolution: the new agoras of the 21st Century", 45th International Conference International Society for the Systems Sciences, July 8th-13th 2001.
Banathy, B., 2003, "Self-Guiding Evolution of civilisation", *Systems Research and Behavioural Science*, 20:4309-323.
Baruma, I. and Margalit, A., 2004. *Occidentalism: The West in the Eyes of its Enemies*, Penguin, New York.
Bateson, G., 1972, *Steps to an ecology of mind: A revolutionary approach to man's understanding of himself*, Ballantine, New York.
Bausch, K., 2001, *The Emerging Consensus in Social Systems Theory*, Kluwer/Plenum.
Beck, U., 1992, *Risk Society Towards A New Modernity*, Sage, London.
Beck, U., 1998, *Democracy Without Enemies*, Polity, Cambridge.
Bogue, R., 1989, *Deleuze and Guattari*, Routledge, London.
Bourdieu, P., 1986, "The forms of capital", Reprinted from Richardson, J. in: *Education, Culture, Economy and Society*, A.H. Halsey, ed., 1997, Oxford University Press, Oxford.
Buchanan, I. and Colebrook, C., 2000, *Deleuze and Feminist Theory*, Edinburgh University Press.
Checkland, P. and Holwell, S. 1998, "Action research: Its nature and validity", *Systemic Practice and Action Research*, **11**(1):9-21.
Chomsky, N., 2003, *Hegemony or Survival: America's Quest for Global Dominance*, Allen and Unwin, Crows Nest, NSW.
Churchman, C. West, 1971, "The design of inquiring systems", *Basic Concepts of Systems and Organization*, Basic Books, New York.
Churchman, C. West, 1979a, *The Systems Approach*, Delta, New York.
Churchman, C. West, 1979b, *The Systems Approach and its Enemies*, Basic Books, New
Churchman, C. West, 1982, *Thought and Wisdom*, Intersystems Publications, California.
Edgar, D., 2001, *The Patchwork Nation: Rethinking Government-Rebuilding Community*, Harper, Sydney.
Edwards and Gaventa, J., 2001, *Global Citizen Action*, Colorado, Boulder.
Fine, Michelle, 1994, "Working the hyphens: Reinventing Self and Other in Qualitative Research" in: *Handbook of Qualitative Research*, Denzin & Lincoln, Sage.
Firkin, R., 1999, Alice Springs Public Library Internet Project, Unpublished Report.
Fivaz, R., 2003, "Unidimensional life", *Systems Research and Behavioural Science*, 269-287.
Fukuyama, F., 2004, *State Building: Governance and World Order in the 21st Century*. Cornell University Press, New York, Ithaca.
Gaventa, J., 2001 "Towards participatory local governance: Six propositions for discussion", Paper presented to the *Ford Foundation*, LOGO Program with the Institute of Development Studies, June.

Gaventa, J. and Valderrama, C., 1999, "Participation, citizenship and local governance", Background note for workshop on *Strengthening participation in local governance*, Institute of Development Studies, June.

Greenfield, S., 2002, *The Private life of the Brain: Emotions, Consciousness and the Secret of the Self*, John Wiley and Sons, New York

Habermas, J., 1995, "Reconciliation through the public use of reason: Remarks on John Rawls's Political Liberalism", *The Journal of Philosophy*, XC11 (3):109-131.

Habermas, J., Derrida, J, and Giovanna, B. 2003, *Philosophy in a Time of Terror Dialogues with Jurgen Habermas and Jacques Derrida*, interviewed by Borradori, Giovanna, University of Chicago Press, Chicago, pp, 16-18, 72-74.

Haraway, D., 1991, *Simians, Cyborgs, and Women. The Reinvention of Nature*, Free Association Books, London.

Harding, R., 1993, Opportunity costs: Alternative Strategies for the Prevention and Control of Juvenile Crime", in: *Repeat Juvenile Offenders: The failure of selective incapacitation in Western Australia*, R. Harding, Research Report No 10, University of Western Australia, Crime Research Centre, page 141.

Hindess, B., 2003, "Responsibility for others in the modern system of states", *Journal of Sociology*, **39**(1):23-30.

Hofstadter, D., 1979, *Godel, Escher, Bach an Eternal Golden Braid*, Basic Books, New York.

Jamrozik, A., 2004, *The Chains of Colonial Inheritances: Searching for Identity in a Subservient Nation*, University of New South Wales Press, Sydney.

Jones. B., 1990, *Sleepers Wake! Technology and the Future of Work*, 3rd.ed., Oxford University Press, Melbourne.

Laslow, K. and Laslow, A. 2004, "The role of evolutionary learning community in evolutionary development: the unfolding of a line of inquiry", *Systems Research and Behavioural Science*, **21**:269-280.

Laszlo, E., 1991, *The Age of Bifurcation: Understanding the Changing World*, Gordon and Breach. Philadelphia.

Laszlow, A., 2001, "The epistemological foundations of evolutionary systems design", Systems Research and Behavioural Sciences, **18**:307-321

Long, M., 2002, "Beyond traditional boundaries: government in the information age", *Australian Journal of Public Administration*, **61**(1):3-12.

Longo, M., 2004, "Hostile receptions: dilemmas of democracy: legitimacy and supranational law", *Australian Journal of Politics and History*, 50(2):211-228.

Maruyama, M., 1980, "Mindscapes and science theories", *Current Anthropology*, 21:589-608.

McIntyre-Mills, J., 2000, Global Citizenship and Social Movements: Creating Transcultural Webs of Meaning for the New Millennium, Harwood, Netherlands.

McIntyre Mills, J., 2003, Critical Systemic Praxis for Social and Environmental Justice: Participatory Policy Design and Governance for a Global Age, Kluwer, London.

McIntyre, J., 2002, "A community of practice approach to knowledge management and evaluation re-conceptualized and owned by Indigenous stakeholders", Evaluation Journal of Australasia, 2(2): 57-60.

McIntyre, J., 2003a, "Participatory democracy: Drawing on C. West Churchman's thinking when making public policy", Systems Research and Behavioural Science, 20:489-498.

McIntyre, J., 2003b, "Yeperenye dreaming in conceptual, geographical and cyberspace: a participatory action research approach to address local governance within an Australian Indigenous Housing Association", Systemic Practice and Action Research, **16**(5):309-338.

McIntyre, J., 2003c, Critical Systemic Practice for Social and Environmental Justice, Kluwer. New York.

McIntyre, J., 2004, "Facilitating critical systemic praxis by means of experiential learning and conceptual tools to enhance our capacity to design systemic policy and praxis that promotes understanding through better governance", Systems Research and Behavioural Science, **21**:37-61.

McIntyre, J. 2005a, Working and re-working the conceptual and geographical boundaries of governance and international relations *Systemic Practice and Action Research* Vol. 18, No 2 157- 220.

McIntyre, 2005b,. Critical and systemic practice to address Fixed and Fluid identity and politics at the local, national and international level *Systemic Practice and Action Research*. 18 , No3 223-258

Midgley, G. and Ochoa-Arias, A., 2001, "Unfolding a theory of systemic intervention", Systemic Praxis and Action Research, 14(5): 615-649.

Midgley, G., 1992, "Pluralism and the legitimation of systems science", *Systems Practice*, 5(2):147-172.

Midgley, G., 1997, "Dealing with coercion: critical systems heuristics and beyond", *Systems Practice,* 10(1):37-57.

Midgley, G., 2000, "Systemic intervention: philosophy, methodology, and practice", Contemporary Systems Thinking, Kluwer, New York.

Midgley, G., 2003, Second Order Cybernetics, systemic therapy and soft systems thinking, Sage, London.

Otto, D., 1999, "Everything is dangerous: some poststructuralist tools for rethinking the universal knowledge claims of Human rights law", *Australian Journal of Human Rights*, accessed 9/22/2004

Pilger, J., 2002, *The New Rulers of the World*, Verso, London.

Putnam, R., 1995, "Bowling alone", *Journal of Democracy,* **6**(1):65-78

Singer, P., 2002, *One World: The Ethics of Globalisation*, The Text Publishers, Melbourne.

Veitch, J. 1977, *Descartes: A Discourse on Method: Meditations and Principles*, Everyman, London.

Yeatman, A., 2003, "Global ethics, Australian citizenship and the 'boat people': a symposium", *Journal of Sociology*, **39**(1):15-22.

Warren, E., 1999, *Democracy and Trust*, Cambridge University Press, Cambridge.

Wadsworth, Y., 2001, "The mirror, the magnifying glass, the compass and the map: Facilitating participatory action research", in *Handbook of Action Research*, P. Reason and H. Bradbury, Sage, London.

White, L., 2001, "'Effective Governance' through complexity thinking and management science", *Systems Research and Behavioural Science*, 18(3):241-257, Wiley.

Chapter 14

EDUCATION FOR ENGAGED CITIZENSHIP

HAMMOND, D.

Abstract: In the midst of rapidly shrinking support for the public sector, coupled with increasingly prescriptive state and federal mandates for meeting narrowly defined content and assessment standards, the mission of education is being called into question, as is the integrity of the democratic process. With market forces trumping all other values, such apparently laudable initiatives as the "No Child Left Behind" legislation in the United States become a means of centralizing control and reinforcing what Paolo Friere has called "the 'banking' concept of education." In this paper I examine a variety of pedagogical practices that foster an alternative approach, which Friere refers to as "problem-posing education". Key elements in such practices include an interdisciplinary curriculum, dialogue and critical inquiry, and community-based learning opportunities. Given West Churchman's life-long concern with cultivating ethical practices in management, an important starting point is the educational system, a fact not lost on those who would direct that system toward their own more limited ends.

Key words: Education, Ethics, Interdisciplinary Studies, Critical Inquiry, Dialogue, Learning Communities, Community-Based Learning

14.1 CULTIVATING DEMOCRACY AND SOCIAL JUSTICE IN A CORPORATE WORLD

In defining the systems approach, West Churchman suggests, "on the one hand, we must recognize it to be the most critical problem we face today, the understanding of the systems in which we live." He goes on to point out, on the other hand, that the problem of the "appropriate approach to systems" has not been solved. Rather, "the nature of systems is a continuing perception and deception, a continuing re-viewing of the world, of the whole

system, and of its components. The essence of the systems approach, therefore, is confusion as well as enlightenment." He then goes on to articulate some principles of a "deception-perception approach to systems" (1968, 230-231):
1. The systems approach begins when first you see the world through the eyes of another.
2. The systems approach goes on to discovering that every worldview is terribly restricted.
3. There are no experts in the systems approach.

When I first met West Churchman in 1993, while enrolled in a graduate program in the history of science at the University of California in Berkeley, his greatest concern was with the number of children in the world who were dying each day from starvation and lack of basic care. During the mid-1990s he was working with a number of students and community members on developing a framework for Global Ethical Management (GEM). He was deeply concerned about the suffering and injustice that he observed in the world, and passionately devoted to finding effective ways of addressing these problems. His principles for the deception-perception approach provide a useful starting point for articulating a systemic approach to education. In order to understand the complexity of the systems in which we live, it is important to be able to integrate insights drawn from a wide variety of disciplinary perspectives – economics, politics, sociology, psychology, and the natural sciences (especially biology and ecology), as well as the humanities. In addition, educational practice that is truly aimed at cultivating a democratic society needs to instill in its students the ability to communicate effectively, to ask meaningful questions, and to listen to points of view that challenge their own.

Unfortunately, as Ron Miller points out, the dominant trends in contemporary education seem to be moving in the opposite direction:

> "Public education as it developed in the twentieth century became a mechanized process of inducting young people into the culture of modernity. By "mechanized" I mean carefully managed and controlled by a central authority, so that personal differences of style, desire, and aspiration were blurred by the need to conform to standards and pre-established roles. At the time, and still very much in our time, this definition has made sense; in a culture that values efficiency, competition, and the production and consumption of material goods above all else, what other purposes could schools serve?" (Miller, 2000, 5)

Far from the open-minded inquiry that Churchman's argument would support, the model being increasingly imposed on educators from kindergarten to the university is grounded in a commitment to the certainty of objective knowledge, and fails to address the more complicated ethical dimension of life. As Churchman suggests, "ethics is not a body of theory substantiated by facts. Instead, it is a process of continuously – and I think eternally – discussing and debating and occasionally fighting over the issues. In short, ethics is a dialectical process in which all humanity, past, present, and future, must take part" (1979, 118).

14.2 INTERDISCIPLINARY APPROACHES TO LEARNING

In a 1980 report "intended for all who care about the quality of our common life," the newly formed Commission on the Humanities described the significance of the humanities as follows:

"The essence of the humanities is a spirit or an attitude toward humanity. They show us how the individual is autonomous and at the same time bound ... to humankind across time and throughout the world. The humanities are an important measure of the values and aspirations of any society. Intensity and breadth in the perception of life and power and richness in the works of the imagination betoken a people alive as moral and aesthetic beings, citizens in the fullest sense. They base their education on sustaining principles of personal enrichment and civic responsibility. ... They can approach questions of value, no matter how complex, with intelligence and good will. They can use their scientific and technical achievements responsibly because they see the connections among science, technology and humanity." (Report of the Commission on the Humanities, 1980, 3)

Acknowledging that they were "deeply concerned about the serious social deficiencies of perception and morale," they identified the primary roots of these problems: "Our society has increasingly assumed the infallibility of specialists, the necessity of regulating human activity, and the virtues of material consumption." Suggesting that "[t]hese attitudes limit our potential to grow individually and to decide together what is for the common good," they ask the following questions, which are highly pertinent to Churchman's concerns, as well as my own: "When does specialization suffocate creativity, denigrate the critical judgment of nonspecialists, or undermine the idea of leadership? When does regulation become regimentation? At what point does materialism weaken the will to conduct our lives according to spiritual and moral values?"

In the fall of 1997, I had the good fortune of joining the faculty of the Hutchins School of Liberal Studies, at Sonoma State University (one of 23 campuses in the California State University system). Founded in 1969 as an interdisciplinary and seminar-based liberal studies program, it embodies the essence of the humanities as articulated above. It has gained a national reputation for its integrated and interactive approach to learning. A recently drafted mission statement for the program reads as follows:

> "The Hutchins School is committed to innovative pedagogy and interdisciplinary inquiry into vital issues, through seminar-based learning, thematic and cross cultural courses, small classes, and collaborative learning. We graduate students who are life-long learners, skilled in analysis, synthesis, and multiple modes of communication."

Inspired by the groundbreaking work of Robert Maynard Hutchins in fostering interdisciplinary and collaborative initiatives in both teaching and research as Chancellor of the University of Chicago from 1929 to 1951, the Hutchins School is only one of many institutions devoted to interdisciplinary inquiry and collaborative learning. Churchman himself was actively supportive of interdisciplinary initiatives at UC Berkeley, where he helped to foster the establishment of student-designed interdisciplinary majors. And the breadth of his own career - as a student of philosophy, founder of operations research, professor in the business school at Berkeley and, into his 80s, professor of ethics in the Peace and Conflict Studies program - provides an impressive example of cross-disciplinary synthesis.

For the past decade, the Hutchins School has been actively engaged in the Learning Community movement, which fosters interdisciplinary pedagogy and the creation of opportunities for students to integrate knowledge from a variety of perspectives in connection with specific themes or problem areas. At the 1999 Teacher-Scholar Summer Conference of the California State University's Institute for Teaching and Learning, faculty from the Hutchins School presented a workshop on "Developing Effective Curricular Learning Communities in the CSU." In her opening presentation on "Learning Community Models," Jean McGregor (Director of the National Learning Communities Project of the Washington Center for Undergraduate Education at Evergreen State College) defined learning communities as follows:

> "A variety of approaches that link or cluster classes during a given term, often around an interdisciplinary theme, that enroll a common cohort of students. This represents an intentional restructuring of students' time, credit and learning experiences to foster more explicit connections between students, between students and their faculty, and between disciplines." (MacGregor, 1999)

Indeed, the key dimension of the learning community model is the cultivation of the ability to make meaningful connections, both with faculty and peers, and across disciplinary boundaries. As a result, this model emphasizes active and collaborative learning, incorporates value-based knowledge, and encourages civic engagement as part of the curriculum.

14.3 DIALOGUE AND CRITICAL INQUIRY

In 1986, the American Association for Higher Education sponsored the development of a statement of "principles for good practice in undergraduate education." The principles as articulated were "based on a view of education as active, cooperative, and demanding". According to the first three of these principles, good educational practice "encourages student-faculty contact, encourages cooperation among students, [and] encourages active learning" Hatfield, 1995, 1).

The cultivation of skills in dialogue and collaboration is key to the development of participatory decision-making processes, as well as the emergence of a more truly democratic society. Further, if individuals are to become more actively engaged in the decisions that shape their lives, they need to have a sense of ownership in the process, instead of passively deferring to the "expertise" of those in leadership positions. Unfortunately, the dominant educational model tends to reinforce passivity, dramatically restricting the potential for what Paolo Friere (1970) refers to as "liberatory education".

Using categories identified by John Broadbent, in his paper on Participatory Action Research (2004), the dominant pedagogical approach is generally positivistic (i.e. "objective", linear, predictive, expert-based, seeking simplicity and the domestication of time). As a result, students are trained to think that there are simple, pre-determined answers to complex problems, and they are less likely to question the simplistic, short-sighted, and self-serving presentations that pass for public discussion of challenging issues in most of today's media.

In contrast, the Hutchins School, like an increasing number of similar programs committed to critical inquiry, offers an interactive, student-centered pedagogy, in the spirit of Freire, integrating elements of a more constructivist and critical orientation. It is constructivist in the sense that it is concerned with the meaning that students themselves find in the material they are reading. It is dialectical, pluralistic, participatory, and seeks to cultivate mutual understanding. Students often report that they learn the most from hearing the various perspectives of their peers; at the end of her second year, one student reported that the most important thing she had learned was to listen to something she disagreed with, without immediately reacting.

The Hutchins pedagogy is critical, in that issues of power are not only an integral part of the curriculum, but are also addressed through the structure of the learning environment itself. By allowing students to share the power of the traditional authority figure, they learn to have confidence in their own ideas and to challenge received wisdom when their own perceptions support alternative views. Furthermore, the interactive learning approach also provides a context for the creation of "transformative spaces" for envisioning alternative futures (Broadbent). In addition, it enhances skills in collaborative problem solving, as well as such values as mutual respect and inclusion, which are essential to the realization of a more equitable and sustainable future.

14.4   COMMUNITY-BASED LEARNING

In addition to the integrated curriculum and the interactive, seminar-based approach to learning, perhaps the most important element in this alternative educational model is the commitment to community-based learning. Through research and service grounded in the local community, students begin to realize that learning is more than homework and books; rather it connects them to something greater than themselves. By becoming more actively engaged in the larger world, they can more clearly see the connections between what they are learning in the classroom and what is going on in the "real world."

In May 2002, three members of the Hutchins faculty teamed up with two representatives from Sustainable Sonoma County (a local non-profit) in a five day "Engaged Department Institute," co-sponsored by the California State University system and Campus Compact. Although already familiar with the ideal of civic engagement as an important dimension of a more enlightened curriculum, I gained a much greater appreciation for the necessity of involving community partners in the curriculum development process, as well as the potential for community building inherent in the concept of civic engagement, which the Hutchins team defined as follows:

> "Direct involvement in household, classroom, workplace, and local, national, and global communities, sustained by critical analysis, ethical reflection, coalition-building, and action for social change. Hutchins fosters civic engagement through curriculum development, dialogue, service with multiple communities, public policy analysis, and social action."

As defined in resource materials prepared by Campus Compact,

"Service learning means a method under which students learn and develop through thoughtfully organized service that: is conducted in and meets the needs of a community and is coordinated with an institution of higher education, and with the community; helps foster civic responsibility; is integrated into and enhances the academic curriculum of the students; and includes structured time for students to reflect on the service experience." (Campus Compact, 2000, 15)

It is perhaps no longer appropriate for the university to maintain its traditional isolation from the surrounding community. Greater support for higher education might be cultivated if the university were to provide resources and services through more active engagement of its students in community-based projects. Stronger connections between the university and the K-12 community could provide exciting opportunities for collaboration between students and faculty at both levels, while also serving the needs of the local community.

14.5 EMBODYING THE CONCEPT

An example from my own teaching that embodies all three of these dimensions is an upper division seminar, "The Global Food Web", which I offered in the fall of 2004. The idea for this seminar emerged out of a series of public forums on global climate change that I helped to organize with other Sonoma State faculty and members of the local community. At the time, I heard Helena Norberg Hodge (of the International Society for Ecology and Culture) comment in a radio interview that the single greatest thing we could do to reduce greenhouse gas emissions would be to rebuild the local food system. She noted that the average food item travels roughly 1500 miles and that local agricultural systems throughout the world are being destroyed by the influx of subsidized products from the increasingly centralized agribusiness industry.

In addition, I was inspired by the work of Fritjof Capra's Center for Ecoliteracy in Berkeley, California, which has been working to integrate ecological awareness into the public schools through supporting the development of school gardens and "farm to school" projects, bringing locally grown food into school lunches. The Center has developed a "Fertile Crescent Network" of educators, farmers, school food service directors, and other community members throughout northern California to facilitate these projects, which are helping to preserve small family farms at the same time that they are addressing the growing epidemics of obesity and diabetes among young children, resulting from the ever increasing consumption of fast food (the devastating results of which can be seen in the recent film, "Supersize Me").

Since many of the students in the Hutchins School are planning to become teachers, this seemed like an ideal topic for a seminar that would lend itself quite well to an interdisciplinary approach. Understanding the global food system entails an appreciation for the economics and politics of food production, ecological considerations including the importance of soil conservation and land use planning, the chemistry of food and digestion, and the psychological dimensions of eating. Further, the topic provided excellent opportunities for community engagement, which they could then carry into their own teaching. Drawing on connections cultivated through the Fertile Crescent Network, I brought in members of the local community to talk about various aspects of the local food system and to present a variety of service learning opportunities. Many of the students worked in school gardens, some worked in the local food bank, one woman did a survey of the local schools, and a couple of students became involved in a local campaign to ban genetically engineered food.

This was one of the most successful seminars I have ever taught; students unanimously agreed that they learned more from this course than they had in any other course they had taken in their college career. I believe this is because it allowed them to understand the complex interactions between economic, ecological, biological and psychological dimensions of life in the context of a topic that was directly relevant to their own lives, and their engagement in the local community was an experience that they all found enormously valuable. And, in relation to Churchman's concerns, students were able to contribute in small but significant ways to the care and feeding of the next generation, and to see how their efforts might be applied in a larger context.

14.6 FINAL REFLECTIONS

One of the greatest challenges plaguing our culture is the fragmentation we experience in our lives, both individually and collectively. The creation of opportunities for re-establishing

> "connections between the various aspects of our lives is perhaps one of the most critical tasks confronting our educational institutions today. And yet, as Miller points out, the dominant model tends to reinforce fragmentation and 'mechanization'."

Many of us are convinced that school systems, as they were purposely designed and as they currently function, are inhumanely mechanical, undemocratic, damaging to personal growth and community health, and in a word, obsolete. We argue that they reflect a mechanistic view of the world that denies the noblest and best qualities of the human spirit. (2000, 7)

Appealing to such crusaders for educational reform as Emerson, Thoreau, Dewey, Montessori, Steiner, Goodman, and Illich, Miller argues that the current system focuses on "train[ing] young people to fulfill their roles in a vast, impersonal social machine" and to "compete for success in a system that only cares about their skills and credentials." Alternatively, he suggests that the creation of meaningful identity is an important dimension of the educational process that has been seriously neglected. (Miller, 2000, 5)

As Friere suggests,

"Apart from inquiry, apart from the praxis, individuals cannot be truly human. Knowledge emerges only through invention and reinvention, through the restless, impatient, continuing, hopeful inquiry human being pursue in the world, with the world, and with each other." (1970, 349)

The world is ripe for the emergence of a more truly integrated approach in education, extending beyond the institutional context into the surrounding community. Robert Hutchins appealed to the central importance of dialogue in education, the "Great Conversation," the nurturance of which was for him the highest calling of education. (Hutchins, 1952)

More recently, in an eloquent critique of the fragmentation and reductionism inherent in our contemporary understanding of our world, Wendell Berry also underscores the need for "authentic conversation" among the disciplines (2000, 23). Cultivating these three key dimensions – interdisciplinary curriculum, dialogue, and civic engagement – holds a great deal of promise for the renewal of our current educational system, nurturing a more vital and ultimately more socially relevant learning environment.

An interdisciplinary curriculum, combined with an interactive, student-centered approach to learning is particularly well suited to students planning a career in education, providing a pedagogical model that they can implement in their own classrooms. This in turn will help their own future students develop the skills in dialogue and collaboration that are essential to the cultivation of a thriving democracy.

Perhaps the most important dimension of this approach, however, is the opportunity for fostering civic engagement, through enhancing the potential for dialogue between various sectors of the community, including local businesses, government, and non-profits. Indeed, the educational sector is the ideal site from which to renew public dialogue, and I believe it has a responsibility to do so in these increasingly challenging times.

REFERENCES

Berry, Wendell, 2000, *Life is a Miracle: An Essay Against Modern Superstition*, Fellowship Press, New York.

Broadbent, John, 2004, "Participatory Action Research: A Methodology for the Study of Evolutionary Systems?", Proceedings of the 48th Annual Conference of the International Society for the Systems Sciences, Asilomar, CA.

Campus Compact, 2000, *Introduction to Service Learning Toolkit: Readings and Resources for Faculty*. Providence, Brown University, RI:

Churchman, C. West, 1968, *The Systems Approach*, Dell Publishing, New York.

Churchman, C. West, 1979, *The Systems Approach and its Enemies*, Basic Books, New York.

Commission on the Humanities, 1980, *The Humanities in American Life: Report of the Commission on the Humanities*, University of California Press, Berkeley.

Freire, Paulo, 1970, *Pedagogy of the Oppressed*, Herder and Herder, New York. (Page numbers reference excerpt, "The Banking Concept of Education," in *Ways of Reading: An Anthology for Writers*, D. Bartholomae and A. Petrosky, 1999, Bedford/St. Martin's, New York.

Hatfield, Susan Rickey, ed., 1995, *The Seven Principles in Action: Improving Undergraduate Education*, Ankar Publishing, Bolton, MA.

Hutchins, Robert M., 1952, The Great Conversation: The Substance of Liberal Education. Chicago: Encyclopedia Britannica, Inc.

McGregor, Jean, 1999, "Learning community models," presentation in workshop on *Developing Effective Curricular Learning Communities in the CSU* at CSU Teacher-Scholar Summer Conference, California.

Miller, Ron, ed., 2000, *Creating Learning Communities: Models Resources, and New Ways of Thinking About Teaching and Learning*, Foundation for Educational Renewal Brandon, VT.

Chapter 15

DIALOGUE FOR CONSCIOUS EVOLUTION

CHRISTAKIS, A.N.

Abstract: The chapter is based on a forthcoming book titled Dialogue for the Information Age Democracy. It focuses on the critical role of dialogue in the evolutionary guidance of social systems. It presents findings from grounded theoretical research starting with the Club of Rome prospectus on The Predicament of Mankind in 1970 (http://www.cwaltd.com/pdf/clubrome.pdf). This research led to the discovery of a disciplined dialogue paradigm called Structured Design Process (SDP). The SDP is suitable for engaging stakeholders in designing their social systems in the context of the escalating complexity of the Information Age. The paradigm is founded on a new geometry for languaging, called Archanesian, which postulates three axioms: (a) the escalating complexity of social systems designing, (b) the cognitive limitations of the observer in processing too many observations simultaneously, and (c) the difficulty in determining the relative saliency of observations. These three axioms become the foundation for the construction of terminology, methodology, and architecture for the SDP. In this chapter I will focus primarily on the architecture, which is composed of 35 components, six of which are laws discovered by systems thinkers of the 20th century. The six laws and their interactions in the context of structured dialogue for the production of meaning and community wisdom (DEMOSOPHIA) are presented in the chapter. The SDP has been applied for the last twenty years to address national, international, and inter-organizational design challenges. The results have been extraordinarily satisfactory to the practitioners and participants of these projects. A brief story from the design arena will also be presented.

Key words: Dialogue, structured dialogue, social systems, designing, CogniScope, democracy, Information Age, globalization, stakeholders, design arena

15.1 INTRODUCTION

People all over the world aspire to participative democracy, and yet the democratic planning and design of any social system, from cities to national health care programs, is threatened by the inability to engage stakeholders in a meaningful and productive dialogue. The fundamental premise espoused in this chapter is that in this critical juncture of the evolutionary process of humanity, there cannot be conscious evolution without the capacity and commitment to excavate through true dialogue the wisdom of the people in the Agoras of the 21st century Global Village.

My intent is not to deprecate traditional democratic and managerial methods. There is, however, sufficient empirical evidence that leaders and managers, even when they espouse the democratic ideal, they cannot implement it in the situations they are confronting. One of the root causes for most of the failures of participative democracy is the erroneous presupposition that a dialogue approach for the practice of democracy in the Information Age could be founded on a paradigm that is similar to the one used in the Agora of Athens 2,500 years ago.

This observation led me to commit thirty years of my professional life to developing, in collaboration with others, a new paradigm for participative democracy that is applicable to all people independent of their educational or power status. The Structured Design Process (SDP) paradigm is suitable for boundary-spanning dialogue across disciplines and cultures (Christakis and Brahms, 2003). It has been practiced all over the world in a variety of settings ranging from the emancipation of oppressed indigenous people (Christakis and Harris, 2004) to the development of safer pharmaceutical products (Vogt, 2002) and health care services for patients (Parker et al., 2004).

Bela H. Banathy in his book Guided Evolution of Society (2001) captures very elegantly and succinctly the critical role of dialogue in societal evolution. He wrote:

"Dialogue is a vehicle for understanding cultures and subcultures in organizations. And organizational learning depends upon such cultural understanding. It facilitates the development of a common language and collective mental models. Thus, the ability to engage in dialogue becomes one of the most fundamental and most needed human capabilities. Dialogue becomes a central component of any model of evolutionary transformation."

While dialogue is a central component of the conscious evolution of society, I will argue in this chapter that the conventional approach to engaging people in dialogue, even though it is necessary, it is not sufficient for enabling stakeholders to address democratically, systemically, and collaboratively the complexity of Information Age issues. It must be complemented with a new scientific paradigm. I will describe the axioms, terminology, methodology, and architecture of this new paradigm. For a deeper philosophical, theoretical, and methodological discussion the interested reader is referred to a forthcoming book titled Dialogue for the Information Age Democracy (Christakis, 2005).

15.2 THE GEOMETRY OF THE SDP PARADIGM

In the beginning of the 20th Century it became necessary to complement classical mechanics with relativistic mechanics in order to explain certain physical phenomena that were observed at that time by experimental physicists. The theory of classical mechanics, which was discovered in the 18th century by Newton, was not able to explain accurately some new observations. Newton's theory used as its mathematical foundation the axioms espoused in the geometry of space postulated over twenty centuries earlier by a Greek mathematician named Euclid. The theory of relativistic mechanics, which was discovered by Einstein in the early parts of the 20th century, employed axioms espoused by a different geometry of space. This new geometry of space was invented in the 19th century by a German mathematician named Riemann. The axioms of the Riemannian geometry differ drastically from those of the Euclidean geometry. For example, Euclid postulated that it is possible from a point to draw only one parallel line to another line. Riemann, in his geometry of space, rejected this axiom and postulated that is possible to draw an infinite number of parallel lines from one point. It was this new Riemannian geometry of space that was instrumental in Einstein's development of the theory of relativity. The message here is that in developing the theory of relativity Einstein had to adopt a new geometry of space. Without the adoption of the Riemannian geometry, the development of the theory of relativity was not possible or conceivable.

Building on this historical analogy of the need for a paradigm shift from classical to relativistic theories in the world of physics, I believe that the discovery in 1956 by Miller of the physiological phenomenon of cognitive limitations of humans, namely that human beings cannot process simultaneously more that seven observations (Miller, 1956), necessitates the adoption of a new dialogue paradigm for addressing responsibly, systemically, and collaboratively the escalating complexity of the Information Age issues. The new paradigm is based on a different "geometry for languaging," so to speak. In other words, the theory of the SDP paradigm is founded on a new geometry, just like the theory of relativity was founded on the new geometry of Riemann. I will call this new geometry Archanesian, in honor of my favorite Cretan village of Archanes, which has a history of 6,000 years. Furthermore, it should be pointed out that the theory of SDP is responsive to the contemporary requirement of Third Phase Science (de Zeeuw, 1996) to discover and apply a new paradigm for the inclusion of the observers in the construction of high quality observations for the issues facing them in the Information Age.

The development of the theory for the SDP paradigm is axiomatic, just like the theory of relativity is axiomatic. As is well known in science, the utility of any scientific theory should be determined on the basis of the results of observations obtained from its applications in the arena of practice. In my book (Christakis, 2005), I elaborate on the philosophy, theory, and methodology of the SDP, and present some stories of its application in the arena. The book is composed of five parts, the last four of which are strongly connected. Part I is dedicated to the history of the past and the future. It sets the context and offers background information on why and how the new dialogue paradigm was developed over a thirty-year period. Part II discusses the philosophical foundations on which the paradigm is founded. The foundations of Part II inform the espoused theory and its practice in the arena, presented in Part III. The foundations are also informed in a cyclical mode by the lessons learned in the arena, which is Part V, thus grounding the theory. The theory presented in Part III informs and guides the development of methodology, which is discussed in Part IV. Finally, the methodology provides the roadmap for the practice of the theory in the arena of applications, which is presented in Part V. The cyclical model of connections among Parts II, III, IV and V closes by making sure that the lessons learned in the arena, i.e., Part V, are transferred to the foundations and the theory, i.e., Parts II and III, so that the theory evolves while being grounded in the arena of applications. This cyclical model connecting the domains of foundations (Pat II), theory (Part III), methodology (Part IV), and arena (Part V) has been referred to in the systems sciences literature as the Domain of Science Model (DSM). In other words, any science should be composed of four domains with the domain of applications in the arena feeding and grounding the domains of foundations, theory, and methodology in a continuous evolutionary spiral.

In light of the axiomatic position advocated for the development of the SDP theory it is appropriate to stipulate the three axioms of the Archanesian geometry from the outset. These axioms are:
- Complexity Axiom: Complex social system designing situations are multidimensional. They require that observational variety should be respected in the dialogue among the observers, in an effort to strive for

comprehensiveness. Comprehensiveness, however, is an objective not easily attainable by human observers, even through the practice of a disciplined dialogue paradigm, such as SDP.
- Cognitive Limitations Axiom: Observers are subjected to cognitive limitations during social system designing dialogue, which must be explicitly recognized and avoided during the dialogue. Cognitive limitations demand that designing teams: (a) control the pace of knowledge generation and assimilation, and (b) control the number of observations and relationships that observers must manage simultaneously during the dialogue.
- Saliency Axiom: During social system designing dialogue, understanding of relative saliency of observations can be brought into play only when the observer's authenticity, learning, and appreciation of variety are ensured so that the observers are able to construct categories of observations before assessing the relative saliency of individual observations.

Readers interested in elaboration and explanation of these three axioms are referred to (Christakis, 2005).

15.3   TEN KEY DEFINITIONS

The three axioms inform and require the definition of ten key terms for the practice of the paradigm. Terminological clarification is very critical for the Archanesian geometry. The ten key terms are:

1. **Dialogue**: The capacity of observers to engage in creating meaning, wisdom, and action through communicative and collaborative interaction.
2. **Culture**: The capacity of a group of observers to interact through dialogue and to build a community of stakeholders founded on a consensual language.
3. **Conscious Evolution**: The capacity of a culture to be futures-creative in designing social systems with the engagement of the stakeholders.
4. **Future**: The state of a social system that is not a mere extrapolation of the past and present.
5. **Complexity**: A state of being of the observer in which he/she are not able to comprehend fully the multi-dimensionality of a social system designing situation they are embedded in, and hence are confused by it.
6. **Triggering question**: A question framed carefully by a social system designing team for the purpose of delineating the relevant context of a designing situation.
7. **Observation**: The authentic, concise, and content-specific response by an observer/stakeholder to a triggering question.

8. **Explanation**: The elaboration by an observer/author of the meaning of his/her observation for the purpose of making it transparent to other stakeholders participating in a social system designing situation.
9. **Problem**: The discrepancy between the belief by an observer of "what ought to be" and his/her observation of "what is," for a particular social system designing situation.
10. **Issue**: The discrepancy between the belief by an observer of "what ought to be" and his/her observation of "what is," coupled to an appreciation that the pluralities of beliefs is an integral component of the social system designing situation.

15.4 THE ARCHITECTURE AND EVOLUTION OF SDP

The Architecture of SDP is composed of thirty-five components, which have been grouped into seven modules. The seven modules, and the approximate time period of their development, are:
- The 7 Consensus Methods: (1) Nominal Group Technique, (2) Interpretive Structural Modeling, (3) Ideawriting, (4) DELPHI, (5) Options Field, (6) Options Profile, and (7) Trade-off Analysis (Years 1972-1982);
- The 3 Key Role Distinctions: (1) Context, (2) Content, and (3) Process (Years 1982-1985);
- The 3 Application Phases: (1) Discovery, (2) Design Dialogue, and (3) Action (Years 1989-2001);
- The 4 Stages of Inquiry: (1) Definition or Anticipation, (2) Design of Alternatives, (3) Decision, and (4) Action Planning (Years 1989-1995);
- The (5 Cs) of SDP, namely:(1) Community of Stakeholders, (2) CogniScope™ Team, (3) Consensus Methods, (4) *CogniSystem* Software and (5) Collaborative Facility (Years 1985-1995);
- The 7 Graphic Language Patterns: (1) Elemental observation, (2) Problematique, (3) Influence tree pattern, (4) Options field pattern, (5) Options profile/scenario pattern, (6) Superposition pattern, and (7) Action plan pattern (Years 1970-1989); and
- The 6 Dialogue Laws: Requisite: (1) Variety (Ashby), (2) Parsimony (Miller), (3) Saliency (Boulding), (4) Meaning and Wisdom (Peirce), (5) Autonomy (Tsivacou), and (6) Evolutionary Learning (Dye) (Years 2001-2003)

In the following sections I will elaborate on the last three of the seven modules of the SDP architecture presented above.

15.4.1 Elaboration of Module (e)

15.4.1.1 The Five "C's" of SDP

There are five components that must co-exist during the conduct of the design dialogue phase, i.e., module (c) of the Architecture. The five components are graphically shown in Figure 15.1: (1) the Community of stakeholders (observers) to be engaged in SDP for the social system designing situation; (2) the *CogniScope*™ inquiry design and facilitation team; (3) Consensus methods, selected from the universe of available problem-solving and design methods, on the basis of technical and behavioral criteria for productive dialogue leading to communicative action; (4) *CogniSystem* software, for recording observations and meanings, exploring pair-wise relationships among observations, producing efficiently graphic language patterns of relationships among observations, and displaying them on a large screen, with the flexibility to amend the observations and the patterns continuously; and (5) a Collaborative facility which promotes the comfort of the participants, and has the capability to display visually the graphic patterns constructed during the structured dialogue. The facility has been called "Observatorium," in honor of Harold Lasswell, the founder of policy sciences, who over thirty years ago invented the concept of the "Social Planetarium" (Warfield, 1994).

Figure 15-1: The Five C's of SDP

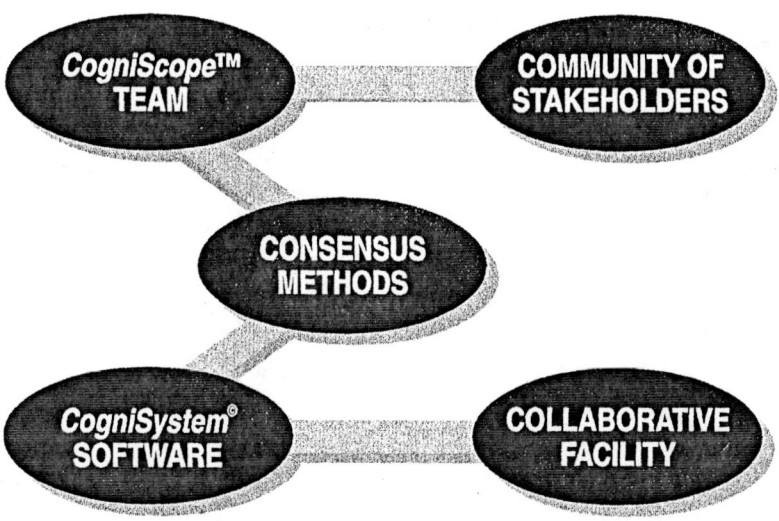

The community of stakeholders includes representatives of all the parties who are affected by a situation and should be involved in the design of an action plan to improve it. These stakeholders are primarily responsible for generating content by means of the structured dialogue.

The *CogniScope*™ team consists of (1) the inquiry design team that collaboratively designs the context of the situation and the triggering questions, and (2) the facilitation team. During the design dialogue phase of Module (c), the facilitation team requires at a minimum a lead facilitator, a recorder, and a computer software operator. After the sessions, one or more members of the *CogniScope*™ team produce a report that describes what happened and presents the results in a transparent manner to the participants.

The consensus methods used in SDP are: (1) Nominal Group Technique, (2) Interpretive Structural Modeling, (3) Ideawriting, (4) DELPHI, (5) Options Field, (6) Options Profile, and (7) Trade-off Analysis

The *CogniSystem* software is a refined version of the software that was worked on from the very beginning of the development of SDP back in the early 70s. From the start of the theory development it was felt necessary to use interactive software. The goals of the software development were: (1) to lessen the cognitive demands on the designing participants; (2) to generate better designs; (3) to speed the designing process; and (4) to maintain participant-driven rather than expert-driven deliberations. The *CogniSystem* keeps track of participant observations, recording them, displaying them by means of the graphic language patterns of the Archanesian geometry, arranging them in accord with participant decisions, and organizing the efficient delivery of products to participants.

The last component (see Figure 15.1) of a well-designed SDP application is a facility with the following physical characteristics:
- Comfortable chairs;
- Proper seating arrangements;
- Sufficient wall space to display the results of participant observations and decisions;
- A screen or wall on which developing information such as triggering questions, definitions, observations, influence patterns, and option profiles can be projected; and
- Computers, projectors, and printers.

A floor plan displaying a configuration of the collaborative facility with thirty-six participants is displayed in Figure 15.2

15. DIALOGUE FOR CONSCIOUS EVOLUTION

Figure 15-2. An Example of a Floor Plan for a Collaborative Facility Accommodating Thirty-Six Participants

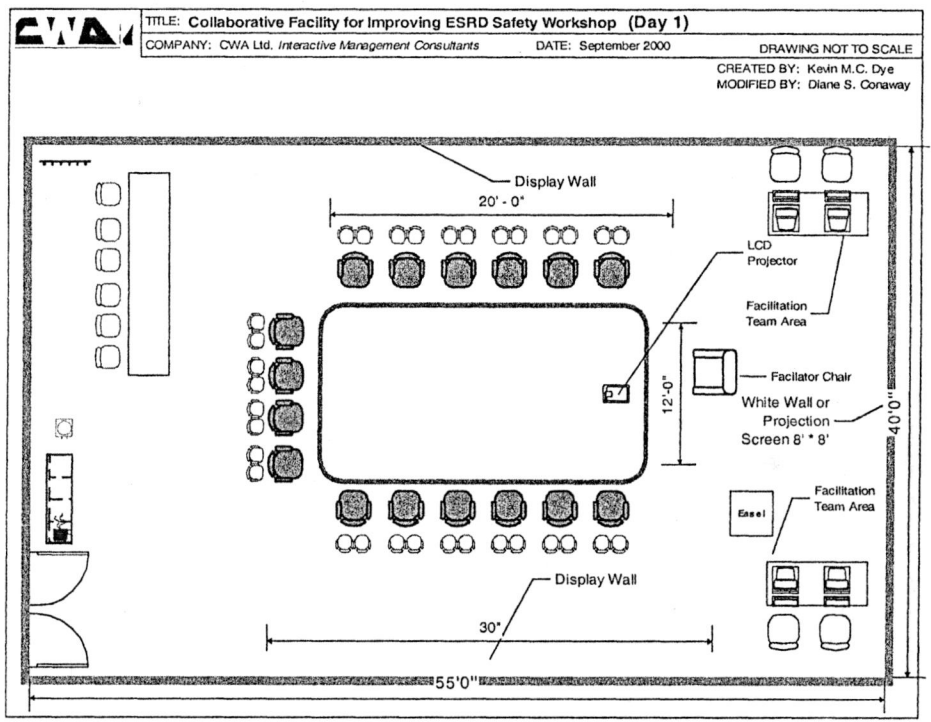

15.4.2 Elaboration of Module (f)

15.4.2.1 The Seven Graphic Language Patterns

In accordance with the three axioms of the Archanesian geometry, special attention was paid in the development of the theory of SDP to the adoption and utilization of suitable graphic language patterns. By 1989, seven language patterns were identified and utilized. These are:
- Elemental observations, which are employed by the participants as they express their well-developed or nascent ideas about a design situation in response to a triggering question.
- The Problematique is the preliminary structural model collectively generated by a structured dialogue focusing on paired relationships among a set of observations and using computer assistance.
- The Influence Tree Pattern (also called the Tree of Meaning) is the hybrid graphic/textual interpretive construct visually displaying and interpreting the relationships discovered in the Problematique dialogue.
- The Options Field Pattern is the graphic/textual representation of the dimensionality of a design situation.
- The Options Profile/Scenario Pattern is a graphic representation within the context of the Options Field Pattern that identifies the most salient options that the participants decide to pursue in their action plan.
- The Superposition Pattern combines the Influence Tree of Meaning generated in the Problematique dialogue with the action options presented in the Options Profile to graphically display the impact of selected actions on the roots and branches of the Influence Tree by means of a color-coded master Influence Tree Pattern (see Figure 15.6).
- The Action Plan Pattern embodies the assigned responsibilities, time lines, interaction patterns, and monitoring activities decided on by the participants.

Six of the seven language patterns of the Archanesian geometry are displayed graphically in Figure 15-3.

15. DIALOGUE FOR CONSCIOUS EVOLUTION

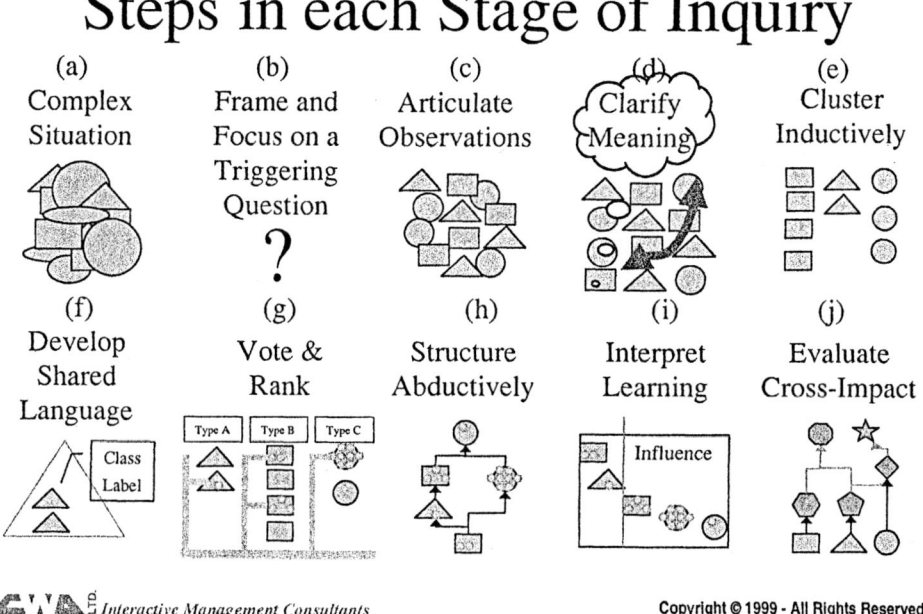

Figure 15-3. Six Graphic Language Patterns of the Archanesian Geometry

15.5 THE DIALOGUE GAME

Structured dialogue differs from many other methodologies that poll stakeholders to find out what problem they think is most important, or what action they deem most effective, but do not probe deeper, proceeding directly to prescribing action plans to meet problems and effect change. Action plans designed by other methodologies usually fail because they fail to identify the crucial leverage points. By performing the extra steps of constructing graphic language patterns, structured dialogue enables designers to go the extra miles that can spell the difference between merely talking about and actually effecting needed evolutionary change.

To point up the importance of these extra steps, consider the following example. It demonstrates how structured dialogue distinguishes between importance, as we intuitively perceive it, and influence or leverage, as it plays out in a social system designing situation.

Module (g) of the SDP architecture identifies the six laws proposed by systems scientists during the 20th century, which play a critical role in the construction of observations by the participants of an SDP application. Enforcing the implementation of all six laws is essential for successful structured dialogue. They do not, however, as we will see, have equal importance and influence. Here they are in a concise description stated as the six principles of structured dialogue:

Table 15.4. Six Principles of Structured Design Process Paradigm

A.	(____) APPRECIATION OF THE **DIVERSITY OF THE PERSPECTIVES** OF OBSERVERS IS ESSENTIAL IN MANAGING COMPLEX SITUATIONS. (Ashby's Law of Requisite Variety).
B.	(____) **STRUCTURED DIALOGUE** IS REQUIRED TO AVOID THE COGNITIVE OVERLOAD OF OBSERVERS. (Miller's Law of Requisite Parsimony).
C.	(____) THE **RELATIVE IMPORTANCE** OF OBSERVATIONS CAN ONLY BE UNDERSTOOD THROUGH COMPARISONS WITHIN A SET. (Boulding's Law of Requisite Saliency).
D.	(____) **MEANING AND WISDOM** ARE PRODUCED IN A DIALOGUE ONLY WHEN THE OBSERVERS SEARCH FOR RELATIONSHIPS OF SIMILARITY, PRIORITY, INFLUENCE, etc. WITHIN A SET OF OBSERVATIONS. (Peirce's Law of Requisite Meaning).
E.	(____) DURING DIALOGUE IT IS NECESSARY TO PROTECT THE **AUTONOMY AND AUTHENTICITY** OF EACH OBSERVER IN DRAWING DISTINCTIONS. (Tsivacou's Law of Requisite Autonomy in Distinction-Making).
F.	(____) **LEARNING** OCCURS IN A DIALOGUE AS THE OBSERVERS SEARCH FOR RELATIONSHIPS AMONG THE MEMBERS OF A SET OF OBSERVATIONS. (Dye's Law of Requisite Evolution of Observations).

At this point of the discussion, I would like to ask the reader to take a minute and try to rank these six principles. To do this game most effectively, ponder each principle and then rank the most important principle by putting a "1" in the blank before the principle you think is most important, a "2" in that blank for the next most important, etc.

When you have finished ranking the principles in terms of their importance, switch your mindset to consider the systemic influences these principles have on each other. To do that you would have to ask yourself questions like the following:

"Supposing that in a group dialogue the participants were able to implement:
(PRINCIPLE # A: DIVERSITY OF PERSPECTIVES)

15. DIALOGUE FOR CONSCIOUS EVOLUTION

> Will this SIGNIFICANTLY enhance their capacity to implement:
> (PRINCIPLE # B: STRUCTURED DIALOGUE)
> In the context of participating in a structured dialogue session to design a social system?"

Dialogue game participants typically answer NO to this particular question. To methodically determine the influence relationships among the six dialogue principles, a group playing the game would have to answer 30 pair-wise questions such as the one shown above. With the support of the CogniScope groupware, players need to answer only 15 questions instead of thirty to generate an influence tree. This represents a factor-of-2 efficiency gain for just six observations. In actual designing situations, which usually include a large number of observations, the efficiency gain might be a factor of 10 or more. This efficiency gain in group-designing can reduce the work of months to that of just days when people need breaks from interminable meetings. It also reduces the likelihood of groupthink, which often occurs when people accept premature closure due to other pressing needs.

As a result of this pair-wise inquiry among the six principles, a tree-like pattern, like the one shown in Figure 15.4, emerges displaying graphically how the influences of the principle(s) at the root of the tree work together to take the form of a "tree of meaning."

The interpretation of the Tree implies that in order to produce meaning and wisdom through dialogue, namely Principle # D at the top, we must ensure that all the principles appearing along the trunk of the Tree are enforced. Principle # B, Structured Dialogue, is the most influential principle because it is located at the root of the tree. On the other hand, Principle # D, Meaning and Wisdom, is the least influential, even though it is the most important and desirable outcome from the conduct of dialogue.

This simple example makes it clear that a very important idea, such as "meaning and wisdom" is not necessarily influential, when a system of ideas is constructed by exploring and mapping a relationship among the ideas, like the tree of meaning shown above. Furthermore, the prerequisite to the accomplishment of important ideas or ideals is the requirement to enable the stakeholders by means of true dialogue to discover and agree on the influential ideas, and then to empower them to implement those fundamental ideas in the spirit of a collaborative leadership model (Magliocca, Christakis, 2001). The discovery of the roots of the tree of meaning by the collaborative engagement of the stakeholders avoids the pitfalls of acting on the "erroneous priorities" as determined by aggregating individual participants' votes on relative importance among a set of ideas. Unfortunately this form of determining priorities by means of group voting on relative importance is a mistake practiced by many decision-makers. The discovery of the truly effective priorities, however, cannot be accomplished without a systemic and systematic exploration of the interdependencies among the ideas/observations. Discovering effective priorities is one of the challenges of the escalating complexity of the Information Age, and why we need to apply the new democracy paradigm.

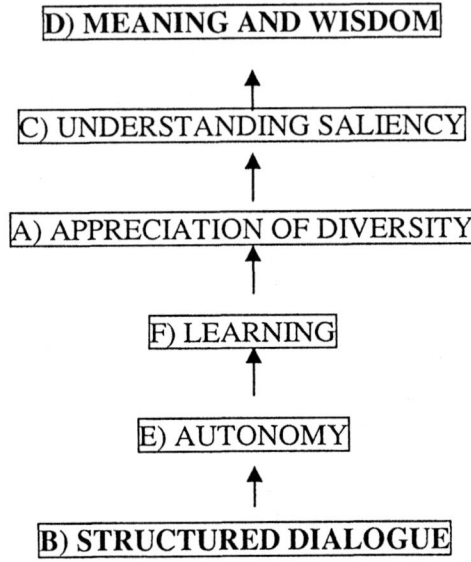

Figure 15-5. A Tree of Meaning

15.6 A STORY FROM THE DESIGN ARENA

15.6.1 Context

A social system designing team from the field of special education allocated about three months participating in a series of workshops employing the SDP paradigm for the purpose of designing a Continuous Improvement Focused Monitoring (CIFM) process. The participants to the workshops were representatives from the educational community with expert knowledge in the field of CIFM. The participants were initially engaged in a series of workshops for the purpose of re-designing the CIFM process, which they will then have to implement in the field with school districts throughout the state of Michigan.

15. DIALOGUE FOR CONSCIOUS EVOLUTION

After the designing team completed the re-design of the CIFM process, it was decided to conduct a "root cause analysis workshop" with the engagement of the CIFM designers as workshop participants. The purpose of this particular workshop was to try to anticipate any factors that might inhibit the successful implementation of the CIFM process in the arena of practice. The intention was to conduct an anticipatory root cause analysis before the launching of the CIFM, as opposed to the traditional root cause analysis, which usually focuses on identifying the root causes of existing systemic problems.

In revisiting the findings of the work of the CIFM designing team it is important to draw a distinction between "observer-independent" and "observer-dependent" root cause analyses.

15.6.2 Observer-Independent Root Cause Analysis

Generally speaking, observer-independent root cause analysis is a data-based procedure for ascertaining and "analyzing" the causes of problems in an effort to determine what can be done to solve or prevent them. The goal of root cause analysis goes beyond merely "fixing" the problem. It seeks to actually prevent it from happening again.

15.6.3 What Root Cause Analysis Needs to Include

Over 30 years of research and development teach us that effective and reliable root cause nalysis must provide three essential qualities:
1. **The process in addition to facts must take advantage of people's knowledge while preventing the biases of experts from controlling the direction of the investigation.**
   Methods which allow or even encourage the specialists/analysts themselves to choose which aspects of a problem to focus their search for solutions run a strong risk of failing to identify the best solutions. Knowledgeable stakeholders are better equipped to determine which solutions are the best, so it is desirable for them to have visibility of all of the available avenues toward prevention.
2. **The process must depict the facts of the case so that the causal relationships are clear and the causal relevance of those facts can be verified.**
   Root cause analysis needs a process which validates our thinking so that we can be sure we have included all of the relevant facts, and at the same time, only the relevant facts. It is the strong dependency on facts or data that makes this form of root cause analysis observer-independent, namely it minimizes dependence on "subjectivity" in favor of "objectivity."

3. **The process must also help us understand what actions must be taken to implement potential solutions and who in the organization needs to take those actions.**

   Once every possible avenue toward prevention is identified, the root cause analysis must identify what specific actions need to be taken. Is a new policy needed? If a policy already exists, then why wasn't it effective, and what steps do we need to take to make it effective in the future? And who in our organization needs to take those steps?

15.6.4 Observer-Dependent Root Cause Mapping

In the case of the CIFM root cause analysis, it was recognized from the outset that this particular root cause analysis, in addition to being anticipatory, it also had to be "observer-dependent" and not "observer-independent." The distinction between observer-dependent and observer-independent is founded in the context of the evolution of science from First Phase Science, e.g., Newtonian physics, to Second Phase, e.g., quantum physics, to Third Phase, e.g., second order cybernetics (de Zeeuw, 1996).

Newtonian physics was deliberately constructed in the 18th century to be observer-independent, and to be invariant in terms of space and time. For example, an observer seeing in 2000 an apple falling in London from a British tree, and a different observer seeing in 2004 an apple falling in Los Angeles from an American tree, will report the same phenomenon. However, in the arena of education, where the CIFM process will be applied to monitor progress of educational policies, the observers are observing different phenomena in terms of school performance indicators at different locations at different times, all of which are equally valid in the context of their particular situation. It follows that the challenge for the conduct of observer-dependent root cause analyses in the field of **education is to enable the observers/stakeholders to construct high quality observations collectively, collaboratively, and systemically**. This is exactly what was accomplished in the CIFM workshop by employing the SDP paradigm.

It appears that this very fundamental distinction between first and third phase sciences is ignored in a number of root cause analyses dealing with phenomena in the educational systems design arena, leading to erroneous results. In order to distinguish this third phase science form of root cause analysis from the traditional one founded solely on objectivity and facts, we have named it Root Cause Mapping (RCM).

15.6.5 Root Cause Mapping for the CIFM

In the case of the CIFM application of the RCM, the stakeholders were first engaged in a generative dialogue in response to the following triggering question:

"What factors do we anticipate will emerge as inhibitors to the successful implementation by OSE/EIS of the CIFM cycle of activities in its field applications?"

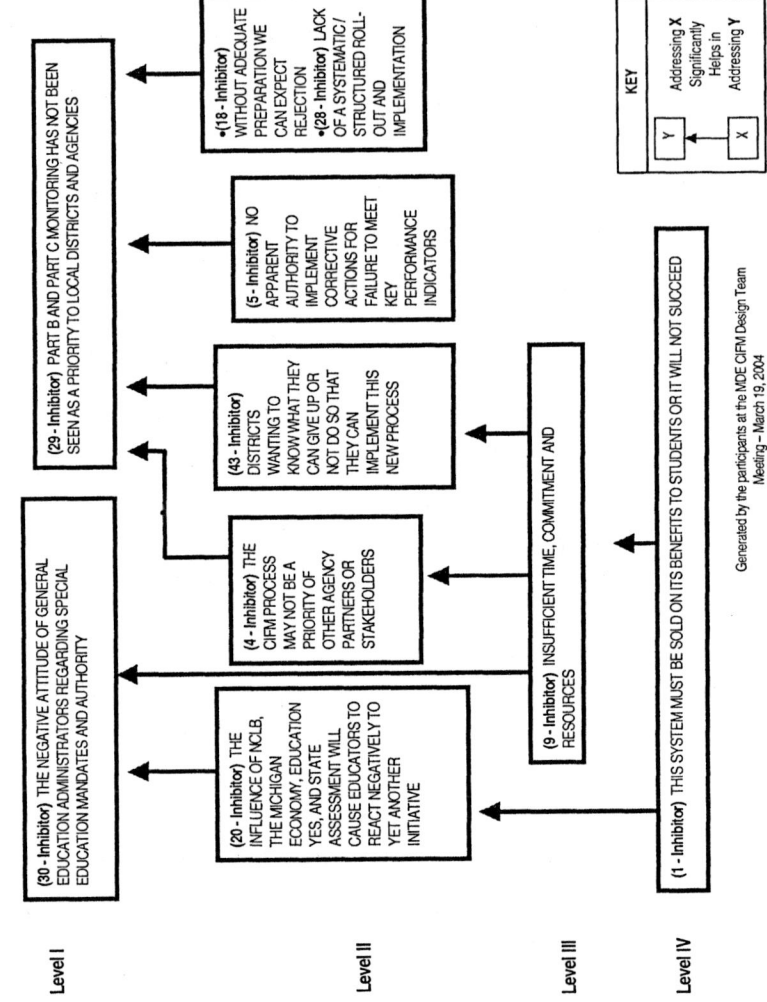

Figure 15-6. Amended Root Cause Map of Inhibitors to the Successful Implementation of CIFM

15. DIALOGUE FOR CONSCIOUS EVOLUTION   297

Figure 15.7: Superposition of Action Options onto the Root Cause Map of Inhibitors to the Successful Implementation of CIFM

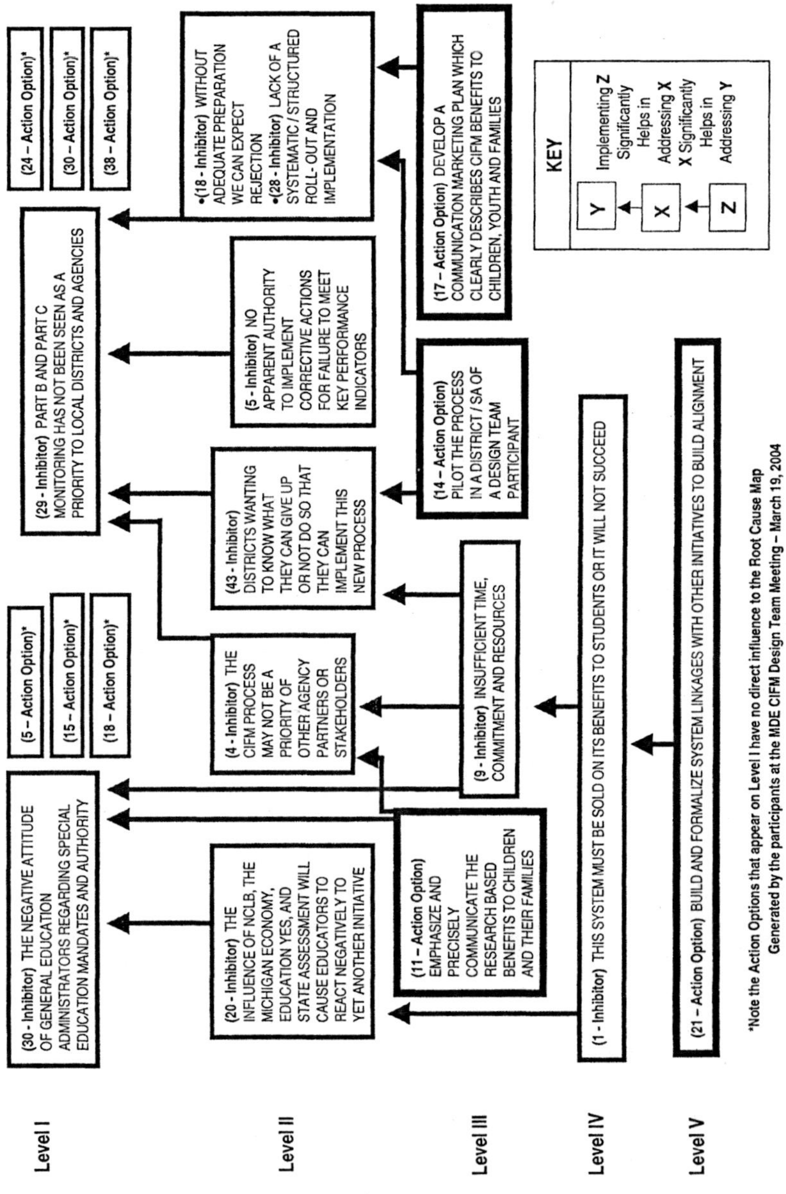

Level I
- (30 - Inhibitor) THE NEGATIVE ATTITUDE OF GENERAL EDUCATION ADMINISTRATORS REGARDING SPECIAL EDUCATION MANDATES AND AUTHORITY
- (29 - Inhibitor) PART B AND PART C MONITORING HAS NOT BEEN SEEN AS A PRIORITY TO LOCAL DISTRICTS AND AGENCIES
- (24 – Action Option)*
- (30 – Action Option)*
- (38 – Action Option)*

Level II
- (20 - Inhibitor) THE INFLUENCE OF NCLB, THE MICHIGAN ECONOMY, EDUCATION YES, AND STATE ASSESSMENT WILL CAUSE EDUCATORS TO REACT NEGATIVELY TO YET ANOTHER INITIATIVE
- (4 - Inhibitor) THE CIFM PROCESS MAY NOT BE A PRIORITY OF OTHER AGENCY PARTNERS OR STAKEHOLDERS
- (5 - Inhibitor) NO APPARENT AUTHORITY TO IMPLEMENT CORRECTIVE ACTIONS FOR FAILURE TO MEET KEY PERFORMANCE INDICATORS
- (18 - Inhibitor) WITHOUT ADEQUATE PREPARATION WE CAN EXPECT REJECTION
- (28 - Inhibitor) LACK OF A SYSTEMATIC / STRUCTURED ROLL- OUT AND IMPLEMENTATION
- (5 – Action Option)*
- (15 – Action Option)*
- (18 – Action Option)*

Level III
- (11 – Action Option) EMPHASIZE AND PRECISELY COMMUNICATE THE RESEARCH BASED BENEFITS TO CHILDREN AND THEIR FAMILIES
- (9 - Inhibitor) INSUFFICIENT TIME, COMMITMENT AND RESOURCES
- (43 - Inhibitor) DISTRICTS WANTING TO KNOW WHAT THEY CAN GIVE UP OR NOT DO SO THAT THEY CAN IMPLEMENT THIS NEW PROCESS
- (14 – Action Option) PILOT THE PROCESS IN A DISTRICT / SA OF A DESIGN TEAM PARTICIPANT
- (17 – Action Option) DEVELOP A COMMUNICATION MARKETING PLAN WHICH CLEARLY DESCRIBES CIFM BENEFITS TO CHILDREN, YOUTH AND FAMILIES

Level IV
- (1 - Inhibitor) THIS SYSTEM MUST BE SOLD ON ITS BENEFITS TO STUDENTS OR IT WILL NOT SUCCEED

Level V
- (21 – Action Option) BUILD AND FORMALIZE SYSTEM LINKAGES WITH OTHER INITIATIVES TO BUILD ALIGNMENT

KEY
- Implementing Z Significantly Helps in Addressing X
- X Significantly Helps in Addressing Y
- Y ← X ← Z

*Note the Action Options that appear on Level I have no direct influence to the Root Cause Map
Generated by the participants at the MDE CIFM Design Team Meeting – March 19, 2004

In response to the above question the stakeholders identified forty-six inhibitors (red ideas). After voting individually and subjectively on the inhibitors of higher relative importance, the participants constructed, through a strategic dialogue focusing on the relationships between pairs of inhibitors, a Root Cause Map (see Figure 15.5). This figure displays graphically the four inhibitors that were located at the roots of a tree-like pattern. This group work was completed during the first day of the workshop.

On the second day of the workshop, the stakeholders engaged again in a generative dialogue in response to a different triggering question, namely:

"What are Action Options for overcoming the Inhibitors by focusing at the roots of the Root Cause Map?"

They proposed over forty preventative action options (blue ideas) for addressing the inhibitors. Voting individually and subjectively on the relative importance of these preventative measures, they identified fifteen that received three or more votes, with thirty people voting. Ten of the fifteen important preventative actions were subsequently superimposed onto the Root Cause Map, by engaging the group in a strategic dialogue of relational voting in order to discover the effectiveness of the actions in addressing the root causes of the Map.

The deeper a preventative action is located in the superposition map of actions onto inhibitors, the stronger is its effectiveness in overcoming the inhibitors to the successful implementation of the CIFM process. On the basis of supermajority voting, it was discovered that only four of the important preventative actions exerted strong leverage in overcoming the root cause inhibitors. The stakeholders must give preferential consideration in implementing those four effective preventative actions (See Figure 15.6)

15.7 CONCLUSIONS

From the discussion of the SDP paradigm presented in this chapter we can draw the following conclusions:
- Structured dialogue is foundational in enhancing conscious evolution of society;
- Designing social systems with the use of structured dialogue ensures that the six systemic principles of: (1) variety, (2) structure, (3) understanding, (4) meaning, (5) authenticity, and (6) learning, are practiced in accordance with the tree of meaning (see Figure 15.4);
- Structured dialogue enables the emergence of a consensual language, which is the prerequisite for the sustainability and evolution of a culture;

- Structured dialogue facilitates the development of a designing culture in a community of stakeholders;
- Structured dialogue ensures higher quality designs for improving products and services for complex social system designing situations.

The Information Age and the trend of globalization (Friedman, 2000) have had adverse impacts on the practice of participative democracy by people all over the world. However, these trends should not preclude us from taking steps toward consciously guiding the evolution of social systems. It is our responsibility to shape this evolutionary process so that all people on the planet may benefit. Participative democracy is seen as the most practical way toward this ideal, a return to the experience of the agora, where ordinary people have the opportunity to shape the issues of governance pertinent to them. It is not enough, however, to only bestow the responsibility of citizenship upon those who were granted such a privilege in ancient Athens. Needless to say, our modern agoras must also include women and those who were relegated to the slave class. We must not return to the ancient agora but bring it forward with the benefit of our current wisdom and experience. And we must do this by recognizing the importance of applying the new paradigm of SDP founded on the three axioms of the Archanesian geometry, both of which were discussed in this chapter and are more elaborated in the forthcoming book by Christakis (2005).

REFERENCES

Alexander, G.C., 2003, "Interactive management: An emancipatory methodology", *Systems Practice and Action Research*, **15**:111-122.
Apel, K., 1981, *Charles S. Peirce: From Pragmatism to Pragmaticism*, University of Massachusetts Press, Amherst.
Ashby, R., 1958, "Requisite variety and its implications for the control of complex systems", *Cybernetica*, **1**(2):1-17.
Banathy, B.H., 1996, *Designing Social Systems in a Changing World*, Plenum, N.Y.
Banathy, B.H., 2001, *Guided Evolution of Society: A Systems View*, Plenum, N.Y.
Bausch, K., 2001, *The Emerging Consensus in Social System Theory*, Plenum, N.Y.
Bausch, K., 2000, "The practice and ethics of design", *Systems Research and Behavioral Science*, **17**(1):23-51.
Boulding, K., 1966, *The Impact of Social Sciences*, Rutgers University Press, New Brunswick.
Christakis, A.N., 1973, "A new policy science paradigm", *Futures*, **5**(6):543-558.
Christakis, A.N., 1987, "High technology participative design: The space-based laser", in: *General Systems*, John A. Dillon Jr., ed., International Society for the Systems Sciences, **XXX**:69-75.
Christakis, A.N., 1988, "The Club of Rome revisited", in: *General Systems*. W.J. Reckmeyer, ed., International Society for the Systems Sciences, **XXXI**:35-38.

Christakis, A.N., 1993, "The inevitability of demosophia", in: *A Challenge for Systems Thinking: The Aegean Seminar*, Ioanna Tsivacou, ed., University of the Aegean Press, Athens, Greece, pp. 187-197.

Christakis, A.N., 1996, "A people science: The CogniScope system approach", *Systems: Journal of Transdisciplinary Systems Sciences*, 1(1).

Christakis, A.N., 2004, "Wisdom of the people", *Systems Research and Behavioral Sciences*, **21**:317-330.

Christakis, A.N., 2005, *Dialogue for the Information Age Democracy*, Information Age Publishing, Inc., Greenwich, Connecticut

Christakis, A.N. and Warfield, J.N., 1987, NSF DTM Ohio.

Christakis, A.N., Warfield, J.N., and Keever, D., 1988, "Systems design: Generic design theory and methodology", in: *Systems Governance*, Michael Decleris, ed., Publisher Ant. N. Sakkoylas, Athens-Komotini, Greece, pp. 143-210.

Christakis, A.N. and Dye, K.M., 1999, "Collaboration through communicative action: Resolving the systems dilemma through the CogniScope", *Systems: Journal of Transdisciplinary Systems Sciences*, 4(1).

Christakis, A.N. and Brahms, S., 2003, "Boundary-spanning dialogue for 21st-century Agoras", *Systems Research and Behavioral Sciences*, **20**:371-382

Christakis, A.N. and Harris, L., 2004, "Designing a transnational indigenous leaders interaction in the context of globalization: A wisdom of the people forum", *Systems Research and Behavioral Sciences*, **21**:251-261.

Churchman, C.W., 1971, *The Design of Inqiring Systems: Basic Concepts of Systems and Organizations*, Basic Books Inc. Publishers, New York.

Churchman, C.W., 1979, *The Systems Approach and its Enemies*, Basic Books Inc. Publishers, New York.

de Zeeuw, G., 1996, "Second order organizational research", *Working Papers in Systems and Information Sciences*, University of Humberside, Hull, England.

Dye, K.M., Feudtner, C., Post, D., and Vogt, E.M, 1999, Developing Collaborative Leadership to Reframe the Safe Use of Pharmaceuticals as a National Health Priority, *Final Report*, CWA Ltd. Paoli, PA.

Dye, K.M. and Conaway, D.S., 1999, Lessons Learned from Five Years of Application of the CogniScope™ Approach to the Food and Drug Administration, *CWA Ltd. Report*, Paoli, PA.

Friedman, T.L., 2000, *The Lexus and the Olive Tree*, Anchor Books, New York.

Habermas, J., 1984, *The Theory of Communicative Action*, Vols. I and II, Polity Press.

LaPointe, G., 1999, (Doc # 99141), "A Socio-Psychological View of Social Systems and Design", *Proceedings of the 43rd Annual Conference of the International Society for the Systems Sciences*, J.K. Allen, M.L.W. Hall, and J. Wilby, eds.

Lopez-Garay, H., 2001, "Dialogue among civilizattions: What for?", *International Journal of World Peace*, **XVIII**:15-33.

Mcintyre, J., 2004, "Facilitating critical systemic praxis (CSP) by means of experiential learning and conceptual tools", *Systems Research and Behavioral Science*, **21**:37-61

Magliocca, L.A. and Christakis, A.N., 2001, "Creating a framework for sustainable organizational leadership: The CogniScope system approach", *Systems Research and Behavioral Science*, **18**:259-279.

Miller, G.A., 1956, "The magical number seven, plus or minus two: Some limitations on our capacity for processing information", *Psychology Review*, **63**:81-97.

Murthy, P.N., 2000, "Complex societal problem solving: A possible set of methodological criteria", *Systems Research and Behavioral Science*, **17**:73-101.

Parker, T.F., 2004, "The chronic kidney disease initiative", *Journal of the American Society of Nephrology*, **15**:708-716.

Roberts, N., 2002, *The Transformative Power of Dialogue*, Elsevier, New York.

Simon, H.A., 1974, "How big is a chunk", *Science*, **183**:482-488.

Tsivacou, I., 1997, "The rationality of distinctions and the emergence of power: A critical systems perspective of power in organizations", *Systems Research and Behavioral Science*, **14**(1):21-34.

Taylor, J.B., 1976, "Building an interdisciplinary team", in: *Perspectives on Technology Assessment*, S.R, Arnstein and A.N. Christakis, ed., Science and Technology Publishers, Jerusalem, Israel, pp. 45-63.

Turrisi, P.A., ed., 1997, *Pragmatism as a Principle and Method of Right Thinking*, State University of New York Press.

Vogt, E.M., 2003, "Effective communication of drug safety information to patients and the public", *Drug Safety*, **25**:313-321.

Warfield, J.N., 1988, "The magical number three, plus or minus zero", *Cybernetics and Systems*, **19**:339-358.

Warfield, J.N., 1994, *A Science of Generic Design: Managing Complexity Through Systems Design*, Iowa State University Press, Ames, Iowa.

Warfield, J.N. and Christakis, A.N., 1987, "Dimensionality", *Systems Research*, **4**:127-137.

Warfield, J.N. and Cardenas, A.R., 1994, *A Handbook of Interactive Management*, Iowa State University Press, Ames.

SECTION F

THE USE OF NEW METAPHORS TO REFOCUS THE MANAGERIAL DISCOURSE

Chapter 16

MAKING FRIENDS OF ENEMIES
From critical systems ethics to postmodern ethics

Yu, J.E.

Abstract	The paper evaluates the contributions of contemporary philosophies namely critical philosophy and poststructuralism as a means for facilitating communication and participation in practice. On the one hand, this paper examines the contributions that critical systems thinkers have made to 'critical systems ethics'. The approach seeks to improve social practice. On the other hand, the paper outlines the idea of a poststructuralist philosophy, by drawing on Foucault and Deleuze to reconstruct 'postmodern ethics'. To open up new forms of possibility within the postmodern ethics, an alternative mode of thinking is proposed.
Key Words:	Critical systems ethics; poststructuralist philosophy; postmodern ethics; images of thought.

16.1   INTRODUCTION

In this paper, I wish to evaluate the contributions that contemporary Western philosophies namely critical theory (philosophy) and poststructuralist philosophy have made to develop a systems approach that tends to be governed by ethical and moral judgements in a participative manner. Understanding a systems approach and its rationale from Churchman's philosophy, I identify a great achievement in 'critical systems heuristics', which is characterized by efforts to make what might be called the 'emancipatory systems approach'. On the other hand, I examine an alternative mode of thinking that Foucault and Deleuze have directed at poststructuralist philosophy focusing in particular on his concerns about the complex and dynamic relations of knowledge, power and ethics within which decision makers have engaged in social inquiring process in a participatory manner. Following Foucault's concept of problematization, Deleuze visualises an event that encounters randomness and unpredictability of chaotic behaviours which we experience in organizational and social contexts. This provides us with a new discourse to pose a set of questions and problems to open new possibilities to organize the collective thought to bring about social transformation through the 'postmodern ethics' that is likely to happen within the 'post-industrial society'.

16.2 CHURCHMAN'S IDEA OF A SYSTEMS APPROACH

West Churchman (1968, 1979) believes that knowledge comes from "others" or "enemies" as we remember what he said, 'The systems approach begins when first you see the world through the eyes of another'. In this famous phrase, Churchman suggests that people can step outside a system they are in and mentally try to consider it through the lenses of other people's values.

To Churchman, the "enemies" of systems approach provide a powerful way of learning about the systems approach, precisely because they enable the rational thinker to step outside the boundary of a system and to look at it. It means that systems thinkers are not necessarily involved within a system but are essentially involved in the "outside" of systems rationality. In churchman's systems approach, decision makers or system designers admit the freedom of individual thinking and our world view is very restricted and that it is possible to develop an alternative 'mode of thinking'. Philosophically, Churchman (1968, 1971)'s systems approach is influenced by the dialectics of Kant and Hegel and is an example of a Singerian inquiring system. Kant (1988) believes that underlying any fact-network there will be taken-for-granted assumptions which cannot be questioned on the basis of that fact-network. In this sense, Kantian notion of categorical imperatives is developed from the taken-for-granted assumptions and it constructs the moral code of conducts that aims at a universalistic orientation in human life (Yoo, 1991, p. 54). Kant (1956) tries to consider the human will as if we are acting from duty or inclination by trying to apply our goals to the rest of society. By starting from the premise that man(sic) is rational and that he recognizes himself as a *means* and as *an end*, Kant expects this form of rationality to enable actors to treat everyone as equals. That is what Kant proposes in the form of a moral maxim. On the other hand, Hegel's notion of the dialectic follows a trilogy of thesis, antithesis and synthesis. The description of a fact on the basis of the set of assumptions forms a thesis. It then follows that to challenge one set of assumptions by another set of assumptions in the form of an antithesis may produce a new thesis which will be a synthesis of theses, and so process continues. Singer was trying to find a suitable design for a human inquiring system. A Singerian Inquiring System is a never ending process of inquiring in the form of Thesis-Antithesis-Synthesis established by building up a fact-network at the outset and then challenging it with an antithesis in a continual process of change. For Churchman, the Hegelian notion of objectivity is that objectivity is developed through the confrontations of diverse subjectivities. Based on the works of Kant and Hegel, Churchman (1968, 1979) developed the participative decision-making process of social planning that happens through the 'sweeping in' process of diverse perspectives and 'unfolding' a wide range of social, political, economic and environmental issues of real-world situations. In Churchman's dialectical approach, different evaluations are possible. This makes a systems approach more practical for social systems design, which tends to highlight issues of organizational transformation and social change (Ulrich, 2004, p. 203). Churchman (1979) argues that conflicts are inherent within social systems. These conflicts mainly stem from the four enemies of the systems approach, namely, politics, morality, religions and aesthetics, which belong to the domains of irrational thoughts and 'non-scientific' kind of discourse. Churchman's dialectical approach deals successfully with these inherent conflicts as his approach pays sufficient attention to the existence of irrational thoughts, differing interests, values, beliefs and philosophies, that is, 'others', 'enemies' of Churchman's systems approach. Appreciating Churchman's dialectical approach, Ulrich (1987, 1988) developed 'Critical Heuristics of Social Planning' as he takes the contribution of a critical philosopher, Habermas on the development of 'critical and discursive ethics' that is based on the 'communicative rationality' and 'discursive ethics' (Habermas, 1984, 1987, 1993).

In the next section, we will discuss the contribution of critical systems thinking that have made to improve the field of rational inquiry to social planning with the relevance of moral consideration for management of organizations, and the improvement of social conditions in our society.

16.3 CRITICAL SYSTEMS THINKING AND SOCIAL SYSTEMS DESIGN

According to Habermas (1972), necessary conditions for the constitution of knowledge exist. From the concepts of work and interaction, technical and practical interests have been established in order to realize defined goals under given conditions. However, the exercise of power inhibits open and free social interaction that is, communicative action or symbolic action which is defined and governed by norms, rules, values of the existing social systems.

In order to produce the genuine social scientific knowledge, the understanding of the other social actors is necessarily a third cognitive interest, namely, an emancipatory interest.

Habermas argues that the task of social scientists is to establish self-reflective knowledge aimed at understanding "free choice" that is in freedom to behave in a certain way without social restrictions of domination and power through the social process called collaborative democracy.

Following Habermas, Ulrich argues that some sort of participative debate is necessary to explore the value judgements of all stakeholders on the development of social planning in order to achieve an emancipatory interest towards a civil society.

In other words, Ulrich (1983) seeks to uncover the normative social planning of systems designs for the involved clients, decision makers and planners, and the affected witnesses or citizens.

This approach to social systems design is regarded as a new development in the systems movement. Critical Heuristics of Social Planning is a great achievement in emancipatory systems (Jackson, 1991, p. 187). In this sense, Ulrich's 'Critical Systems Heuristics' (CSH) is called as the emancipatory systems approach (Jackson, 2003, p. 23). Ulrich suggests that CSH is useful in that it forces us to accept our inability to comprehend the 'whole' system. CSH, which based upon the concept of a practical reason in terms of Kant (1788), is supposed to help us achieve a 'solution' or explain the ways we have found a 'solution' or 'improvement' in social practice.

Kant distinguished the phenomenological world as "appearances" and noumenal world as "things in themselves", but Ulrich (1983, p. 184) argued that these are two sides of the same thing.

The first being concerned with the phenomenal "surface" as it appears to our senses or sensibility (as it is the source of our knowledge) from a certain standpoints.

The other is concerned with its noumenal "essence" as we comprehend it by reflection.

Using semanticist Alfred Korzybski's words, "A map is not the territory", Ulrich makes clear that the 'territory' is the noumenal reality that we cannot grasp as such. On the other hand, 'map' represents the phenomenal reality that appears to us from a certain viewpoint. For instance, when we deal with the social reality, that is, phenomenal reality is only the 'map' (i.e. empirical and analytic science), we must reflect on the 'normative' implications of our own standpoints *vis-à-vis* the 'territory', which belongs to our living society which we cannot objectively understand as such. Then the 'territory' implied that historical, moral and political reality exist behind the given 'map' (i.e. social reality to be mapped as a social system). Thus, the metaphysical and normative presuppositions flowing into the maps of "appearances" in terms of which we "see" the territory of the noumena (Ulrich, 1983, p. 185).

Put differently, the 'boundary judgement' inevitably flows into social maps and systems designs as 'synthetical *a priori* judgements' remain theoretically problematic but are practically indispensable. In this sense, the boundary judgement obeys the notion of 'correspondence' that means the source of *a posteriori* concepts (social practice) and our *a priori* concepts and judgements (systems rationality) should be mapped and reproduced with one another (Ulrich, 1983, p. 286).

From a critical point of view we assume that we are dealing not with dualistic thinking (in terms of either p or q) or dichotomies (i.e. subject vs. object, idealism vs. realism, reason vs. practice, etc.), but with third possibility or alternative 'pluralism' that practical reason requires a mediating 'third' between reason and practice in critical systems practice (Ulrich, 1983, p. 163; pp. 272-274).

For Ulrich (1983, p. 228), what is considered as a 'problem' is what belongs to the section of social reality that is "to be mapped."

A 'solution' is that "there is no objective solution but only a critical solution to the problem of boundary judgements" (Ulrich, 1983, p. 230).

In critical heuristics of social systems design, therefore, the systems concept is needed when we define a 'boundary criterion', that is, everything within the systems boundary can be said to belong to the 'system', while everything outside this boundary refers to the "environment" (Ulrich, 1983, p. 191).

When we apply the boundary judgements of CSH into the relevance of moral and ethical considerations for management of organizations and an improvement of social practice, 'critical systems ethics' is developed to reveal the emancipatory interest.

Critical systems ethics tends to secure the improvement of life conditions and comprehensive rationality by considering the mutual understanding and agreement which are made up on the basis of the communicative ethics and ethics of the whole systems (Yoo, 1991, p. 172). In addition, CSH depends on the critical power of the practical reason for its emancipation from the given objective facts. This presupposes what Ulrich's concept of "polemical employment of reason" is about (Ulrich, 1983, p. 279; 1987, pp. 281-282). Being critical systems practitioners, thus, implies that we must question all the assumptions or *a priori* judgements that are presupposed in one's standpoints. Being critical or emancipatory systems thinking, it is beyond the viewpoints of ontological complexity, it questions the claims of scientific discourse and reveals the limitation of scientific statements (Ulrich, 1987, 1998a).

The practical process of CSH should be followed by the internalisation of boundary questions. Bearing in mind the 12 boundary questions the practitioners should apply these boundary questions systematically in the 'problem situations' to see what the normative consideration should be achieved (Ulrich, 2000). In discussing the epistemological grounds for Ulrich's CSH, Jackson (1985) argues that a position such as Ulrich's sometimes seems to suggest taking a "common sense" view of materialism, which is inconsistent with his epistemological position that is based either on the synthetical *a priori* subject or on cogent intersubjectivism (Ulrich, 1983, 1987, 1993, 1996).

Mingers (1992) also argues that in Ulrich's CSH it is accepted that there are socially "disadvantaged" or the oppressed without equal distribution of power, knowledge and education under a coercive power structure, and it could be extended so that its practical implications is fully to explore in the case of helping a key client (i.e. "the socially disadvantaged") to bring about change in an existing structure.

Responding to these critics, Ulrich (1995) reconstructs his scheme in second-order concepts of boundary critique. For instance, where the legitimacy lies in using systems rationality in order to make the competent citizens in a civil society? Ulrich (1998) argued that 'reflective practice' should be grounded not only in the use of critical systems methods but also in heuristically proper use of an adequate notion of a civil society. Second-order concepts of boundary critique is based on the suggestion that the practice of CSH can be associated with "pragmatization" of critical systems thinking (CST), and further, the idea of critique can be reinterpreted with the Foucauldian notion of 'problematization' (Foucault 1984).

In considering "pragmatization" of CST, target group needs to understand the concerned situation and help to rescue a systemic sense of 'improvement' that is eventually to improve the "Larger System". On the other hand, the basic *a priori* assumption of the critical employment of boundary judgements can be shown its appropriate relevance in a real-world context so that it explains how communication, debate and decision taking on the social planning happen in everyday situations (Ulrich 1995).

Thus, this recent development of boundary critique as its second-order concepts can bring in the different rationalities of the involved planners and all those "responsible citizens" together, so as to reach "the equality of citizens as citizen" that is needed if we are pursuing an ideal of the polemical employment of boundary judgements (Ulrich 1995, pp. 10-13). It is then social systems design to be implemented as a participatory approach to problem solving or decision making for social planning that respects all perspectives of decision makers, planners or systems designers, "enemies" and others, based on the belief that diversity is the basis for developing critical systems ethics. In critical systems ethics, in order to create a condition of making the critical employment of boundary judgement, social planners should work with others, or at least encouraged to listening to others as the basis for generating good dialogue (Yoo, 1991, pp. 40-75). Then, critical systems ethics is indispensable knowledge which is based on the communicative rationality by securing the improvement of life conditions through the problem-solving approach that tends to uphold on the whole systems ethics.

If we take an example of CSH, in the case of improving quality standards in the National Health Service (NHS) in the U.K., CSH can provide a dialogic arena in which problems of communication associated with the diverse groups of people might be bridged (Gregory, *et al.* 1994). Through a series of workshops used a number of guiding questions as the basis on which facilitators hoped to facilitate the discussion, participants used a number of boundary questions to improve the process of discussion which could provide opposition to one another's viewpoints that enabled them to generate greater 'comprehensiveness' in the ongoing search for understanding and ethical awareness of social planning. In this way, using CSH, it gives responsibility to the facilitator in the process of discussion in order to encourage participants to consider options for thinking and action which exceed the boundaries of their initial perspectives. It is that CSH gives ethically justifiable responsibilities to the facilitators who intend to intervene on questions relating to the truth, sincerity, intelligibility and moral acceptability of statements made in communities that such a systemic intervention is guided by a process of critical appreciation. It clearly shows that the practical aspect of CSH assists a pragmatized version of the Habermasian validity-claims(Gregory and Romm 1994). Midgley (2000, p. 151) argues, "if value and boundary judgements (of CSH) are intimately linked, then it is contradictory to claim that any moral idea can be universally applied" (bracket is mine). In this sense, being critical systems practitioners, implies that we must notice that some practitioners accept the internalization of boundary questions during the process of social systems design, whilst others resist the principles of CSH that are based on 'pragmatization' of the Habermasian validity-claims.

16.4 THE SEARCH FOR A NEW MODE OF THINKING FOR THE POSTMODERN AGE.

16.4.1 Introduction

William and May (1996) have shown that our society has already moved to a new era s known as the "Fourth Epoch" (Mills, 1970, p. 184). According to new demands from this transition, the social scientist needs to search for new values, identities and the alternative ways of life. The Modern Age is being replaced by a postmodern period (William and May, 1996, p.158). In his book, "The Coming of Post-Industrial Society, which was written in 1973, the American sociologist Daniel Bell foresaw new development in the 21st century, now known as a 'Postmodern Age' or 'Post-Industrial Society', which is characterized by a global change that moves towards what is called the 'knowledge-based and digital economy'.

In the context of the knowledge-based , digital economy, three major trends shape our postmodern future, that is, globalization, information technology and managerial innovation (Cummings and Worley, 2001, pp. 4-5). These trends made us challenge the taken-for- granted assumptions and values that underlined, what has come to be known as the 'project of Enlightenment', that is, the pursuit of social progress through the search for 'scientific' or objective knowledge, which can be obtained through the application of human reason and the scientific method (Hancock and Tyler, 2001).

In systems and organizations sciences, many writers argue how multiple perspectives, different modes of thinking and metaphors are useful to open up powerful ways of thinking to deal with the complexity of organizational, social, economical and political situations (Hatch, 1997; Midgley, 2000; Jackson, 2003; Yu, 2004a, b).

In the context of systems thinking, however, little attention is paid to the poststructuralist, Gilles Deleuze, who has the eye of the 'outsider'. We will discuss Deleuze's 'image of thought' in the next section.

16.4.2 Being open to the unknown: the image of thought

Gilles Deleuze's thought of the unthinkable leads to 'natural' phenomena, which can only be studied by the 'science of the sensible'. The French philosopher, Deleuze, creates a new understanding of what it means to 'think' about desire, life and the ethics of social systems.

Deleuze shows how it is possible to approach the 'unthinkable' by means of the 'image of thought'. Our contention is that Deleuze's philosophy points

16. MAKING FRIENDS OF ENEMIES

beyond itself to another mode of reasoning which it called 'transcendental empiricism' (Patten, 1996, p. 11).

In the transcendental empiricism of Deleuze, philosophy is no longer a question of interpreting, evaluating and judging, but finding the 'image of thought' which moves and transforms the transcendental nature of thought in experimentation.

For Deleuze, the image of thought explores the peculiar and unpredictable natures of events, which challenges the established way of thinking towards understanding the nature of the 'natural' world, chaos and life. It brings devaluation of reason and consciousness; thus, thought can no longer be subjected to any guiding and grounding hypothesis and questions, but rather must begin by posing the new sorts of questions and problems, and a creation of new values[1].

It creates a new mode of unconscious thought and existence, and new possibilities of life (Goodchild, 1993, p. 39; Yu, 2001, pp. 172-174). The thought is defined as an attempt to grasp the 'whole', either as an underlying substance, unconscious nature of transcendental reality, an inner ground of all corporeal entities, and the exterior movement. Transcendence is presupposed as being encountered in the whole. The exterior movement inspires the image of thought.

When Deleuze and Guattari (1988, p. 5) claim our "thought lag behinds nature" an unthought operates alongside thought which is still the element 'outside' the system. When we say is 'outside' refers to the areas of unconsciousness, nature and life, transcendental being or becoming rather

---

[1] It produces real movement or actions, any movement in life or the joy, any experience or "becoming imperceptible" or "molecular unconsciousness" according to the Law of Nature (Deleuze and Guattari, 1977, pp. 273-296; 1988, pp. 232-307). In the 'transcendental empiricism', the 'immanent plane' reveals itself in the form of three determinations which represent the understanding of an unconscious and transcendental reality of creative thought, desire and life. As we will argue throughout the following sections, the transcendental reality has essentially duplicity and subsists between the two extremes of mind and body, the 'heights' and 'depths', subjects and objects. In actual fact, these extremes are coexistent and correspondent to a parallelism wherein actual states and virtual image of thought are conceived in terms of two aspects of an unconscious transcendental reality. Put differently, the pure event (in the terms of Deleuze) is grasped in the twin dimension of the imperceptible and unknown 'simulacra' (Deleuze, 1990). In the *Logic of Sense*, Deleuze (1990) is concerned with constructing a surface of thought, characterized as 'chaotic dynamics'. The image of thought represents the 'unthinkable' within a 'plane of immanence'; it appears like a collection of simulacra and phantasms which refer to a problematic field. Upon the plane of immanence, an effect of corporeal bodies has appeared in the present moment that can only be realised through 'sense'. This effect is called as an event, which is a pure event which does not have any physical or material aspects of it.

than meaning literally an outside environment which the human consciousness can appreciate.

Thought happens as an effect of movement and changes 'outside' thought. There is a 'transcendental', in other words, the higher power of decision-making which operates when decision makers look for all the possibilities and 'affirm all chances'.

In experimentation, all transcendent presuppositions have to be renounced to refuse use of reason or rationality alone in the service of particular human interests.

This allows thought to be free and to go beyond the boundary of 'territorization'. It proceeds to the unthinkable of chaos and complexity. Here, the image of thought encounters the chaos from the concepts and ideas being created in the form of 'sense'.

In experimentation, we receive questions from 'outside' and posing a series of new questions and problems, look for all chances and possibilities and construct assemblage and collage which produce a multiplicity of alternatives and solutions. In experimentation, it does not merely look for acceptable and satisfying solutions, but develops the questions in its own problematic field. In this process of experimentation, the questions become the reflexive thought where it considers the unthinkable that forces decision makers to think and act, and it carries thought to a 'transcendental' exercise in which unthinkable force of life acts within thought. In this way, thought becomes a reflexivity that it returns back to the thought itself.

16.4.3 What is an event?

Deleuze (1990) introduces the notion of an event. *The Logic of Sense* in Deleuze's work, aims to articulate an event as the point of contact between bodies and languages, and hence life and thought. An event extracts from the state of affairs, Deleuze (1990, pp. 9-11) shows how an event is related with bodies, thought of life and language through neither designation, manifestation, nor signification, but an event emerges from the 'border' between subject and object (or "state of affairs"). The pure form of the event is the determination of the body, which affects our thought. A pure event inheres or subsists not in an empirical mode of existence, but in a transcendental mode of existence of the one who lives it, or in whom it is actualised. Thus an event has power to change or affect our bodies, our speech or statements and our way of life. Having extracted an event from bodies, their mixtures, their actions, and passions, Deleuze (1990, p. 182) shows that language is related to bodies, and thought of life. Deleuze's vision of an event is to create a 'sense' for concepts; sense appears on the 'metaphysical surface' of propositions within language, as a membrane that

16. MAKING FRIENDS OF ENEMIES

overcomes the independent functions of denotation, manifestation and signification at once. An event has an autopoietic entity of "seeing and speaking" (Deleuze, 1990, p. 194; Patten, 1996, p. 4). The bodies can serve as a ground for the event and meaning, but the relations amongst bodies, thought of life, and concepts are themselves conditioned under one another to form a ground for the event and meaning (Deleuze, 1990). Unlike the Kantian assertion to the *a priori* concept of an idea that is developed from the sensibility, which would be a condition of possible experience of individuals, Deleuze (1990, p. 19) suggest a fourth, immanent dimension of the proposition, which he calls it 'sense'. Sense is not at all the attribute of the proposition, thus it is separate from the other dimensions of the proposition. Sense does not emerge or exist in designated physical states of affairs, nor is it manifested in subjective understanding or belief, and neither does it inhere in the signification of concepts, rather it inhere in the form of the event that is expressed by the verb.

Deleuze comes to discover his notion of the 'sense' that it is a condition of possible experience of images, simulacra or phantasm that which is an immediate and ultimate sources of meaning which is created from the relations of languages, bodies and thought of life. Sense flows into a 'pure event' that is the event without any physical or material aspect of it. For instance, in the incident of a snowy weather, we say, "it snows" it means an event has happened. For this example, the pure logical significance of "it snows" without the concept of the physical aspect of "snow" becomes a 'pure event'. As a result of the production of the 'singularities' (*sigularité*), which refers to "one" of impersonal and pre-individual singularities, the "one" of the pure event creates wherein *it* snows in the same that *it* dies. As is claimed in the genesis of the pure event, then, everything is singular if *it* is both collective and private, particular and general, neither individual nor universal (Boundas, 1993, p. 81). Consequently, we may say that any event happening in the world is a kind of repetition of the same pattern of its logical structure of singularities within the 'space' or 'field' of discourses. And as every event has the personal aspects related to a particular subject or 'individual selves' (Deleuze, 1990, p. 111) and it has "impersonal and pre-individual" singularities' aspect of its logical context at the same time (Deleuze, 1990, p. 103). "Cleopatra fells in love with Caesar," "Napoleon comes to the throne of the empire when he wears the crown in the ceremony of coronation." These are concrete examples of the 'singularities' that is the splendor of events (Lee, 1999, p. 92). Deleuze calls these kinds of events as the 'pure events' which occur within the spatio-temporal time of the 'singularities' (for further details of 'singularities', see Deleuze, 1990, pp. 100-108). It is very clear what Foucault (1970, pp. 349-350) gives us a good example of the (pure) events,

"Marc Antony is dead" *designates* a state of things; *expresses* my opinion or belief; *signifies* an affirmation; and, in addition, has a *meaning*: "dying." An intangible meaning with one side turned toward things because "dying" is something that occurs, as an event, to Antony, and the other toward the proposition because "dying" is what said about Antony in a statement. To die: a dimension of the proposition; an incorporeal effect produced by a sword; a meaning and an event; a point without thickness or substance of which someone speaks, which roams the surface of things" (italics are original).

The pure events can be experienced only under the logical dimension of the eternal time as it known as *Aeon*, in contrast to the instant time (*Chronos*) that we usually regard as time that happens within a material body. Since *Aeon* occurs within the 'surface', 'cranny' or 'border' between the bodies and thought of life, through the motions of a body (Deleuze, 1990, p. 61), the 'singularities' or 'singular points' appear that it is only under the conditions of the 'metaphysical fields' which individuals turns their experiences of events into mental process of transformation that created by its 'surface effects' (Deleuze, 1990, pp. 77-81). Thus, the pure events affirm an 'ontological' duration of the movement-image and time-image of the pure becoming in terms of 'an impersonal transcendental field' (Deleuze 1989, 1992). In his metaphysics of the virtual reality (that is, the 'surface effects reality', see Deleuze 1994, pp. 209-211), an "event by itself is problematic and problematizing" (Deleuze 1990, p. 54). Deleuze addresses an event as the particular phenomena that is based on the 'problematic' or that of problems. The term 'problematic' is a central concept for his metaphysics. For Deleuze, a 'problem' is determined by the 'singularities' or singular points as he focuses on the process of 'problematization' that leads us into the 'logical sets of singular points' in the 'transcendental fields'. For example, in terms of mathematics, we just consider the function $X^3-X=f(X)$. Within this function, there are two 'problems' (let us assume that X is either A (1/2) or B (-1/2)). Put differently, there are two singularities or singular points. On this function $X^3-X=f(X)$, other points on this function except A or B are considered as quiet normal, continuous and homogeneous. In this example, if we suppose that a whole function $X^3-X=f(X)$ is a system or logical set of singularities, then it is clear that there are two possibilities that is considered as "a 'problem' in terms of a set of singularities or singular events" (Lee 1999, p.146). Thus, the mode of the event is the problematic, and we conceive a 'problem' in a new way such as producing various events on the conditions of a 'problem' that is, a multiplicity of dispersed points (Deleuze, 1990, p. 55; Foucault, 1970, p. 354; Foucault, 1984). Now, we can speak of events in the context of the problem whose conditions are determined (i.e. if X>0, in the above example, the 'solution' becomes A

(1/2)). If the distributions of singularities correspond to each series being emerged from the fields of problems, how are we to characterize the paradoxical element which runs through the series of singularities? Deleuze introduced the notion of an aleatory point which means 'problems' arise out of a 'question'. He raised himself for identifying the paradoxical element, which traverses a series of singularities. Deleuze (1990, p. 113)'s own terms,

> "It behoves the *event* to be subdivided endlessly, and also to be reassembled in one and the same Event; it behoves the *singular points* to be distributed according to mobile and communicating figures which make of every dice throw one and the same cast (an aleatory point), and of this cast a multiplicity of throws" (italics are original).

In this way, the aleatory point represents the virtual and dynamic order out of which reforming of a series of events within a parallel structure between and through, which the event is displaced and communicated in a coherent manner. The aleatory point brings us to identify three different syntheses of events, which functions as 'transcendental' components of aesthetic. These are connection, disjunction and conjunction (Deleuze, 1990, pp. 23-25; 95-96). When connection is projected as continuity, "a series linked to an erogenous zone appears to have a simple form, to be homogeneous, to give rise to a synthesis of succession which may be contracted as such, and which in any case constitutes as simple connection" (Deleuze, 1990, p. 225). Time is also a dimension of asymmetrical passage, which distributes the disjunctive series ('or'). The disjunction posed a differentiation of series, which creates the new modes of existence which are incompatible with each other. In conjunction or the conjunctive series ('and'), the series is inseparable from a complex form which subsumes under it 'heterogeneous' series, now, the memory of events of the past, there is bodily resurrection, in which a spiritual 'phallus' as the image imposed in the genital zone, the spirit became an event, a becoming, a condition of continuity, which may find a new actualisation in the present. "It gives rise to a synthesis of coexistence and coordination and constitutes a conjunction of the subsumed series" (Deleuze, 1990, p. 225). Based upon this kind of the production of the reorganized series of events happens through the aleatory point which is said to subsist below the conscious threshold. The aleatory point itself is neither idea which corresponds to the language, nor body which corresponds to the "state of affairs," rather it represents the virtual order out of which ideas and bodies are engendered. Thus, the event must engender out of the multiple dimensions of the proposition such as 'denotation' (an indication of the state of affairs), 'manifestation' (a representation from the subject), and 'signification' (the symbolic transformation in language) being generated under the natural phenomena (Figure 16.1).

16. MAKING FRIENDS OF ENEMIES

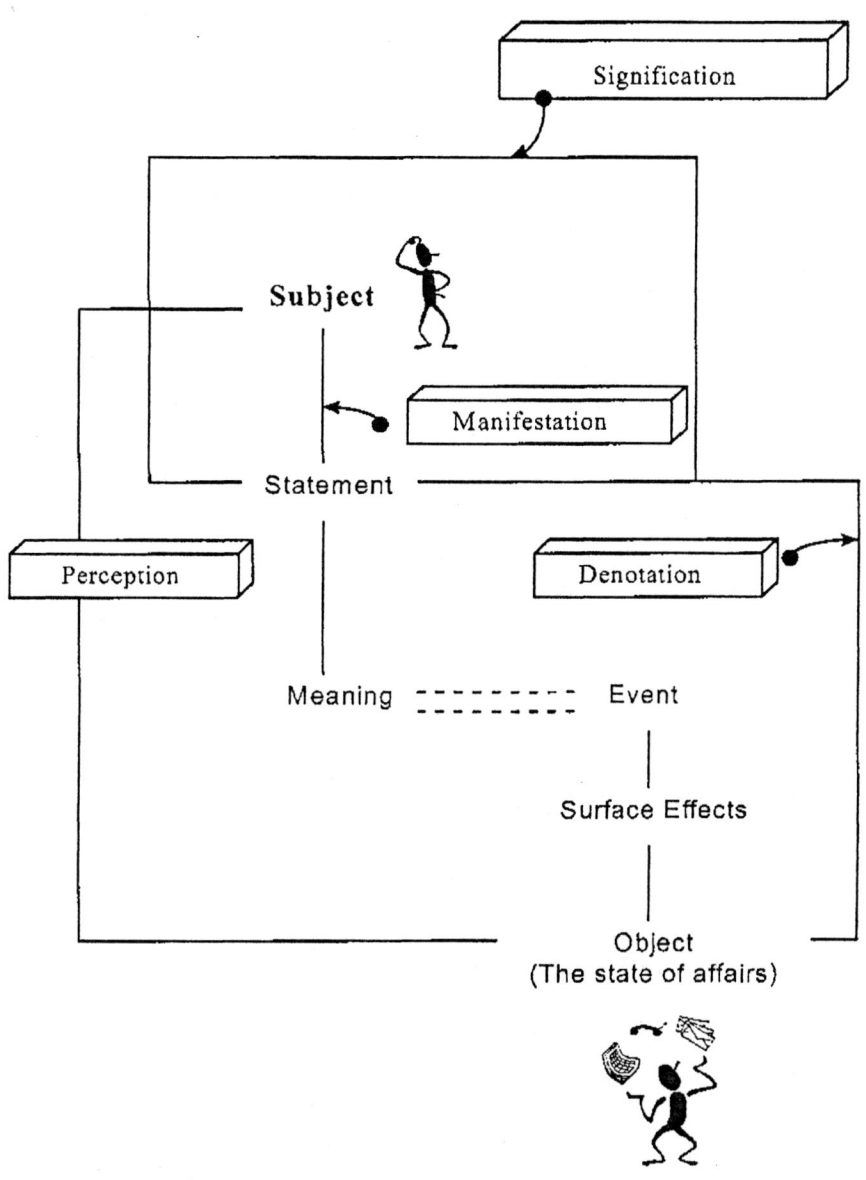

Figure 16.1 A genesis of the event in relation to signification, manifestation and denotation (Deleuze, 1990, pp. 23-25; 95-96)

16.4.4 An immanent methodology for problematization: Dealing with randomness and unpredictability

From these three relationships of denotation, manifestation, and signification, the event does not simply represent the ('moving') image of the world but be proactive or intervene upon it in a way of fulfilling its 'desire'. In this sense, the event functions as a 'vector of deterritorialization' (Patton, 1996, p. 13). Assuming a situation has continuously changing in unpredictable way, the event can be made as a possibility that leads to the multiple choices and alternatives. In this way, a problem-solving process is irrelevant, because the dichotomy of problem-solution is not acceptable in the continuously changing situations which are constantly evolving and transforming into the totally different contexts from the previous situations (Tsouvalis, 1995; Yu, 2001). Dealing with randomness and unpredictability that evolve from the continuously changing situations, following Foucault, Tsouvalis (1995) argues that we should deal with the question of 'what can I (or we) do?' about the concerned situations, which are both the object and subject of enquiry of appreciating the unpredictable contexts. In other words, dealing with unpredictable and continuously changing situations in local and contingent contexts, we need a new mode of thinking what Foucault (1981, p. 13; 1984) aims to instigate the process of 'problematization'. In process of problematization, a form of a critique has to be developed for a new type of the conceptualisation of both objectifying and subjectifying forms of knowledge, actions and their effects. For Deleuze (1994, pp. 70-77), the emergence of the condition of a problem is inscribed within the impetus in the form of the potential tendencies immanent within the unforeseeable variety of forms of a series of events. Then, "something new" is extracted which produces 'expectation' or anticipation for what could happen in the future according to 'ontological fields' that constitutes 'living present', that is "what happens" within the formation of pure spatio-temporal dynamics. Put differently, the continuous process of problematization is happened, 'sense' flows into the 'transcendental' dimension of pure becoming (that is a pure event) which always extends from a pure past to future both at once in experimentation. This allows us to appreciate that an open system of a series of (pure) events can produce as a formation of 'chaotic dynamics' within the problematic fields.

In the example of a real-life situation, the process of problematization occurs when the facilitator (or researcher) acts as an 'observer as participant' who engages in discussions with others (i.e. decision makers, members of staff in the organization) through formal/informal meetings and symposia to assess how people interact and create a series of events within the 'problematizing field'. During the process of making a series of events, the facilitator should bear in mind that there is no 'right', 'true' way of seeing problem situations that fulfil what should be done. Rather, the participants (including the researcher-facilitator as participant) encourage people with an alternative way of thinking to release subjugated ideas and knowledge from minority groups, low-level employees and unpopular 'voices' in the organization (Yu, 2004a). In this process of problematization, the collective thoughts become knowing-in-action that makes connection between power and knowledge that uncovers ways in which knowledge becomes constituted within power relations in specific organizational and social contexts (Yu, 2001, pp. 192-216).

16.5 THE SEARCH FOR A NEW MODE OF ETHICS

16.5.1 Introduction

In postmodern thinking, the power relations give rise to a possible corpus of knowledge and knowledge extends and reinforces the effects of this power (Foucalut, 1977). Whilst Habermas made a 'critical' reaction against the shortcomings of interpretive standpoints on the nature of social reality, Foucault (1971, 1977) seeks to advance a conception of power on the basis of constitution of knowledge through a relation of truth and power. His work is concerned with a new concept of power as it conceives as a whole series of the positive effects through a complex social function. This differs considerably at the substantial level with Marx's inversion of Habermasian concept of 'economic' power (Tsouvalis, 1995, pp. 199-209). Foucault criticised that the 'economical' concept of power is inadequate to understand its nature of social operation. To understand the concept of 'non-economical' power, the awareness of the importance of human body built of explicitly upon the analysis of *Discipline and Punish*. The reason for this is that the interrelation of the truth and power can be manifested through the analysis of human body from punishment. The physical torture and its pain of the condemned man and this visible characteristics of the punishment evokes the public effects that 'signification' of all the crime-punishment slip into the mind of the common spectators in the society. This historical background to various studies of the genesis of power and formation of knowledge can be introduced in modern society as well (Foucault, 1977). Foucault used the Bentham notion of panopticism that shows us that power should be visible, but it is anonymous and thus unverifiable. It functions as the 'apparatus of production' in the form of a network or strategy. It becomes an ultimate source of creating social reality so that "a discursive formation is an ordering of discourse is *both historically specific and socially contingent*" (Tsouvalis, 1995, p. 170, emphasis is added). In this social reality, for Foucault (1972), in his investigations on unknown 'field' of 'statements' *(énoncé),* he rejects man as a free agent with self-consciousness or individually meaning-giving subject (Foucault, 1974). Thus, an individual must situated in his/her position within the field of 'statements' in order that he or she will utter a particular 'statement' that contains his/her particular experience with the *singulier* of discursive events. In his pursuit of the Genealogy, he observed that the relation of power and knowledge to contrast to the 'modernist' concept of power that was believed as negative forces or even "commodity" in social and economic systems (Habermas, 1984a,b, 1987). Dealing with the social realty that evolves from complex and dynamic relations of knowledge, power that is concerned with how people constitute themselves as ethical subjects, Deleuze (1994, p. 6) believes that everything changes around the will to power in accordance with the law of Nature. Nietzsche's notion of the will to power is to ground the repetition in eternal return on both the death of God and dissolution of the "knowing" self (Deleuze, 1994, p. 11). Deleuze (1994) offers his concept of *difference* and *repetition* based on Nietzsche's notion of eternal return which conceived as formless being of *sense,* that is, the image of thought. Sense is in turn located in the complex processes of posing the set of problems and questions. Deleuze's version of Nietzsche's eternal return unfolds as pure movement which always subsumes a plurality of objects which is itself indefinite (Deleuze, 1994, p.13). It operates as determination of difference, as such movement is disparate and distinctive. However, difference can be no more than a predicate in the comprehension of the diversity of the concepts. Such understanding opens up the possibility of difference having its own concept of univocal being, rather than being understood under the Same (as it maintains as identical) (Deleuze, 1994, pp. 40-41). Eternal return 'makes' the difference because it creates new values from the point of view of the conservers of old values. This 'ontological' difference corresponds to questioning, which become problems or problematizing, producing the virtual determinant fields of existence. Such field appears at the level of "simulacra" and maintains them in that state with the power of formless being (Deleuze, 1994, p. 67). It is this dimension of constant movement which constitutes the general form of difference to the eternal return. It is true that the powers of difference and repetition could be reached only by putting into question the traditional or dogmatic image of thought (Deleuze 1994, p. xvi).

16.5.2 Postmodern Ethics: Subjectivization and schizoanalysis

Following Foucault (1988), Flynn (1985, p. 534) argues, the ethics (or "subjectvization") "denotes neither the conscious of phenomenologists nor the atomic individual of empiricism." Deleuze developed Foucault's hint on the subjectivization that humans tend to be have a 'desire' towards 'liberation' that makes people free or emancipate from the repressive structures of overt institutionalisation (Ruiz, 1997, chapter 4). If Foucault's predilection for ethics leads us to liberate from both the production of the obligatory rules for power and the existing 'moral codes' which seems ontologically or even conceptually linked to knowledge and power, then, how do we fulfill a *desire* for the liberation that refers to 'something new' as a means of struggle to 'something old and present'? Deleuze (1988, pp. 107-109) gives us the following clear statements,

> "Everything takes place as if the modes of subjectivization had a long life... The folding or doubling is itself a Memory: the 'absolute memory' or memory of *the outside*, beyond the brief memory inscribed in strata and archives, beyond the relics remaining in the diagrams... Time becomes a subject because it is the folding of the outside and, as such, forces every present into forgetting, but preserves the whole of the past within memory: forgetting is the impossibility of return, and *memory is the necessity of renewal*... as long as we remain on the level of things and states of things we can believe in a 'savage' experience that lets the thing wander aimlessly through consciousness. But *if phenomenology 'places things in parenthesis' as it claims to do, this ought to push it beyond words and phrases towards statements (énoncés), and beyond things and states of things towards visibilities*" (emphasis is added).

The above statements indicate that a new process of subjectivization is possible to develop the 'postmodern ethics', which Deleuze argues, the process of subjectivization is certainly preconditioned by a thought of a 'dice-throw' or it is more like a series of draws in a lottery (see Deleuze, 1988, p. 98). To think means to problematize, says Foucault, it is a 'history of thought' that refers to the games of repetition, of difference, and of the triple (i.e. knowledge, power and the self) root of a problematization of thought (Deleuze, 1988, p. 116). It is interesting that Foucault talked about critique, invention and imaginations about the changing shape of the 'something old and present' in our culture through the notion of problematization. In order to make individuals to participate effectively in the process of decision-making, the concept of subjectivization is crucial for offering individual actors to participate creatively in the process of self-reformation. The creativity of the individuals and self-production of individuals can actively place in a better situation to be involved freely and actively in the process of organizational transformation and social change. Considering the notion of subjectivization as the ethical aspect on a social inquiring process within organizational and social contexts, we should pay attention to the collective and holistic thoughts (and action) on psychoanalysis what Deleuze and Guattari (1977) called as 'schizophrenia' or 'schizoanalysis' (Holland, 1999). Schizoanalysis in *Anti-Oedipus* associates the judgements of the Ethics defined in terms of "Good" and "Bad," rather than "Good" and "Evil". This criterion of "Good" and "Bad" corresponds to the Ethics which presupposes both action and knowledge (Boundas, 1993, p. 70). There is, then, a passion brings out those of a *joy* in favour of "Good," and thus our power of acting is subsequently increased. It may be said that our power of acting is enhanced as if a power of being acted upon insofar as it is filled by passions. In such relation, a *joy* initiates the process of a *desire* that emanates from the "body without organs" (Boundas 1993, chapter 7). Now then, how does one arrive at a maximum of the ethical 'joy' or 'joyful' passions in the process of producing a desire in our daily life? This question corresponds to the aspects of the 'nomadic subjects' who has the pervert conduct which corresponds to "traversing" the 'social codes'. The Oedipal complex has always-haunted parental and familial pathways, which makes 'man' as having traditional figuring of subjectivity with the rational self-consciousness. By taking the Oedipal complex and the fear of castration as unavoidable in the Freud's theory of sexuality, Freud's analysis of the development of the psyche contributes to make the relation between the subject and repressive social structure (Deleuze, 1990). Deleuze and Guattari (1977) consider that oppressive social mechanisms are maintained not only physical and dynamic structure but also by the creation and reproduction of the oedipal personality which adjusts into the given social reality. They claim that this oppressive social existence once again takes place through secondary repression by the interjection of the oedipally structured ego into the unconscious (Howie, 1992, chapter 3). The presence of the social coercive repression appropriate to make the appearance of what Deleuze and Guattari (1977, pp. 10-15) called as the 'social codes' (see also, Holland 1999). Deleuzian Ethics is to show that the 'nomadic subject' who is capable of traversing across the 'social codes' by means of pure becoming according to the principles of rhizomatic systems (Deleuze and Guattari, 1977, pp. 16-17) Pure becoming is expressed in terms of the notion of aleatory or the aleatory point which corresponds to the nomadic subject (Bogue, 1989). By postulating the concept of the nomadic subject, it corresponds 'paradox' of a pure becoming which means the acting against a taken-for-granted assumption and rules that are manifested in habitual ways of thinking and practices in every-day life at that particular time and space. In this sense, Deleuze calls it as 'paradox' of a pure becoming, whose characteristic is to elude the present, which refers to the paradox of infinite identity (of both directions at the same time-of future and past, of active and passive, and of cause and effect) (Boundas, 1993, p. 40).

In this sense, the nomadic subject is understood in terms of denial of the common sense and the Freudian assumption of the oedipal personality, attitudes and beliefs in the 'history of present'.

Thus, Deleuze's 'paradox' focused on a new mode of behavior and life based on the 'nomadic subject' who rejects the rational and organizational character of capitalism (Deleuze and Guattari, 1977; Deleuze, 1990, pp. 77-81).

The 'nomadic subject' recognizes the 'exit' and is open to the alternative possibilities which constitutes the extended moment of a 'paradox' (Deleuze, 1990, p. 103; see also Holland, 1999, p. 36). Deleuze and Guattari (1977, pp. xx-xxiii) argue,

> "To be anti-oedipal is to be anti-ego as well as anti-homo, wilfully attacking all reductive psychoanalytic and political analyses that remain caught within the sphere of totality and unity, in order to free the multiplicity of desire from the deadly neurotic and Oedipal yoke...*Groups must multiply and connect in ever new ways, freeing up territorialities for the construction of new arrangements.* Theory must therefore be conceived as a toolbox, producing tools that work...A politics of desire would see loneliness and depression as the first things to go. Such is the anti-oedipal strategy: if man is connected to the machines of the universe... *The life that's in him will manifest itself in growth, and growth is an endless, eternal process. The process is everything. It is this process – of desiring-production –* that Anti-Oedipus sets out to analyse" (Emphasis is added)

Deleuze and Guattari's *schizoanalysis* can be used as an approach that is based on a new 'collective subjectivity', and not by professionals and experts. Thus we proposes the participatory approach to encounter 'rhizome and traversing' in organizations and communities (Deleuze and Guattari, 1977, pp. xxi-xxii).

Schizoanalysis requires a discovery of the 'outside' by returning to a more unconscious and natural forms of a 'savage' experience (Deleuze and Guattari, 1977, chapter 4, see Deleuze and Guattari, 1988, pp. 7-12 for the details of principles of the rhizome). The nomadic subject follows the principles of rhizomatic systems, in order to liberate a *desire* from the given social reality. Ethical subjects (or changing agents) can have an impact on the power-knowledge networks through strategic action (Midgley, 2000, p.

168). What does an ethical subject need to do in order to become a new collective subjectivity, to able to think and act more ethically, politically and practically in a group formation?

As Yu (2001, 2004a) has shown, the facilitator as an ethical subject must be prepared to accept any challenges on the grounds of the contexts-based justification (based on the criterion of 'Good' and 'Bad') which may arise from the specific culture and history in the given situations in order to bring out a joy in favour of 'Good' through the process of subjectivization and schizoanalysis (see Yu, 2001, pp. 198-216).[2]

16.6 CONCLUSION

Contemporary Western philosophy of critical theory and poststructuralism have made contributions to the development of a systems approach. . Firstly, the critical systems approach (i.e. CSH) insists on the application of rationality and scientific knowledge to the development of ethical and moral judgements for improving the process of social inquiry. Secondly, it has contributed an alternative mode of thinking about rationality and discursive ethics.

These contributions enable us to open up an alternative way of life that allows the nomadic subjects to traverse the existing social codes.

In this way, there would be another possibility for 'making friends of enemies' in order to enhance our restricted world view into an alternative mode of thinking that is not only governed by scientific knowledge but also created by the image of thought.

The image of thought will succeed in relating thought about life, chaos, and the lived experience of people, not representation.

This allows thought to approach the 'unthinkable' which forces us to think about our thinking.

Thought will relate to the movement, in return 'becoming', an 'ontology', as an abyss which only a transcendent value could come to fill.

---

[2] In order to fulfill a *desire* within organizations and communities, research into a new 'collective subjectivity' is required by means of participatory approach into the *ethical* and *political* processes in our daily life. Having considered a new collectivity, in Deleuze and Guattari's framework, the nomadic subject should reject a tradition model that presupposed by a taken-for-granted assumption and the hypothesis. Then, the notion of nomadic subject will create the new kinds of lines of thought, values, beliefs and ethical judgements (Lee, 1997). Being an ethical subject means one has to reflect on the mode of existence through the image of thought. In this sense, an ethical subject does not rely on subject-centred reason. In other words, ethical subjects can be created through the complex and dynamic relationship of power and knowledge in the contingent and local contexts.

The exterior movement inspires 'thought', an unthought operates alongside 'thought' which is still the element of 'outside'.

This is a matter of empiricism, experimentation, lived experience and life. The experimentation does not merely look for solutions, but develops the question in its own problematic field.

The reflective question carries thought to a 'transcendental exercise' where it considers the 'unthinkable' which forces us to think. In experimentation, we receive questions and problems from 'outside' of thought, and construct collage which produces a multiplicity of solutions. However, here we must leave these matters to be pursed in more detail in further research within the specific cultural contexts.

REFERENCES

Bogue, R., 1989, *Deleuze and Guattari*, Routledge.
Boundas, C.V., eds., 1993, *The Deleuze Reader*, Columbia University Press, New York.
Churchman, C.W., 1968, *The Systems Approach*, Dell Co., New York.
Churchman, C.W., 1971, *The Design of Inquiring Systems*, Basic Books, New York.
Churchman, C.W., 1979, *The Systems Approach and its Enemies*, Basic Books, New York.
Cummings, T.G. and Worley, C.G., 2001, *Organization Development and Change*, 7th ed. South-Western College Publishing/ Thompson Learning, Ohio, USA.
Deleuze, G., 1988, *Foucault*, Athlone Press, London.
Deleuze, G., 1989, *Cinema 2: The Time-Image*, (trans. Tomlinson, H. and Galeta, R.), Athlone, London.
Deleuze, G., 1990, *The Logic of Sense*, (trans. M. Lester with C. Stivale), Althlone, London.
Deleuze, G., 1992, *Cinema 1: The Movement-Image*, (trans. Tomlinson, H. and Habberjam, B.), Athlone, London.
Deleuze, G., 1994, *Difference & Repetition*, (trans. Patton, P.), Athlone, London.
Deleuze, G. and Guattari, F., 1977, *Anti-Oedipus: Capitalism and Schizophrenia*, (trans. Hurley, R., Seem, M. and Lane, H.) University of Minnesota, Minneapolis.
Deleuze, G. and Guattari, F., 1988, *A Thousand Plateaus: Capitalism and Schizophrenia*, (trans. Massumi, B.), The Athlone Press, London.
Flynn, T., 1985, "Truth and subjectivation in the later Foucault", *The Journal of Philosophy*, pp. 531-539.
Foucault, M., 1970, "Theatrum Philosophicum", in *Michel Foucault: Essential works of Foucault 1954-1984*, Vol. 2, Faubion, J. D., eds., Penguin, London.
Foucault, M., 1971, *L'ordre du discours: Lecon inaugurale au College de France Prouncee lee*, 2 December 1970, Gallimard, Paris.
Foucault, M., 1972, *The Archaeology of Knowledge*, Routledge, London (originally published as *L'Archeologie du Savoir* in 1969)
Foucault, M., 1974, The Order of Things: An Archeology of the Human Science, Tavistock, London.
Foucault, M., 1977, *Discipline and Punish: The Birth of the Prison*, Allen Lane, Penguin Books, London.
Foucault, M., 1981, Questions of Method, Ideology and Consciousness, **8**:3-14.

Foucault, M., 1984, "Polemics, politics and problematizations", in: *Michel Foucault: ethics, essential works of Foucault 1954-*1984, P. Rabinow, 1994, vol. 1. pp. 111- 119, Penguin, (the interview conducted by Paul Rabinow in May 1984).
Foucault, M., 1988, *Technologies of the Self*, in L. Martin, et al, eds., q.v., pp. 16-49.
Goodchild, P.S., 1993, *Chaos and Eternity: Gilles Deleuze and the question of philosophy*, PhD thesis, University of Lancaster, U.K.
Gregory, W. and Romm, N., 1994, "Developing Multi-Agency Dialogue: The Roles(s) of Facilitation", *Working Paper* No. 6, University of Hull, U.K.
Gregory, W., Romm, N., and Walsh, M., 1994, "The Trent Quality Initiative: A multi-agency evaluation of quality standards in the National Health Service", *Research Report*, University of Hull, U.K.
Habermas, J., 1972, *Knowledge and Human Interests*, Heinemann, London.
Habermas, J., 1984, *The Theory of Communicative Action, Volume One: Reason and the Rationalisation of Society*, Polity Press, Cambridge.
Habermas, J., 1987, *The Theory of Communicative Action, Volume Two: Lifeworld and System: A Critique of Functionalist Reason*, Policy Press, Oxford.
Habermas, J., 1993, *Justification and Application: Remarks on Discourse Ethics*, Ciaran P. Cronin (tran.), Polity Press, Cambridge.
Hancock, P. and Tyler, M., 2001, *Work, Postmodernism and Organization: A critical introduction*, Sage, London.
Hatch, M.J., 1997, *Organization theory: Modern, symbolic and Postmodern Perspectives*, Oxford University Press, Oxford.
Holland, E.W., 1999, *Deleuze and Guattari's Anti-Oedipus: Introduction to Schizoanalysis*, Routledge, London and New York.
Howie, G.O., 1992, *Fragmented Subjectivity: A critical study of the Anti-Oedipus: Capitalism and Schizophrenia*, Ph.D.Dissertation, Faculty of Social and Political Science, University of Cambridge.
Jackson, M.C., 1985, "Book review on W. Ulrich: The itinerary of a critical approach", *Journal of the Operational Research Society*, **36**(9):878-881.
Jackson, M.C., 2003, Systems Thinking: Creative Holism for Managers, Wiley, Chichester.
Kant, I., 1956, *Critique of Practical Reason* (1788: origin), Bobbs-Merrill, Indianapolis.
Kant, I., 1988, *Critique of Pure Reason* (1781: origin), J.M. Dent and Sons Ltd.
Lee, J.W., 1997, *The Traversing* (Korean version), Min Em Sa, Seoul Korea.
Lee, J.W., 1999, *The Age of Simulacre: Deleuze and his Philosophy of Events*, Keorum, Seoul, Korea
Midgley, G., 2000, *Systemic Intervention: Philosophy, Methodology, and Practice*, Kluwer Academic/Plenum Publishers, New York.
Mills, C.W., 1970, *The sociological imagination*, Originally pubulished in 1959, Penguin, Hatmondsworth.
Mingers, J., 1992, "Recent developments in critical management science", *Journal of the Operational Research Society*, **43**(1):1-10.
Patton, P., eds., 1996, *Deleuze: A Critical Reader*, Blackwell, Oxford.
Ruiz, M., 1997, *Psychophyical Parallelism in the Philosophy of G. Deleuze*, PhD Thesis, University of Warwick, U. K.
Tsouvalis, C., 1995, *Agonistic Thinking in Problem-Solving: The Case of Soft Systems Methodology*, PhD thesis, University of Lancaster.
Ulrich, W., 1983, *Critical Heuristics of Social Planning*, Wiley, Chichester.
Ulrich, W., 1987, "Critical heuristics of social systems design", *European Journal of Operational Research*, **31**:276-283.

Ulrich, W., 1988, "Churchman's 'Process of Unfolding'- Its significance for policy analysis and evaluation", *Systems Practice*, **1**:415-428.
Ulrich, W., 1993, "Some difficulties of ecological thinking, considered from a *Critical Systems Perspective*: A plea for critical holism", *Systems Practice*, 6(6).
Ulrich, W., 1995, "Critical system thinking for citizens: A research proposal, *Research Memorandum*, No.10, Centre for Systems Studies, University of Hull, 28th November, 1995.
Ulrich, W., 1996, "A primer to critical systems heuristics for action researchers", *Research Memorandum*, Centre for Systems Studies, University of Hull.
Ulrich, W., 1998, "Systems thinking as if people mattered: Critical systems thinking for citizens and managers", *Working Paper* No. 23, Lincoln School of Management, University of Lincolnshire and Humberside, Lincoln.
Ulrich, W., 2000, "Reflective practice in the civil society: The contribution of critically systemic thinking", *Reflective Practice*, 1(2):247-267.
Ulrich, W., 2004, "In memory of C. West Churchman (1913-2004): Reminiscences, retrospectives, and reflections", *Journal of Organizational Transformation and Social Change*, Intellect, 1.2 & 1.3, pp. 199-219
Williams, M. and May, M., 1996, *Introduction to the Philosophy of Social Research*, UCL, London.
Yoo, B.Y., 1991, *A study on Critical Systems Ethics for Democratic Citizenship Education*, PhD Dissertation, Seoul National University, Korea.
Yu, J.E., 2001, *Towards rhizomatic systems thinking in management science*, DPhil Dissertation, University of Lincoln, U.K.
Yu, J.E., 2004a, "Reconsidering participatory action research for organizational transformation and social change", *Journal of Organizational Transformation and Social Change*, Intellect, 1.2 &1.3, pp. 111-141.
Yu, J.E., 2004b, *An Invitation to the Management of Difference: Systems thinking for knowledge creation and learning*, Samsung Economic Research Institute, Seoul, Korea.

Chapter 17

THE APPLICATION OF AN EPISTEMOLOGICAL INQUIRY TO INCREASE OUR UNDERSTANDING OF COMPLEX ISSUES
Enlarging Churchman's "Circle of Enemies"

VAN GIGCH,J.P.

Abstract: In this paper we explore a special aspect of C. West Churchman's epistemology which is concerned with a method that he pioneered to increase our knowledge of a problem. In the **Systems Approach And Its Enemies**, Churchman (1979), identified four main enemies of the Systems Approach (SA) namely: Politics, Morality, Religion and Aesthetics. In doing so he pioneered the use of the *epistemological inquiry* an approach which can be used to enlarge the scope of available knowledge and to increase our understanding of the underlying concepts of these issues. Whereas Churchman worked with four enemies, there are many more enemies lurking in the horizon and hindering the application of the Systems Approach to any problem. We will illustrate an extension of Churchman's approach by using the examples of Terrorism, Globalization and Democracy. In so doing we show an aspect of Churchman's multifaceted epistemology.

17.1 SEARCHING FOR ENEMIES: AN ILLUSTRATION OF AN EPISTEMOLOGICAL INQUIRY

Churchman (1979) in his *Systems Approach and Its Enemies* provided the original idea to identify the enemies/adversaries of the Systems Approach. It takes many readings to comprehend what Churchman meant. What is not obvious is that, by urging us to search the so-called 'enemies', Churchman provides a glance at the depth of own his mind. Without overtly stating so, he illustrated the application of an *epistemological inquiry* which can be used to increase our knowledge base. We will illustrate this approach by applying it to three different problems.

Extending the search for "enemies" is in keeping with the message he gave in the conclusion of another book, The Systems Approach (Churchman, 1968), where he refused adamantly to state what the Systems Approach consists of, except to make the following four statements:

1. "The Systems Approach begins first when you see the world through he eyes of another."
2. "The Systems Approach goes on to discovering that every worldview is terribly restricted."
3. "There are no experts in the Systems Approach," and
4. "The Systems Approach is not a bad idea."

We would forgive a neophyte who would consider the above four points cryptic. We hope that the interpretation provided in this essay may begin to answer some of the questions that the above statements leave unanswered.

17.2 A MATTER OF DEFINITIONS: ADVERSARIES VERSUS ENEMIES - ENEMIES/ADVERSARIES/ANTAGONISTS

We first note the similarity/difference between the words 'enemy' and 'adversary.' Other authors use the word adversaries to denote all the forces which impede the progress of a discipline.

In The Concise Oxford Dictionary (1976), the concepts of adversaries and enemies are considered synonyms. However other sources show a distinction between these two concepts. An *enemy* is seen as a hated opponent and is usually considered a person who hates another and eagerly seeks his/her defeat. Other words used in lieu of 'enemy': 'opponent'; 'hostile army' or 'hostile nation', an 'alien.' On the other hand, an *adversary* is an opponent who is *not* necessarily hated. An adversary is someone who is 'in front of', opposed, coming from another direction, averse. Adversary indicates one against the other without necessarily the intent to harm.

As an example, in the conflict between Ronald Reagan and Gorbachev during the Cold War, they called themselves adversaries but not enemies. Reagan acknowledged that he considered Gorbachev a worthy adversary: they faced one another but did not necessarily hate each other. Finally, the word "antagonist" is also a synonym of adversary and these concepts can be used interchangeably.

17.3 TERRORISM

To search for "enemies," in the Terrorism Problem, we refer to Ackoff and Strümpfer's article "Terrorism: A Systemic View," **Systems Research and Behavioral Science, 20: 287-294, 2003.**

Ackoff and Strümpfer (2003) first refer to the fact that *"a major factor in the success of terrorism lies in the fear and paralysis it induces."* **The Enemy**: Fear induced by physical insecurity.

"Through violence, terrorism conducts what is primarily a psychological war directed at affecting the mind and behavior of the public." A war against terrorists is not what can be called a "conventional war." The enemy does not represent a country in the traditional sense, its commanders do not have command centers, they are not even clearly recognized or identifiable. **The Enemy**: Not conventional, cannot be identified, "does not have a face," it is invisible.

One of terrorism's aims is *"to obtain acceptance of its value system and aims,"* through threat and coercion. In democracies, this acceptance if gained through representatives who vie for votes through persuasion and politics. **The Enemy**: Induces fear rather than persuasion.

One side in a conflict sees terrorists, the other side sees heroes and/or freedom fighters. **The Enemy**: Depending who gives the support, they are considered heroes/martyrs or misguided warriors.

Motivation: It is difficult to discern the enemy's motivation. There is no question but that terrorism's breeding ground are areas of the world where people live in poverty and in dire conditions. What starts as nationalistic movements is rapidly transformed in pools of desperate individuals who have no jobs, no housing and whose living conditions are dismal. Under these circumstances, out of desperation they resort to terrorism to stake their claims. In the extreme, life is not worth fighting for and death is a symbol of their plight. **The Enemy**: Desperation brought about by destitution and despair.

Terrorist acts cannot be analyzed rationally. In using violence, the adversary retaliates with violence and this endless vicious circle vitiates any possibility of determining where right or wrong lies. Both sides in a conflict are in the grips of an action-reaction loop where no reason prevails. **The Enemy**: Vicious circle of violence and revenge.

What is fact and what is faith? We recall Churchman's words who referred to the enemy's ideology as a genuine "worship." Churchman did not pretend that this worship had always any religious foundation. However, in terrorism, beliefs are grounded in a religious-like faith, where reasoning in the traditional sense does not exist. **The Enemy:** Ideology worshipped as religious faith.

Those who must fight terrorism must shed all conventional wisdom, traditional norms of combat, obsolete institutional forms which cannot face the apparently disorganized chaos inflicted by terrorists. **The Enemy:** Believing in conventional wisdom and accepted practices.

Countries that try to fight terrorists must change the mindsets and world views of its population as well as that of the individuals in charge of organizations whose responsibility is to fight against this new mode of attack. It is appropriate to refer to a paradigm shift: the old paradigms which spelled how to fight a conventional enemy are obsolete and must be replaced. As Ackoff and Strümpfer (2003) state: *Competently dealing with emerging dilemmas requires a paradigm shift, a shift in the way we think, in ways of conceptualizing reality and our value framework.* **The Enemy:** Entrenched paradigms.

Means to combat terrorism. As Ackoff and Strümpfer (2003) note, until now countries which have been attacked by terrorists or which have suffered terrorists attacks use "*violence against violence.*" It is patently obvious that this is the worse antidote for this kind of phenomenon. Violence escalates to the point of no return. Heroes are made martyrs and on both sides of the equation, protagonists justify their violent acts as a viable and effective defense. It never works. **The Enemy:** The mindset that considers that violence can only be countered with more violence and that violence can only be remedied with more of the same.

Ackoff and Strümpfer state that "We have to think like terrorists." Indeed. Unless we gain a minute understanding of the motives and modes of action of the terrorists we will unable to neutralize them or thwart their unconventional behavior. The Enemy: Our inability of understanding the opponent's mindset or point of view.

Modes of organization to combat a complete new form of threat are utterly obsolete. It would take volumes to explain why the present organization to gather intelligence in any of our countries is obsolete and incapable to handily the new threats posed by foreign speaking enemies who use high speed communications and the media to communicate among themselves and to send their message to the world. At the time of writing, the US has not been able to secure its borders, to find ways to track down terrorists around the globe and worst of all, security in the country is still in the same state of disarray as on 9/11/2001. It is difficult to move an "elephant"....and the inertia and shear size of the US government coupled with the barriers of politics and infrastructure makes it impossible to provide a nimble response to the new threats. New terrorists attacks are a potential which cannot be discounted because of the time it takes for institutional arrangements to change and to adapt to new circumstances. **The Enemy:** The inertia of obsolete institutional arrangements and procedures cannot be changed fast enough when pitted against light and mobile tactics.

Global Problem. According to Ackoff and Strümpfer: "Terrorism is a problem that cannot be addressed successfully within one nation. It is a global problem that can destroy society faster than pollution or the exhaustion of resources." **The Enemy**: To believe that terrorism is an evil which can be fought only locally or that any country can "go-it-alone."

Development as an Antidote to Terrorism. Ackoff and Strümpfer argue that the promotion of development in societies that breed terrorists would be desirable as an equalizer of wealth, improvement in the quality of life and in the living conditions of entire population deprived of jobs, and the mere necessities of water and food. Furthermore, instead of working to shore up these population centers they are inextricably the targets of war and violence. Violence breeds violence and a vicious circle ensues out of which none of the adversaries can extricate themselves. It further deepens the big hole of chaos in which they are all lost and involved at infinitum. **The Enemy**: Dire economic and social conditions that breed extremism.

Enemies Within and Enemies Abroad

The risks of terrorism can be subsumed under Politics or under Religion, two of the main enemies of the Systems Approach singled out by CWC.

Because of its scope, Terrorism must be considered on an international basis. The perpetrators—whoever they may be—operate international networks and can penetrate national borders and commit their awful deeds. Indeed only a concerted 'Systems Approach' can begin to comprehend the multifaceted dangers involved. And yet, each nation must in addition coordinate international action with national measures. As an example, the US, three years after the biggest terrorism attack on the scale of what happened on September 11, 2001 on US soil, is still groping around aimlessly asking questions about how this event could have happened, the lapses in security, intelligence and lack of coordination among agencies in charge of protecting the public from such catastrophes.

Of course the application of the Systems Approach at the national level is imperative. However, we submit that failing to take a Systems Approach was not by far the main reason that this disaster took place.

Basically speaking, what we are stating is that –there is no system approach or any approach that can begin to comprehend and unravel who is responsible for what happened—let alone prevent a new catastrophe.

Several US National Commissions headed by Congress or appointed by Congress concluded investigations into the lapses of US security and lack of knowledge of intelligence in relation to terrorism in the US. **(US National Commission, 2004), (US Select Committee, 2004). The Enemies**: Politics and Religion and their old-fashioned bureaucracies, where these concepts must be understood in the context of Churchman's meanings.

Religious Fundamentalism

Many of the dangers of terrorism stem from religious considerations. We consider in particular the dangers of religious fundamentalism. Fundamentalism can be of two kinds: 1) Religious and 2) Scientific. The latter is not relevant to this study and will be overlooked here.Fundamentalism, in the religious sense, is the extreme position which certain religions adopt to assert their claim that they hold the whole "revealed truth." The more extreme the position taken, the more intolerant they seem to be. Occhiogrosso P. (1994) in his *The Joy of Sects* points out that there exists Christian and Islamic fundamentalists. C. West Churchman was quite correct in identifying Religion as a major "enemy" of the Systems Approach. Fundamentalists believers are virulent, intolerant and not liable to yield to reason. Their "worship" is strict and does not allow exceptions. Several religious wars are presently in progress in the name of this extreme worship. The conflict between Catholics and Protestants in Ireland still simmers with hatred and vendettas which to the reasonable mind seem futile and groundless. There is no way that some sort of Systems Approach can make any progress in this inhospitable terrain. The Enemy: The worship of a fundamentalist faith.

Attempting to Modify Culture

Again, to repeat, in the opinion of this author, the so-called "war on terror" cannot be won by fighting it with violence. Violent means engenders violent reactions. Instead of generating goodwill it creates the very justification for more terror and adversaries. Trying to impose the US American-style of democracy or culture on other cultures is counter-productive. It creates more dangers than is worth it. Here the Systems Approach has found natural boundaries which should be warning signs of more dangers and more enemies. It is utter nonsense to believe that our imperialism can translate into transplantation of our culture. Enough antagonism already exists on the ground that the ill-rated tradition of the "Ugly American" dominates many foreign areas and wants to dominate even more. By going after other cultures in a pre-emptive mode and try to change them, we create more enemies rather than defuse the antagonisms and animosities. (See also K, Bausch, in Volume # 1 of this Series, 2005). **The Enemy**: In a sense, we are our own worse enemies. (See Bausch, 2005 and Yu, 2006)

Limitations Of The Systems Approach: Are All Terrorisms Part Of The Same Conspiracy?

In 2004, the US Government and Russian counterparts contended that terrorist attempts in both countries belonged to the same source i.e. they advocated the application of the Systems Approach to both situations alike, regardless of political realities which are entirely different. Another approach with a more limited view could be as effective concerning the situations in the US and in Russia are different. In this view the Systems Approach to terrorism would advocate a limited local or regional approach which contends that the problems in each of those two countries must be treated differently and the approach must be taken country by country. Terrorism in Indonesia is different that that occurring in Chechnya. Here we must refuse a wholesale application of the Systems Approach and take a country by country approach. **The Enemy**: Thinking that the applicability of the Systems Approach is universal.

17.4 GLOBALIZATION

This is a question which by its shear size and scope demands that it be treated by some comprehensive methodology like the Systems Approach. There is no question that the ramifications are huge and almost nameless due to their pervasiveness and intricacies. Named by Churchman and determine what other impediments to the Systems Approach exist. They would then be classified as enemies/adversaries or simply hindrances/barriers to implementation. Basically, Globalization can be classified as an economic question. It started when the non-industrialized and underdeveloped countries of the world wanted to benefit from the accumulated 'riches' of the industrialized countries. It could be considered an example of applying the Systems Let us expand the circle of enemies beyond the four Approach at a global scale.

As an example, when countries such as Mexico realized that its lower wage scales would attract industries and jobs which could compete advantageously against a country like the US where wages and costs were relatively high. Soon through accords like NAFTA, Mexico saw whole industries move in when the relative cost advantage reduced the overall cost of manufacturing in Mexico in comparison when produced elsewhere. Sometimes portions of the manufacturing content was carried out in one country and the other portions in Mexico. **The enemy**: Changing economic conditions which prevent promote local temporary conditions of time and place.

After a few years of this practice, Mexico was "outflank" by other countries. China undercut Mexican costs and the industry which had migrated from the US to Mexico, ruthlessly moved out again to China for instance where it could reap the new cost advantage.

It is not difficult to associate these global economic shifts to upheavals in global trade and the law of comparative economic advantage among trading and competing nations. Do these shifts leave a positive or negative balance? **The Enemy**: Same as above.

As we described, the advantage to Mexico was fleeting. In the end, China is the country which benefits (for a while, shall we say) from offering the lowest wages of any other country on earth. Is it fair? Is it ethical? Will it last? Is it to the credit of the Systems Approach? Mexico had to relinquish its status of country with the lowest manufacturing costs and it saw its manufacturing sector shrink in favor of China. **The Enemy**: Global competition spurs countries to be economical and geo-political adversaries.

Globalization trends have sparked serious and justified complaints from the labor forces which have seen their jobs replaced. It shows that none of the so-called global/total trends can be applied willy-nilly without weighing advantages and disadvantages in the short term/long term. As in all planning, it is the "unanticipated consequences" that cause the most disruptions and invariably, it is the lower socio-economic classes who suffer the consequences of these consequences the most. It is noteworthy to remark that some countries like the US tried to enforce labor laws that would protect the living conditions of the workers in these low wage countries from exploitation (in particular of working children) and decent living conditions.

The Nike Shoe Company moved to reap the economic advantage of lower wages: Can we really expect that it will enforce US labor standards in its new location, regardless of its pious ethical pronouncements? **The Enemy:** Difficulty to enforce human rights in another country.

Globalization provides us with several lessons. The Systems Approach is a process which cannot be applied indiscriminately across the board to all countries and to all circumstances. The large industrialized nations should not be allowed to push the cost of wages to their lowest denominator in order not to sacrifice the human and labor rights of disadvantages workers in far-way countries. US corporations moving abroad should pay for decent living wages without ruining the labor markets where it decides to build a plant abroad. Some claims have been made that these same corporations should compensate the workers in the US who are left unemployed as a result of the move. Is that realistic? **The Enemy:** Economic advantage in a state of flux.

Greed and the biggest profit cannot be the only criteria by which we measure the success of globalization.

Globalization may be an economic remedy which is advantageous to some and rejected by others. Large foreign corporations should not be allowed to dictate the working policies of local companies in particular when they have a tendency to be so efficient as to ruin the local economy instead of becoming a welcome complement to it.

What may be beneficial for Canada may be devastating for India, the Philippines or India. **The Enemy**: Morality of a nation's foreign and domestic policies.

Present discussions of "outsourcing" in the US, (where "outsourcing" is seeking foreign contractors with lower costs to provide some of the services a company needs), may be an indication that even the US may have to renege on its promises of "all out globalization" because "outsourcing" sends the jobs of US workers abroad and increases US unemployment. **The Enemy**: Choice between honest competition or rigging advantage to gain economic gain.

17.5 GLOBAL WARMING

In many respects and strictly from an epistemological point of view, the problem of Global Warming resembles that of Globalization except that to solve Global Warming it is absolutely essential that countries act in consort with one another. No half measure will save the planet and all local interests will have to be subsumed in the total equation if we are to save the planet for future generations. As the name implies this is a climatic effect that affect the whole globe and must be considered from a 'total system' perspective. The question is vast and there is no room in this short essay to justice to it. We mention it as an example which is can only be dealt by and through the Systems Approach. But will local, national parochial, partisan and political other impediments prevent it from ever be solved as "a total system problem?" We leave the subject untreated to jump to another major topic which illustrates our initial intent.

17.6 DEMOCRACY

17.6.1 Democracy and Imperialism

Attempts by a country like the US to dominate the world stage smacks of Imperialism: The US wants to establish dominance either militarily (as exemplified by the invasion of Iraq on the pretext of going to war to impose democracy on another nation); or economically (by conquering a monopoly or establishing same (the US is using too much foreign oil and must therefore secure sufficient supplies no matter what its costs).

Imperialism can be seen as a political stance to dominate the world stage. From a domestic point of view, Imperialism can be seen as an attempt of one country to trample the human rights of another, to dominate its culture or to try to impose its political, religious and economic points of view. (West, 2004)West (2004) sees this dominance as threats to democracy, namely:

- *"The dogma of free markets fundamentalism"* which was explained above under Globalization and its unanticipated effects
- *"Aggressive militarism,"* as exemplified by the so-called US 'pre-emptive' incursion in Iraq, and,
- *"Escalating authoritarianism"* as illustrated by attempts to impose political will as well as dominance of information and media sources. Furthermore, imperialism also extend to other issues such as spurning the international community when issues of international importance such as like Global warming, the environment and other issues are at stake.

The Enemy can be better understood in terms of a), b) and c) above.

17.6.2 Democracy. Identifying the Enemies

We will again attempt to identify the "enemies" of democracy and use this epistemological process as an inquiry to gain expertise of the subject.

The Ideal
- There are many forces which are attempting to water down the original concept of democracy.
- Democracy was always an "ideal" and as Kloppenberg (1995) admits the reason that democracy is a widely accepted idea that receives global appeal and acceptance is because it means different things to different people. May be because the concept is an ideal it has no enemies—that's a novel way of finding acceptance which we must consider as a way to reach consensus. **The Enemy**: Departure from ideal or myth of what democracy could be can be used as a rallying point around to reach consensus.

Authority Versus Freedom/Liberty
- Kloppenberg (1995) refers to the dual imperative of all societies i.e. the necessity to maintain order, equilibrium and civility and the people's aversion of rules, and authority which in the extreme could lead to anarchy. So what shall it be: freedom and liberty or order enforced by

authority? **The Enemy:** Arbitrary authority, political, military of otherwise or at least the fear of domination from the same.

Corruption, Dishonesty and Venality
- It is not clear whether nowadays, business and government is more prone to scandals than it has ever been. Scoundrels who are dishonest have always existed. In most countries around the globe, private business and public corporations as well as governments are rife with corruption of high officials and of schemes which show that many people disregard norms of honesty and use these venues for their own enrichment. **The Enemy:** Tottering between extremes of optimism and pessimism i.e. that either the human kind is basically honest, compassionate and benevolent or it is dishonest, individualistic and egoistic.

Sense of Responsibility and Accountability

It would appear that the original commitment to the ideal of democracy has come to mean "the unbridled pursuit of wealth and power." (Kloppenberg, 1995). Even politicians or in particular politicians who seek the highest positions are overt liars, never admit mistakes and always shift the blame to others. It is rare to find a politician who loses his job due to malfeasance or unethical practices. **The Enemy:** same as immediately preceding.

Natural Virtue Versus Legal Coercion

There was a time when the high calling of working for one's government was considered an honor and even a duty. No longer, serving the public is seen as suspect and connotes the desire to seek private enrichment instead of sacrifice. To be virtuous use to be a matter of private consent and adherence to natural law. Nowadays, it is the adherence to laws and the presence of legal authority which protects most citizens from rogue individuals who want to prey on each other. **The Enemy:** Belief in natural virtue is eroded and is replaced with legal coercion or obligation.

Egoism and Individualism

At least in the US, one of the most important "democratic virtues of democracy" is individualism 'a theory' which allows the individual to pursue whatever objective regardless of any natural impediments and quite often at the expense of competing interests to the point where rivalries end in feuds, trials, and even local wars which may end up in bloody conflicts and even death. Is Egoism stronger in the US than elsewhere or is it a sign of good times/bad times. When conditions are 'tough,' people may have a tendency of looking first after themselves rather than to help others or be generous and altruistic. Individualism is also a very strong American trait which drives people to seek objectives by themselves, to reject welfare or any form of help from the outside.

(Baier, 1991) **The Enemy**: Human being's individualism and greed.

Participation and Representation

Coupled with egoism and individualism the issue of participation and representation co-exist. If everyone wants to "fend for themselves" there is no tolerance for participating in public issues to help others and to spend time which could be 'better spent' pursuing private interests rather than public ones. So it is that US citizen participation in elections is low by Western standards. To serve one's country in public office is not always seen as a virtue, an honor or an obligation.

One important impediment to effective participation and the feeling that representation is not working due to the physical and 'virtual' distance of representatives from the electorate. **The Enemy**: Refusal to participate and trust responsible representatives.

The Ethics Issue

In all of the above we witness a "clash of interests" which reveals the myriad of enemies facing democracy beyond the four identified by Churchman in his epochal The Systems Analysis and Its Enemies (1979). Many of the above could be classified as pertaining either to Politics, Morality, Religion or Aesthetics. However, a more minute parsing of the concept of enemies/ adversaries quickly reveals that the world is populated by enemies/adversaries/antagonists at every turn and that life consists of a constant struggle to reconcile these opponents so that life can resume in peace and in some form of harmony. **The Enemy**: Constant need to remind citizens of the importance of ethical conduct to protect the integrity of the social contract.

17.6.2 Other factors which Affect Democracy and Ability to Govern

Optimum Size of a Country's Population

Countries call themselves "democracies" regardless of their ethical, social and political practices. As a study of the subject will attest 'democracy" is a vast concept whose outline is difficult to grasp regardless of whether the democracy is participative, liberal, deliberative of any other kind. (Goodman, 1993). The size of a country's population is decidedly a factor in this regard.With more than 280 million people, the US has lost its most precious asset which was an "efficient" governance system which was the example of the world. India and China have also due to their respective size become difficult cases in point.

The "American-style of management which was the envy of the world has evolve to where "muddling through" is the best method we can find to govern. Some reasons for this deterioration: Representative government has become too complex to be effective due to lack of popular interest, the force of the media and advertisement which distorts the news in favor of the incumbent party in power, extraordinary influence of the "lobbies" which can through outlandish donations sway representatives and distort public opinion and can easily lead to a visible form of corruption. **The Enemy** : Size of a country's population erodes direct representation and intermediaries between the citizenry and their representatives can lead to overt corruption.

Size of Government

Large countries are becoming difficult to govern. Size complicates the 'business' of organizing an efficient apparatus to ensure governance. Small countries such as Norway, the Netherlands, Ireland, Sweden can build a management system that its population can understand and can staff with responsible people. The example of business and efficient rules of management preclude oversize and overcomplexity. **The Enemy**: Once a certain size and complexity is reached, no Systems Approach can overcome the complexities that come with size and shear intricacies of a government's infrastructure.

Size of organizations.

The scientific discipline of Organization Theory draws up magnificent lists of rules, theories, suggestions on how to best run an organization. It seems that in time of terror or merely unusual circumstances as brought about by a hurricane, an earthquake, wild fires etc. the best plans go awry. Until a few years ago, nuclear accidents were the most "unusual" system failures that could conceivable occur. Nowadays it is the possibility of a terrorism attack against whose fury no 'Systems Approach' is applicable.

Specifically the boundaries of the kind of problems raised in this essay— namely Terrorism, Globalization and the waning of Democracy as we know it, are unknown. Are these problems to be tackled globally, nationally or locally? Among a jittery citizenry, on can imagine a person losing his/her mind on a crowded street due to a deranged mind. It may take the aspect of international proportions, when it may purely be local and inconsequential and hence controllable.

The size of the organization involved in these incidents seems to have a bearing on whether the rational mind can apprehend the gravity of the dangers or not. When an organization like Enron collapses dues to fraud and its own mistakes, we can understand the problem. However, then a terrorist attack such as September 11, 2001 takes place outside the boundaries of normal occurrences, it boggles the imagination, and nothing in our "kit-back of organization tools" makes us ready for such an eventuality. Given that we are not ready now to handle such emerging risks, it is doubtful that we ever will overtake complexity. **The Enemy**: Inertia of over-sized governmental bureaucracies and of large corporate conglomerates, together with their inherent ramifications and complexities are difficult to understand and to unravel to identify misdesigns and possible sources of system failures.

17.7 CONCLUSIONS

17.7.1 The Search of so-called "Enemies of the Systems Approach" as an Illustration of an Epistemological Inquiry.

Why does CWC consider Politics, Religion, Morality and Aesthetics the enemies of the Systems Approach? We attempted to address this question in this essay. Basically, CWC sought to explain that efforts to be rational and use rational thinking for planning and decision making is fraught with hindrances and barriers which apart from the inadequacies of our methodologies is due to the way the world is organized. Human beings congregate in tight communities as a result of their adherence to groups which may be political or religious and to which they owe their allegiance. Furthermore, while seeking to be rational or ethical they naturally impose their own brand of rationality and ethics, because that is what they believe in. Consequently they do not only impose their own political and religious beliefs but also in addition impose their own rationality and morality while interrelating with other human beings. While CWC limited himself to four enemies (Politics, Religion, Morality/Ethics and Aesthetics) it is easy to conjure other enemies.

In a war between two countries we can witness confrontations of culture, technology and many others. Each country tries to find a rational and balanced view of the issues at hand but "the noise of war" obscures objectives, motivations and final outcomes. The world is made up of factions, partisans and enemies. The way CWC describes *the enemies* is original, revealing and certainly not trivial. Having decided that conflicts are the order of the day and that we cannot solve any problem in this world without addressing the notion of enemies, we must suggest that the possible comparison of epistemologies in the direction herein will provide knowledge that may lead to reconciliation of the enemies/adversaries/foes. It is a worthwhile hope that contenders in any conflict should contemplate.

17.7.2 The Comparison of Epistemologies

The crux of the matter has not yet been reached. It consists of comparing the epistemologies between contenders in any conflict. As an example: We need to compare the epistemology of the concept labeled "enemy" with that of the concept labeled "friend." The comparison should lead to revealing the differences in knowledge which an enemy pursues compared with that pursued by a friend. Given that epistemology includes the thinking processes through which knowledge is built, a comparison of these processes will have to be pursued.

Another area of research will compare how friend and foe guarantee the truth of their pronouncements and what is the level of rigor employed by each

17.7.3 On Reaching Consensus: Introducing an Inquiring System at a Higher Level of Logic and Abstraction

Bertrand Russell provided the first hint on how to resolve the conflicting and paradoxical discourses take place at the same levels of logic. He postulated that the conflict would be dissolved if the two conflicting sides of an issue were to be considered at different levels of logic or abstraction, or if the discourses could be discussed from a vantage point of higher logic. CWC used the hierarchy of inquiring systems to show that discourses of divergent logic must of necessity be considered in an inquiring system with a level of logic different from those of the original warring sides. Hampden-Turner (1981) also used Russell's philosophy to confirm that logical paradoxes could be dissolved, absorbed and eliminated by resorting to the metalevel. Later, van Gigch (2003) continued CWC's and Hampden-Turner's account by applying the notion of inquiring systems to any logical conflicts. Another fruitful avenue of research consists of exploring methods by which enemies and foes can be reconciled. The beginnings of this research can also be found in McInyre-Mills (2005).

REFERENCES

Ackoff and Strümpfer, 2003, "Terrorism: A systemic view", *Systems Research and Behavioral Science*, **20**:287-294.

Baier, Kurt, 1991, "Egoism", in: *A Companion to ethics*, Peter Singer, ed., Blackwell, Cambridge, MA.

Bausch, Ken, 2005, "Be your enemy", in: *Rescuing the Enlightenment and Democracy*, Volume 1, McIntyre-Mills, ed., *C. West Churchman's Legacy and Related Works*, Kluwer/Springer, London and New York. See McIntyre.

Churchman, C. West, 1968, *The Systems Approach*, Delacorte Press, New York.

Churchman, C. West, 1979, *The Systems Approach and its Enemies*, Basic Books, New York.

Goodman, Amy, 1993, "Democracy", in: *A Companion to Contemporary Political Philosophy*, R.E. Goodin and Philip Pettit, eds., Blackwell, Cambridge, MA.

Kloppenberg. James T., 1995, "Democracy", "Enlightenment", Entries in: *Companion to American Thought*, R. Wightman Fox and J.T. Kloppenberg, eds., Blackwell, New York.

McIntyre, Janet, ed., 2005, "The Systems Approach and its Enemies," in: *Rescuing the Enlightenment and Democracy*, Volume 1, McIntyre-Mills , ed., *C. West Churchman's Legacy and Related Works*, Kluwer/Springer, London and New York.

National Commission on Terrorist Attacks Upon the United States, 2004, *9/11 Commission Report*, Final Report, Norton, Washington, D.C.

Occhiogrosso, Peter, 1994, The Joy of Sects: A Spirited Guide to The World's Religious Traditions, Image Books, Doubleday, New York.

The Concise Oxford English Dictionary of Current English, 1976, Oxford, Clarendon Press.

Senate Select Committee on Intelligence, 2004, Report on the US Intelligence's Community Prewar Intelligence Assessments on Iraq, Washington, D.C.

van Gigch, J.P. and Janet McIntyre, eds., 2005, *Skepticism And Socratic Wisdom In Modern Holistic Thinking*, Volume # 2, *C. West Churchman's Legacy and Related Works*, Kluwer/Springer, London and New York. (in Process)

West, Cornell, 2004, Democracy Matters: Winning the Fight Against Imperialism, Penguin Press, New York.

Yu, Jae Eon, 2005, "Making Friends of Enemies", in: *Wisdom, Knowledge and Management, Vol2*

EPILOGUE

SYSTEMIC GOVERNANCE AND ACCOUNTABILITY: WORKING AND RE-WORKING THE CONCEPTUAL AND SPATIAL BOUNDARIES THROUGH EMPATHY, LISTENING AND QUESTIONING

MCINTYRE-MILLS, J., Contributor and Collaborator
VAN GIGCH, J., Volume Editor

"The perennial philosopher says that a person who has awakened to the eternal can identify with his or her mortal form, for the eternal simultaneously transcends the world and is present within it. No longer seeking to defend that form at all costs, an awakened person can relate to other people and nonhuman forms of life in compassionate , creative ways...perennial philosophers certainly agree that repressive political-economic structures can drastically impede consciousness evolution, just as those structures can destroy many of the species that have evolved in the past several millions years. Put more positively... perennial philosophers assert that certain noncoercive and materially sufficient political-economic arrangements are necessary if more and more contemporary people are to become aware of further possibilities in the development of consciousness....those arrangements in and of themselves do not "generate" higher consciousness, but instead simply unimpeded the evolutionary tendencies of consciousness itself. Current political-economic arrangements tend to reflect and to reinforce the state of consciousness already achieved, which is why history has been characterized by such violence. Yet within those same arrangements, many people have become aware that there are alternatives...

(Zimmerman, M. 1994: 376-377).

What remains to be said? Our readers will have to admit that the authors who have contributed to this volume spanned a wide range of subjects. The theme that underpins the varied contributions is governance and its potential to enable us to achieve harmonious decisions in a range of arenas (from organisations at the local level to large scale federal decisions within and across national boundaries).

Empowering ordinary people, irrespective of cultural knowledge (language, religion) or racial identity, age, gender, level of education, type of employment, ability and citizenship is an area for systemic governance that is focused on social and environmental justice that cares and advocates on behalf of sentient beings, because we acknowledge the value of diverse species and diverse knowledges. The extent to which we think about others and the environment is a measure of our accountability, our compassion and our ability to manage risk in a sustainable environment. It is a right and a responsibility of good governance.

When we refer to Churchman's Legacy we refer to the inspiration that this scholar (C. West Churchman, 1913-2004) provided in his efforts to look ahead and anticipate how policy makers and managers have to think and practice in order to create connections across areas of concern, in order to enhance our consciousness as care takers and to bridge the gaps between:
- the powerful and the powerless (including all sentient beings)
- the environment and development culture

Governance is bigger than *Management* and implies not only taking into account the traditional administration principles , but 'sweeping in' democratic ideals, equality and fairness regardless of status, identity and economic standing.

Governance addresses ethical concerns which involves much more than comparing behaviour with norms or standards of conduct. It requires a study of the meaning of these norms and determines what is meant by Good/ Bad, Right/ Wrong, Us/Them and other ethical dilemmas dotting the managers' landscape. The ethical dimension reaches also for the aesthetic dimension which seeks to entice the manager to seek solutions which are not only ethically acceptable but which are also beautiful and which reach for the ultimate ideal of quality of life, of the systemic connections across self-other and the environment . The next volume will address some of the ethical issues of accountability and systemic governance.

REFERENCES

Zimmerman, M. 1994. *Contesting Earth's future: radical ecology and postmodernity.Berkley.* University of California Press, pages 376-377)

McIntyre, J. 2005 Working and re-working the conceptual and geographical boundaries *Systemic Practice and Action Research* Vol. 18 , No 2 157- 220.

INDEX

A posteriori concepts, 309
A priori concepts, 113, 135, 309–310, 315
AA. *See* Alcoholics Anonymous
Abduction, 108
Ability to pay model, 152
Aboriginal Australians, 234. *See also* Indigenous people
  children, incarceration of, 240
Academic criticism, countering of, 31–32
Accountability, x, 341
Accounting conventions, 23, 153–154
Accuracy, plausibility v., 82, 85, 95, 100
Ackoff, R., 43, 147, 156, 333–335
Action
  learning, 252, 263
  management as, 170–173
  Plan Pattern, 288
  preparation for, 179–180
  representation of, 56–57
  research, 186, 202, 252
  right, 172, 179
  vision and, 165
Active resources, 188
Activities, 187
Activity Modeling, 192–193
Adaptation
  ideal, 220
  operational, 220–221
Adler, A., 119
Adversaries, enemies v., 332
Adverse Event Management Center, 197
Adverse event reporting (AER), 196–199
  articulation with proposed constructs of, 197–199
  re-formulation of, 199–201
  resolution of problems in, 201
Aeon, 316
Aesthetics, 24, 35, 307, 317, 342, 344. *See also* "Beautiful"
Aether theory, 115
Against Method: Outline of an Anachronistic Theory of Knowledge (Feyerabend), 130
Aggressive militarism, of U.S., 340
Agoras, 280

Agrell, P.S., xxv
Agriculture, 146–147
AI. *See* Artificial intelligence; Ideal adaptation
Aims. *See* Goals; Values
Alcoholics Anonymous (AA), 151, 155–156
Aleatory point, 317, 324
Alice Springs, Australia, 235
American Association for Higher Education, 273
Anaxagoras theorem, 23
Animals
  communication by, 58
  representation by, 53–54
  rights of, 242–245, 260
Animals Australia, 244
Antagonists, 332
Anthropic cosmological principle, 173
Anti-Oedipus (Deleuze & Guattari), 324
ANZUS alliance, 238
AO. *See* Operational adaptation
Applications, domain of, 282
Applied researchers
  indebtedness of, 32
  professional pride of, 31
Approaches, 38, 42
  multiplicity of, 30
  tools v., 36
Archanesian geometry, 282, 286, 288, 299
ARCO, 151
Aristotle, xvi–xviii, xviii, 63, 170, 177
Arjuna, 172
Artificial intelligence (AI), 179
Artigiani, R., 77
Ashby's Law of Requisite Variety, 171, 257
Asylum seekers, 237–239, 243
Asymmetry, 68–69
Athens, 280
Atkin, R.S., 210–211
Audi, R., xiii
Aurobindo, Sri, 177
Australia, 232–245, 253–255
  Aborigines of, 234
  Border Protection Act of, 237, 241
  borders of, 239, 241

democratic culture of, 262
economic disparity in, 253
Elizabeth, 257
Green Valley, 257
Migration Act of, 237
Nationality and Citizenship Act of, 234
New Border regime of, 236
Pacific Solution of, 236–237, 241
Australian Research Council, 230
Authentic conversation, 277
Authority
  liberty v., 340
  restoration of, xi
Autopoesis, 49–50, 70–71, 174, 314
Axiological balance, 220, 223
Axiological profiles, 216–218
Ayer, A.J., 112

Bacon, F., 107, 109
"Bad" research styles, 31
Baguley, P., 210
Banathy, B.H., 280
Baruma, I., 254
Bateson, G., 160, 233
Bausch, K.C., xxv
Bayes, T., 110
Bayes Theorem, 110
Bayesian method, 110
BE. *See* Business efficiency
"Beautiful," 21, 24
Beck, U., 243
Beer, S., xxv, 157–158, 165
Begum, S., 248
Belief
  ethics and, 162–164
  as nonsense, 172
Bell, D., 312
Bell, J., 109
Bell Telephone Labs, 144
Bentham, J., 321
Berlin School, 112
Berry, W., 277
Bertalanffy, K.L., 208
Bhagavad-Gita, 157–158, 171–172, 174, 179–180
Bickerton, D., 50–51, 54, 59, 64
Big Bang Theory, 173
Big Crunch, 173–174
Bits, information as, 66–67
Bohr, N., 77, 111, 124
Bonds, 231
Bone marrow transplant study, 87–88, 91, 95–96

de Bono, E., 38, 129
Boomerang effect, 233, 258, 261–262, 265
Borderless information economy, 231
Borders
  of Australia, 239, 241
  protection of, 239, 241
  subject-object, 314
  thought-life, 316
  of US, 334
Boundaries
  case studies on, 232–241
  conceptual, 231, 257, 263
  geographic, 231, 258, 263
  governance and, xvii
Boundary judgment, 309
Bounded thinking, 245–248
Broadbent, J., 273
Buckley, Y.W., 210
Budgeting, 9, 23. *See also* Accounting conventions
Bunge, M., 213
Bureaucracies, 336
Burkean spiral, 120
Business efficiency (BE), 207, 222
Business, government and, 152

Cameron, K.S., 210
Campus Compact, 274–275
Capacity building, 258
  in strong states, 264–266
  for sustainable future, 263
  in weak states, 264–266
Capra, F., 275
Care, 157
Caretaking, 248–251
Carnap, R., 112
Cartesian thinking, 255, 260
Cassirer, E., 114
Catalyst resources, 188
Categories, 52, 55, 185
Center for Ecoliteracy, 275
A Challenge to Reason (Churchman), ix, 8, 24
Chalmers, A., 133–134
Chaos theory, 255
Chaotic dynamics, 313
Children overboard incident, 236, 240, 249
Children's rights, 240
China, 337–338, 342
Choice, 41
Chomsky, N., 243
Christakis, A.N., xxv
Churchman, C. West, vii–viii

Index

alcoholism of, 155
bibliographic data on, xxv
categories of, 185
ethics, views on, 162–168, 170
grand intent of, 6
honoring of, 25
interview with, 142–160
legacy of, 348
life of, 142
management science/science of
  management, contributions to,
  18–24, 105–106
skepticism of, x, 105–106
CIFM. *See* Continuous Improvement
  Focused Monitoring
Circularity
  of epistemology, 69
  of knowledge, 68, 70
Citizenship, 227, 229, 233–234, 237
  human rights v., 245
  transboundary implications for, 242–245
Client, 185
Closure, 173–174, 233
CogniScope™, 285–286, 291
CogniSystem software, 285–286
Cognition
  knowledge of, 113
  limitations of, 281
  theory of, 67
Cognitive equation, 50, 65
Cognitive Limitations Axiom, 283
Cohen, H., 114
Coherence theory, 114
Collaborative facility, for SDP, 285, 287*f*
Collective subjectivity, 324–326
The Coming Post-Industrial Society (Bell),
  312
Commission on the Humanities, 271
Communication, 170, 231. *See also*
  Language
  animal, 58
  endocrinal, 52
  information and, 62
  open, 249, 251
  personalistic v. nonpersonalistic, 96–97
  two-way, 250
Communism, 152
Community, of stakeholders, 285–286, 299
Community-based learning, 274–275
Companies, 221–222
  BE of, 222
  Company-environment profitability of,
    222
  organization v., 205, 207
  projected actual profitability of, 222
Company-environment profitability, 222
Comparison, 41
Compassion, 233, 238, 248–251, 258–262,
  348
Completion from Without, principle of, 174
Complexity, xxi–xxii, 50, 76, 125
  case studies on, 248–251
  definition of, 283
  management, xviii
Complexity Axiom, 282
Components, 185
Component-systems, 50
Composite enterprise constructs, 189–190
Comprehensiveness, lack of, 184, 283
Computers, 178–179
Comte, A., 111
Conceptual boundaries, 231, 257, 263
Confidence, restoration of, xi
Conjectures and Refutations (Popper), 41,
  130
Conjunction, 317
Connection, 317
Conscious evolution
  definition of, 283
  structured dialogue and, 298
Consciousness, xv–xvi, 259, 260, 262. *See
  also* Mindfulness
Consensus methods, 286–286, 345
Consequentialism, 11
Conservation of nature, 214
Consumers, 152, 187, 197. *See also*
  Customer
Contingency Theory, 212
Continuous Improvement Focused
  Monitoring (CIFM), 292–298
Conventionalism, 115
Core-policy processes, 189, 197–198
Core-support processes, 189, 198, 200
Corporations, public perception of, x
Correspondence, of truth, 116, 120, 132, 309
Corruption, 340–341. *See also* Fraud
Cortina, A., 214
Craft, xvii
Creation myths, 173–174
Creed of Greed, 157, 171–172
Crisis, in science, 127, 129
Criteria
  time-based, 209
  of validity, 75–76
Critical heuristics of social planning,
  307–308

Critical inquiry, 273–274
Critical rationalism, 106, 117–125
Critical Systems Heuristics (CSH), 193, 305, 308–311, 326
Critical Systems Heuristics (Ulrich), 193
Critical Systems Thinking (CST), 191, 193, 202
   pragmatization of, 310–311
   social system design and, 307–311
Critical theory, 305, 326. *See also* Poststructuralism
Critical thinking, 251–254
Crooper, S., 43
Cropper, S., 39
Cross-sector cooperation, 251
CSH. *See* Critical Systems Heuristics
CST. *See* Critical Systems Thinking
Cues, 82, 85, 91
Culture
   in changing world, 228–232
   definition of, 39, 283
   ethics and, 164–166
   in interaction, 231
   modification of, 336
   as molar, 229
   as molecular, 229
Customer, 154. *See also* Consumers
CWC. *See* Churchman, C. West
Cybernetics, 174, 176, 256, 294

Darwinism, 117, 121
Data, 112–113
Dean, B., xxv
Deception-perception approach, 270
Decision makers, 185
Decision-Information-Operations System Model, 186
Deduction, 108, 120–121, 125, 159
Deductive-nomologicalism. *See* Critical rationalism
Definitions, 74, 160, 229, 283–284, 332
Deleuze, G., 305–306, 312–319, 322–326
DELPHI, 286
Delphi Method, 28
Deming, W.E., 144, 154
Democracy, 147–148, 230–231, 246–250, 257, 263
   in corporate world, 269—271
   debasement of, 252, 343
   education and, 270
   enemies of, 340–342
   ethics and, 342
   factors in, 342–343

government, size of, 343
   ideal of, 340
   imperialism and, 339–340
   imposing of, 336
   organizations, size of, 343
   participative, 280, 299
   population and, 342–343
Democratic majority, 241
Denmark, 248
Denotation, 317–319
Derrida, J., 231, 250, 257
Descartes, R., 76
Design, 41
The Design of Inquiring Systems (Churchman), ix, xvi–xvii, 15, 28–31, 162, 185
Desire, 158, 324–326
Despair-hope duality, 10–11
Deutsch, K.W., 213
Development, 206, 335, 348
Dewey, J., 125, 277
Dialectic, xvii, xxi, 258–259, 273, 306
Dialogue, 273–274. *See also* Structured Design Process
   definition of, 283
   game, 289–292
   structured, 298
Dialogue for the Information Age Democracy (Christakis), 281
Differences, 322
   reproduction and, 51
   sorting of, 51–52
Digital economy, 312
Direct experience, 179
Direct opthamaloscope, 91–92
Discipline and Punish (Foucault), 321
Disciplines, differentiation of, 146
Discount factor, 23
Discourse events, 44
The Discourse (Foucault), 35
Disjunction, 317
Dissipative structures, 50, 77
Distributed Singerian Churchmanian Inquiring Systems (DSCIS), 82–84. *See also* SCIS
   further research on, 100–101
   implications for practice of, 101
   process of, 92–95, 99–100
   social context of, 97, 100
   technology and, 91–92, 101
   telemedicine as, 90
   wicked decision problems and, 89, 98, 102
Dithurbide, G., 206

Diversity, 227–228, 254, 256
  managed, 263
Domain of Science Model (DSM), 282
Dror, Y., 34, 36–37
Druker, P., 153
DSCIS. *See* Distributed Singerian Churchmanian Inquiring Systems
DSM. *See* Domain of Science Model
Du mode de'existence des outils de gestion (Moisdon), 40
Dunderdale, P., 210
Duty, Theory of, 11

East Timor, 239–240
Ecology, 207, 209
  efficacy regarding, 219
  of mind, 233
Economic development, 229
Economic disparity, 253
Economic power, 321
Economists, xi
Eddington, A.S., 118
Edison, T., 152
Education
  civic enagement, 277
  democracy and, 270
  dialogue, 277
  interdisciplinary, 271–273, 277
  systemic approach to, 270
Effectiveness, 206–207, 209, 210
  degree of, 219–220
  of priorities, 292
Efficacy, 206–207, 210
  ecological, 219
  level of, 219
  profile, 219
Efficiency, 206–216, 210–211, 342
Egoism, 11, 341
EI. *See* Internal efficiency
Eightfold path, 179
Einstein, A., 111–112, 114, 121, 150, 174, 281
Elemental observations, 288
Elementary enterprise constructs, 187–188
Eliot, T.S., 168
Emerson, R.W., 277
Empathy, xv
Empedocles, 177
Empiricism, xvii, xxi, 110–112, 117, 125, 245, 327
  transcendental, 313
  verificationism and, 108
Enaction, 49–50, 67

Enactment, 85, 95
Endocrine communication, 52
Enemies, 333–339
  adversaries v., 332
  of democracy, 340–342
  friends v., 344–345
  of Systems Approach, 9, 185, 259, 306–307, 312, 331–332, 344
Enemies of the Systems Approach (Churchman), 24
Energy, consumption of, 207
Engaged Department Institute, 274
Engineers, xi
Enlightenment, 107, 246–247, 255, 312
Enron, 9
Enterprise theory, 186
Entropy, 233
Environment, 185
Environmental concerns, 23, 165, 251, 255, 348
Environmental fallacy, 23–24
Environmental justice, 262, 348
Eon Yu, Jae, xxvii
Episteme, xvii
Epistemic humility
  drift towards, 133–135
  introduction to, 105–106
  systemic improvement and, 135
Epistemic modesty, 105
Epistemological inquiry, 331–332, 344
Epistemological totality, 208, 210
Epistemology, 228
  circular, 69
  comparison of, 344–345
  inquiring system of, 4–5, 16
  moral, 10
  natural, 67, 69, 71–72
  program of, 1–2
  projective constructivist, 186
  rationality of, xii
  self-referential, 67–68, 71, 73
  truth and, 12
Eriksson, D.M., xxv–xxvi, 186
Erkenntnis, 112
Ernst Mach Society, 111
ES. *See* Systemic Efficiency
Escalating authoritarianism, of U.S., 340
Essay Towards Solving a Problem in the Doctrine of Chances (Bayes), 110
Ethical behavior, motivation for, 10–12
Ethical imperatives, hierarchy of, 22
Ethicists, xi
Ethics, 145, 152, 154

belief and, 162–164
Churchman's views on, 162–168
culture and, 164–166
democracy and, 342
emphasis on, 156
hierarchy of, 157
legality v., 157, 166
literacy of, xviii–xix
multilevel, 21
new mode of, 321–322
planetary, 164–165
postmodern, 323–326
science and, 145, 153, 155
Science of, 2, 6–9, 12
of war, 159
as well-informed judgment, 166
Euclid, 281
Eudaimonia, xvi, 170
European Union, 264
bounded thinking in, 245–248
federalism of, 242
Events, 314–318
Evidence
falsification of, 118–119
generators of, 92–93
guarantor of, 95
refinement of, 93–94
verification of, 118
Evolution
of SDP, 284–288
social, 280
theory of, 76
Excellence, 206
Expanded pragmatism, 265
Experimentation, 327
Expertise
levels of, 91
technology v., 91–92
Explanation, 284
External dimension, 220–221
External domain, 208

Fact-network, 307
Facts
data and, 112
faith v., 333
value v., 5–6
Factual science, ethical science v., 153
Faith, fact v., 333
Falsification, 118–119, 121–125, 131
abandonment of, 125
accumulation of, 127
of theories, 124

Falsificationism. *See* Critical rationalism
Federalism
subsidiarity and, 242
undermining of, 245–248
Fernández-Ríos, M., 211
Fertile Crescent Network, 275–276
Feyerabend, P., 130–133
Feynman, R., 19
Fiegl, H., 112
Field of statements, 322
First Phase Science, 294
Five "Cs" of SDP, 285–287
FOA. *See* Swedish Defense Research Establishment
van Foerster, H., 174
Form, measure and, 177–179
Formal implications, material implications v., 63–64
Foucault, M., xxi, 35–36, 36, 39, 305–306, 312, 315, 320–323
Foundations, domain of, 282
Fourth Epoch, 312
Fouts, D., 58
Fouts, R., 58
France, 246–247
François, C.O., xxvi
Fraser, M., 240
Fraud, 9, 172
Free choice, 308
Freud, S., 117, 119, 121, 324
Friedman, M., 115
Friends, John, 41–42
Friere, P., 273, 277
Fukuyama, F., xvi–xvii, 264–265
Fundamentalism
free market, 340
religious, 336
scientific, 336
Future, definition of, 283
Future generations, 23
Futures Studies, 28
Fuzzy sets, 74

Gaia theory, 165, 262
Gauss, C.F., 178, 180
GEM. *See* Global Ethical Management
General relativity, 111, 114, 121
Genetic engineering, 153, 165
Geographic boundaries, 231, 258, 263
Germany, 247, 264–265
Gibson, J., 66
van Gigch, John P., x, xxvii, 34, 39, 43–44, 165, 345

Index

Global Ethical Management (GEM), 270
Global food system, 276
Global Food Web, 275
Global good governance, 206–207
Global Utility Function, 211
Global warming, 339
Globalization, 299, 312, 337–339, 343
Goals, 188, 199
Goedelian sentences, 174
Goertzel, B., 65, 69–70, 76
"Good," 21, 24
Goodman, P.S., 210–211
Gorbachev, M., 332
Governance
  boundaries and, xix
  critical perspective on, 251–254
  good, 348
  local, 252, 254
  management v., 348
  regional, 252
  systemic approach to, xix, 251–254
  systemic, inquiring systems for, 258–262
  systemic, process toward, 255–258
  transnational, 228
Government
  business and, 152
  science and, 154
  size of, 343
Graduate School of Business of Chicago, 156
Graphic language patterns, in SDP, 288
Greed, 9, 149, 152, 155, 157, 171–172, 338
Greenfield, S., xx
Gregory, A., 34
Groupthink, avoidance of, 291
Growth, 206
Guarantor, 24, 37–40, 43–44, 95, 185
Guattari, F., 313, 324–326
Guided Evolution of Society (Banathy), 280

Habermas, J., xx, 71, 75, 231, 250, 257, 307–308, 311, 321
Hahn, H., 112
Hammond, D., xxvi
Handy, C.B., 210
Hard formalization, 208
Hard system variables, 5, 7
Hard Systems Thinking (HST), 191–192, 202
Hardin, G., 165
Hardy, G.H., 178
Hartsock, N., xxi
Health sciences centers (HSC), 87, 89–97
Hegel, G.W.F., 18, 114, 306

Heisenberg, W., 77, 111
Hempel, C., 112
Hermeneutic circle, 69
Herschel, W., 123–124
Hickling, A., 42
Hierarchy
  of ethical imperatives, 22
  of inquiring systems, ix, 3–4, 15–16
Hijab, controversy over, 246–248
Hildebrandt, S., 38
Hindess, B., 263
Hitler, A., 145, 153, 156
Holism, 71, 176, 184–185. *See also* Systems thinking
  epistemological, xii
Homer, 60
Hope, 160
Howard, J., 240
HSC. *See* Health sciences centers
HST. *See* Hard Systems Thinking
Human rights, 160, 230, 247, 252, 265
  citizenship rights v., 245
  enforcement of, 338
  Europe and, 265
Human-actor resources, 189, 197–200
Humanities, 271
Hume, D., 109, 111
Humility, 150–151. *See also* Epistemic humility
Husserl, E., 39
Hutchins, R.M., 272, 277
Hutchins School of Liberal Studies, Sonoma State University, 272–274, 276
Hypergame approaches, 192
Hypothetico-deductivism. *See* Critical rationalism

I am Right, You are Wrong (de Bono), 129
ICC. *See* International Criminal Court
Ideal adaptation (AI), 220
Idealism, xvii–xix, xxi, 24, 152, 258
  democracy and, 340
  measurement and, 148
  pragmatism and, 263
Ideas, testing of, xxi–xxii
Ideawriting, 286
Identities
  boundaries and, 232–241
  in changing world, 228–232
  as molar, 230, 240
  as molecular, 230
  sensemaking and, 85, 90
  transboundary, 260

Images of Organization (Morgan), 194
Images of thought, 312–313, 322, 326
Immanent plane, 313
Imperialism, 339–340
Implementation, 16, 143, 147–148, 158, 185
Impossibility Theorem, 211
In-channels, 187, 197
Incongruent perspective, paradox of, 49
India, 342
Indigenous people, 235. *See also* Aboriginal Australians
Indirect opthamaloscope, 92
Individualism, 341
Individuals, characteristics of, 99
Induction, 106–107, 110, 120–121
    deduction v., 159
    proof by, 108–109
Infectious disease study, 87–88
Influence Tree Pattern, 288
Information
    as bits, 66–67
    communication and, 62–63
    dynamics, 77
    knowledge v., 230–231
    meaning and, 61–62
    referential v. non-referential, 64–65
Information Age, 281, 299
In-house domain. *See* Internal domain
Inputs, 187, 197, 216
Inquiring systems. *See also* Distributed Singerian Churchmanian Inquiring Systems; Singerian inquiring system; Singerian-Churchmanian inquiry systems
    definition of, 4
    epistemology, 4–5, 16
    hierarchy of, xi, 3–4, 15–16, 345
    implementation of, 16
    multidiversity of, xi–xii
    real-world, 4
    science, 4, 16
    for systemic governance, 258–262
Institute of Management Science (TIMS), 155, 159–160
Instrumentalities, 22. *See also* Methodologies
Interactive Planning (IP), 192
Interdisciplinary curriculum, 271–273, 277
Internal dimension, 219–220
Internal domain, 208
Internal efficiency (EI), 220–221
International Criminal Court (ICC), 232
International Society for Ecology and Culture, 275

Interpretive Structural Modeling, 286
Interrelations, 168
Intersubjectivity, 250, 312
Introduction to Operations Research (Churchman), 16
Intuition, 76
Intuitive verificationism. *See* Verificationism
IP. *See* Interactive Planning
Iraq, 247, 339
"Is," 9–10
Issue, definition of, 284

Jackson, M., 12, 34, 40, 312
James, W., 125
Jamrozik, A., 238
Jaynes, J., 60
Joy, 324
The Joy of Sects (Occhiogrosso), 336

Kalkaringi Statement, 235–236
Kampis, G., 63
Kant, I., 10–11, 18, 39–40, 113–114, 157–158, 185, 245, 306, 308, 315
Kashyap, A., 179
Katzenstein, P., 264
Killing, 166–167
Kindness, 157
Kloppenberg, J.T., 340
Knowing
    doing and, 48–49, 67, 76
    as process, 250
Knowledge
    autopoesis of, 70–71
    in changing world, 228–232
    circularity of, 68, 70
    of cognition, 113
    craft, xv
    diversity of, 348
    information v., 230–231
    management of, xx
    networked societies and, 253
    objective, 312
    original, 48–49
    power and, xix
    public, 108
    science and, 178
    scientific, xvii
    sources of, 18–19
    validation of, 73–74
Knowledge workers, 158, 252, 263
Knowledge-based economy, 312
Koenisgsberg, Ernest, xxvi
Korzybski, A., 308

Index

Krishna, 172, 174
Kuhn, T.S., 17, 123–126, 128–132

Labor standards, 338
Lakatos, I., 122–123, 132
Language. *See also* Communication
  context of, 74
  definition of, 51
  evolution of, 51–52, 59
  human, 58
  natural, 196
  object, 74
Lasswell, H., 285
Laswell, H., 213
Lateral thinking, 38
Lawyer, xi
Learning
  community-based, 274–275
  interdisciplinary approaches to, 271–273
Learning communities, 272–273
Legal coercion, natural virtue v., 341
Legal rationality, 11
Legality, ethics v., 157, 166
Leibniz, G.W., 18
Lenses, sets of, 255
Levels of thinking, 160, 233–234
Liberatory education, 273
Liberty, authority v., 340
Linear Programming, 191
Linguistics, 61. *See also* Language
Link, P., xix
Linstone, H.A., 31, 34, 39, 43
Literacy, ethical, xviii–xix
Lobbying industry, 342
Locke, J., 18
Logic, xi, xvii, xxi, 258, 345
Logic of Sense (Deleuze), 314
Logical positivism. *See* Positivism
The Logical Structure of the World (Carnap), 112
López, M., 221
Lovelock, J., 165
Luhmann, N., 69–70, 72–73, 75
Lull, R., 19
Lyotard, J., 134

Mach, E., 77, 111–112
Machine-actor resources, 189, 197
Maha Shakti, 176–177
Mahalaksmi, 176
Managed diversity, 263
Management
  as action, 170–173

  authority, restoration of, xi
  complexity, xviii
  governance v., 348
  of knowledge, xx
  perception of, viii
  reflexive approach to, xiii
  risk, xviii
  science of, 149
  science v., 154
  skepticism about, x
  social ills and, 144
  Taylorist Scientific, xi
  Vedantic, 179–180
Management Science, 6–7, 12, 19, 142–143, 155. *See also* Science of Management
  academic criticism of, 31–32
  Churchman's contributions to, 18–24, 105–106
  instrumentalities of, 22
  justifications for, 17–18
  object of study of, 19–20
  questions for, 18
  Science of Management v., 8, 16
  unresolved problems of, 22–24
Management Science, 149
Managers, x–xi
Mandatory sentencing, 240–241
Manifestation, 317–319
Manus, 236, 237
Map, territory v., 308–309
Marburg School, 114
March, J.G., 153
Margalit, A., 254
Maruyama, M., 163
Marx, K., 321
Marxism, 117, 121, 152
Maslow, A., 213
Mason, R.O., 81, 82, 86, 100, 102
Masters of Business Administration (MBA) programs, 143, 153
Material implications, formal implications v., 63–64
Mathematics, 159–160
  information theory of, 66
  model-building with, 149
  representation by, 113–114
  Vedantic, 178–179
Matthews, D., xxvi
Maturana, H., 48–49, 174
MBA. *See* Masters of Business Administration programs
McGregor, J., 272
McIntyre, J., xxvi, 345

Meaning
  information and, 61–62
  tree of, 291
Measurement, 6
  form and, 177–179
  as ideal, 148
  of nature, 77
  of performance, 185
  of science, 77
Medical policies, 199–201
Mega level, 37, 38
Mega policy, 34
Mega systems, 36–38, 43
Memory, 323
Merleau-Ponty, M., 163
Meta policy, 34
Meta system, 38, 40, 42–43
Metadecisions (van Gigch), xii
Metaethics, 10
Metagame approaches, 192
Metaphor, 59–61
Metaphysics, 112, 115–116, 118–119, 316
Metasystem approach, 3–9, 15–16
Methodologies. *See also* Instrumentalities
  domain of, 282
  pluralism of, 131
Mexico, 337–338
Michaelson-Morley experiment, 115
Middle East, 244
Midgley, G., 40, 134, 261, 311
Military, 9
Miller, R., 270, 276–277, 281
Mimetic reproduction, 77
Mind, ecology of, 233
Mind-body, 76
Mindfulness, xvi, xx–xxi, 251, 258–260, 265–266. *See also* Consciousness
Mindscapes, 163–164
Mitroff, I., 28–29, 31, 34, 39, 43, 81–82, 86, 100, 102
Mixed domains, 208
Mobile knowledge workers. *See* Knowledge workers
Modeling Languages, 186
Models, science and, 214
Modus Tollens, law of, 121, 124–125
le Moigne, J., 35, 184, 186, 202
Moisdon, J., 40
Molarity, 229, 240, 250
Molecularity, 229, 250
Montessori, M., 277
Moral Epistemology, 10
Morality, 307, 342, 344

Morgan, G., 17, 193
Morin, E., 35
Motivation, 10–12, 333
Mullins, J., 210
Multidiversity, of inquiring systems, xi–xii
Multilevel ethics, 21
Muslims
  in Denmark, 248
  in France, 246–247
  in the Netherlands, 248

NAFTA. *See* North American Free Trade Agreement
Narrative information, 95, 97
National Economics, 229–230, 252
Nationality, transboundary implications for, 242–245
Natorp, P, 114
Natural epistemology, 67, 69, 71–72
Natural language, presentation in, 196
Natural law, 11
Natural resources, 165
Natural virtue, legal coercion v., 341
Nature, measurement of, 77
Nauru, 236–237, 237, 241
Negations, 56
Netherlands, 248
Networks, 251, 253–255, 263
Neurath, O., 112
Newton, I., 107, 111, 281
Newtonian physics, 109–111, 114, 122–124, 294
NGOs. *See* Non-governmental organizations
Nicomachean Ethics (Aristotle), xvii
Nietzsche, F., 66, 69, 114, 322
Nike Shoe Company, 338
Nobel Prize, 150–151
Nomadic subject, 324–325
Nominal Group Technique, 286
Non-governmental organizations (NGOs), xvii
Nonlinear dynamics, 50
Non-profit organizations, 151. *See also* Non-governmental organizations
Norberg Hodge, H., 275
Normal science, 131–132
Norms, 231
North American Free Trade Agreement (NAFTA), 337
Nuclear weapons, 165
Nussbaum, M.C., xvi

Object language, 74

Index 361

Objectivity, 293, 307, 312
    of observation, 125
    of science, 113
    of truth, 173
Observations, 37–38, 42, 112
    definition of, 283
    objectivity of, 125
    theory-ladenness of, 112, 114–115, 123–125, 128, 135
Observer, 49
Observer-independent root cause analysis
    definition of, 293
    inclusion in, 293–294
Occhiogrosso, P., 336
Occidentalism, 254
Oedipal complex, 324
Ontology, 132–133, 228
Openness, 249–251
Operational adaptation (AO), 220–221
Operations Research, 7, 12, 19, 142–143, 191
Operations Research Society of America (ORSA), 159
Opthamaloscope
    direct, 91–92
    indirect, 92
Opticks (Newton), 107
Options Field, 286–287
Options Profile, 286–287
Organization theory, 146, 153, 186, 190, 193–194, 343
Organizational units, 197–198, 200
Organization-culture resources, 189
Organizations, 219–221
    AI of, 220
    AO of, 220–221
    as brains, 194
    company v., 205, 207
    culture of, 200
    as cultures, 194
    EI of, 220–221
    as machines, 194
    as organisms, 194
    as political systems, 194
    size of, 343
    structure of, 197–198
Organization-structure resources, 189
Orientalism, 254
Original knowledge, 48–49
ORSA. *See* Operations Research Society of America
"Ought," 9–10
Oughtness, 165–166

Out-channels, 187, 197
Outputs, 187, 197, 216, 218, 221
Outside, 313, 323, 327
Outsourcing, 339
Ozdowski, S., 240

Pacific Solution, 236–237, 241
Pacifism, 145–146
Pamanujan, S., 178
Panopticism, 321
Papua New Guinea, 237
Paradigms, 130, 179, 334
    comparison of, 132
    dominant, 133
    of management science/science of management, 17–18
    science as, 126–130
    SDP, 281–282
Parra-Luna, F., xxvi–xxvii, 216
Participation, representation and, 341–342
Participative democracy, 280, 299
Participatory Action Research, 273
Partridge, E., 23
Patriotism, 264
Paul, D.L., xxvi
Pediatric oncology study, 87–88, 91, 96–97
Peirce, C., 74
Perception, sensory, 113
Perennial philosopher, 347
Performance, measure of, 185
Perspectives, taxonomy of, 35–40
Phenomena, definition of, 37
Phenomenalism, 113
Philosophical models, of social systems, 213
Philosophy, discipline of, 144–145
Philosophy of science, 106, 108, 131–132, 134
Philosophy of Science, 106, 169
Phronesis, xvii–xviii
Physical science, social science v., 74
Piaget, J., 186
Planetary ethics, 164–165
Planner, 185
Planning science, 105–106. *See also* Management Science
Plants
    proto-representation of, 52–53
    sensitive, 53
Plato, 158, 160
Plausibility, accuracy v., 82, 85, 95, 100
Pluralism, 309
    methodological, 131
    theoretical, 131–132

POD. *See Prediction and Optimal Decisions*
Polanyi, M., 128
Policy, *mega v. meta,* 34
Political models, of social systems, 213
Politicians, xi
Politics, 307, 336, 342, 344
  boundaries and, 232–241
  molar v. molecular, 250
Popper, K., 41, 117–121, 124–125, 128, 130
Population, size of, 342
Positivism, 106, 111–118, 125, 131
Positivistic modeling, 34
Post-Industrial Society, 312
Postmodern Age, 312
Postmodernism
  ethics in, 323–326
  legacy of, 47–48
  policy implications of, xxi
  practical ramifications of, 78
Poststructuralism, 305–306, 326. *See also*
  Critical theory
Power, 348
  distribution of, 312
  economic, 321
  issues of, 274
  knowledge and, xix
  visibility of, 321
  will to, 322
Practical wisdom, xiii
Pragmatism, xvii–xix, 241, 246
  CST and, 310–311
  expanded, 265
  idealism and, 263
  systemic, xxi
Prana, 177–178
Prediction and Optimal Decisions (POD)
  (Churchman), 2–8, 21, 169
Pre-existing structure, 66
Presentation
  modes of, 95–97, 100
  in natural language, 196
Preston, N., xviii
Prigogine, I., 77
Primary Representational System (PRS), 52,
  54, 57, 59, 64
Principles of Mathematics (Whitehead), 159
Priorities, effective, 292
Probability theory, 110
Probable truth, 109
Problematic, 316
Problematique, 288
Problematization, 312, 320–321
Problems
  definition of, 284
  ordinary v. extraordinary, 19
  space, changing of, 94
  structured, 99
  type of, xi
Processes, 187, 189
Production planning, 174
Productivity, 76, 206
Professional pride, of applied researchers, 31
Profile, 220
Profile efficacy, 219
Profitability, 206, 222
Progress, 3, 116–117, 120, 124–126, 130,
  133, 148
Projected actual profitability, 222
Projective constructivist epistemology, 186
Proliferation, principle of, 132
Prohesis, xv–xvi
Proof, by induction, 108–109
Protolanguages, 58–59
PRS. *See* Primary Representational System
Pseudo-science, 118–119
Public officials, x
Pure becoming, 324–325
Purpose, 185

Quakers, 142, 147–149, 152
Quality control, 144, 154
  statistical, 144–145
  Total, 154
Quality of life, 170
Quantum mechanics, 111, 174, 294
Quine, W.V., 68, 71–72, 112
Quine-Duhem thesis, 121–122, 125

Raja Yoga, 172–173, 180
Randomness, 320–321
Rationality, 306
  epistemological, xii
  legal, 11
  social exercise of, 167
  systemic, xii
  type of, xi
Rau, C., 234–235
RCM. *See* Root Cause Mapping
Reagan, R., 332
Realism, 24
Reality, as map, 308–309
Real-world inquiring system, 4
Reasoning, types of, 30–31
Recursion, of system, 173–174
Reductionism, 112
Redundancy, 70

Reference frames, 128
Referential transparency, 75
Reflection, 67
Reflexivity, xi
Refugees, 237, 239, 242
Regionalism, 228
Regulatory loops, 176
Reichenback, H., 112
Relationships, definition by, 74
Relativism, 130
Relevance, 208
Religions, 162, 166, 180, 307, 333, 336, 342, 344
Remote health care delivery. *See* Telemedicine
Repetition, 322
Representation, xxi–xxii. *See also* Language
  of action, 56–57
  animal, 53–54
  evolution of, 50–51
  level of, 257
  mathematical, 113–114
  participation and, 341–342
  plants and, 52–53
  of society, 56
  of space, 56–57
  of stakeholders, 251
Reproduction
  differences and, 51
  mimetic, 77
Republic (Plato), 158
Research styles, "bad," 31
Resources, 187
Responsibility, 341
Rhizomatic systems, 324–325
Riemann, B., 281–282
Right action, 172, 179
Risk management, xviii
Roles, 197–200
Root Cause Mapping (RCM)
  for CIFM, 295
  of CIFM inhibitors, 296–297*f*
  observer-dependent, 294
Roots, acknowledgment of, 72–73
Ruddock, P., 239
Rules, 188
Rural health care providers, 87, 90
Russell, B., 345

Sales policies, 200–201
Saliency Axiom, 283
Salient cues. *See* Cues
Sampford, C., xviii

Sánchez, J., 211
Saussure, F., 76
Schizoanalysis, 323–326
Schlick, M., 112–115
Schneewind, J.B., 11
Schoorman, F.D., 210–211
Schrodinger, E., 19
Science
  anthropology and, 125
  categorisation of, 134
  crisis in, 127, 129
  cumulative view of, 126, 133
  definition of, 133–134
  demarcation of, 131, 133–134
  ethics and, 146, 155
  factual v. ethical, 153
  First Phase, 294
  foundations of, 74–75
  government and, 154
  history of, 125–126
  inquiring system of, 4, 16
  knowledge and, 178
  management v., 154
  measurement of, 77
  models and, 214
  non-science v., 134
  normal, 131–132
  object language of, 74
  objectivity of, 113
  opinion v., 112
  paradigm of, 126–130
  philosophy of, 106, 108, 131–132, 134
  physical v. social, 74
  positivism and, 111–117
  progress of, 3, 116–117, 120, 124–126
  as public knowledge, 108
  science of, 6–8, 8, 12
  Second Phase, 294
  society and, 154–155
  socio-cultural nature of, 128
  Third Phase, 282, 294
  truth and, 111, 116–117, 119, 120, 124–125, 126, 153
  verificationism and, 106–111
Science of Ethics, 2, 6–9, 12
Science of Management, 149, 155. *See also* Management Science
  Churchman's contributions to, 18–24
  instrumentalities of, 22
  justifications for, 17–18
  Management Science v., 8, 16
  object of study of, 19–20
  questions for, 18–24

unresolved problems of, 22–24
Science of Values. *See* Science of Ethics
Scientific experiments, 121
Scientific Method, 2, 6–7
Scientists, xi
SCIS. *See* Singerian-Churchmanian inquiry systems
Scrafton, M., 249
SDP. *See* Structured Design Process
Second Phase Science, 294
Secondary Representational System (SRS), 52, 57, 59
Second-order cognition, 61
Self-modifying systems, 50
Self-organization, 50, 65
Self-production, 324
Self-referential epistemology, 67–68, 71, 73
Sense, 314–315, 322
Sense datum theory, 113
Sensemaking, 82, 84, 91–92, 95–97, 99–102
   identity and, 85, 90
   intersubjective v. generic, 90–91
   overview of, 85–86
   SCIS and, 86
Sensory perception, 113
September 11, 2001, 238, 247, 257, 335
Service-learning, 275
Sexuality, 180
Shankaracharya, J., 178
Shaping, 41
Shaw, G., 242
Sheep, transportation of, 243–245
Shewhart, W., 144–145
Signals, 57
Significance, 185
Signification, 317–321
Simon, H., 153
Simulacra, 313, 322
Singer, E., xx, 30–31, 148
Singer, P., 18, 39, 81–82, 160, 243, 246
Singerian inquiring system, 306–307
Singerian-Churchmanian inquiry systems (SCIS), 82, 90, 92–93, 99–100. *See also* Distributed Singerian Churchmanian Inquiring Systems
   overview of, 84–85
   sensemaking and, 86
Singularities, 315–316
Skepticism
   Churchman's, x, 105–106
   definition of, x
   about management, viii
Smith, A., 11

Smith, K., 146
Snell, B., 60
Social constructivism, 106, 117, 125–133
Social contract, 11, 342
Social cybernetics, 256
Social domains, 208–209
   external, 208
   internal, 208
   mixed, 208
   values and, 208–209*t*
Social justice, 269–271, 348
Social science, physical science v., 74
Social status, 166
Social systems
   as axiological profiles, 216–218, 223
   critical systems thinking and, 307–311
   general model of, 216
   philosophical models of, 213
   political models of, 213
   sociological models of, 213
   theory of, 210
Society
   representation of, 56–57
   science in, 154–155
Socio-economic well-being, 227–228
Sociological models, of social systems, 213
Socratic approach, xv–xvi
SODA. *See* Strategic Options Development and Analysis
Soft data, 97
Soft formalization, 208
Soft system variables, 5–7
Soft Systems Methodology (SSM), 192
Soft Systems Thinking, 191–193, 202
Solomon, M., 72
Space, representation of, 56–57
Specialization, 271
SRISTI. *See* Vedantic Creation Model
SRS. *See* Secondary Representational System
SSM. *See* Soft Systems Methodology
Stakeholders, community of, 285, 299
Stateness, elimination of, 264
States, strong v. weak, 228, 264–266
Statistical Quality Control, 144–145
Steiner, R., 277
Stereotyping, 254
Stockholm Conference on the Human Environment, 214
Stockholm University, 27
Strange loops, 255
Strategic Choice Methodology, 41

Index   365

Strategic Options Development and
  Analysis (SODA), 192
Stretton, H., 257
Strong states, 228, 264–266
The Structure of Scientific Reasoning
  (Kuhn), 126, 130
Structured Design Process (SDP)
  architecture of, 284–288
  conclusions regarding, 298–299
  dialogue game in, 289–292
  evolution of, 284–288
  five "Cs" of, 285–287
  geometry of paradigm, 281–283
  importance of, 299
  introduction to, 280
  key definitions for, 283–284
  seven graphic language patterns in, 288
  six principles of, 290–291, 298
  story from, 292–298
Structured dialogue. *See also* Structured
    Design Process
  community of stakeholders and, 299
  conscious evolution and, 298
Structured problems, 99
Students, interest of, 29
Subjectivity, 166–167, 293, 325–326
Subjectivization, 323–326
Sub-processes, 189, 197–199
Subsidiarity
  federalism and, 242
  undermining of, 245–248
Success, 206
Superposition Pattern, 288
Superstring theory, 173
Suppliers, 187, 197
Support-of-support processes, 189
Supra-processes, 189
Suroor, H., 248
Sustainability, capacity building for, 263
Sustainable development, 254
Sustainable Sonoma County, 274
Swedish Defense Research Establishment
    (FOA), 28–31
Sweeping-in, of variables, 90, 92, 99, 249,
    254, 259, 307
Synergetics, 50
Synthesis, 78, 82, 84, 246, 258–259,
    306–307, 317
Systemic Efficiency (ES), 209–210, 212,
    215, 219
Systemic Theory of the Enterprise
  case study of, 195–201

composite enterprise constructs in,
    189–190
elementary enterprise constructs in,
    187–188
introduction to, 184–186
organization theory v., 193–194
positioning of, 190–195
summary of, 201–202
Systems thinking v., 191
value of, 190–195
Systemics, 186, 202
Systems
  closure of, 173–174
  creation of, 173–174
  definition of, 207–208
  recursion of, 173–174
Systems Analysis, 191
Systems Approach, 159, 306–307. *See also*
    Systems thinking
  critical theory and, 305–327
  deepening of, 168
  definition of, 269–270, 332
  enemies of, 9, 185, 259, 306–307, 309,
    344
  global warming and, 339
  globalization and, 337–339
  history of, 27
  limitations of, 336–337
  progress of, 12
  universality of, 337
The Systems Approach and Its Enemies
    (Churchman), ix, 30, 165, 184, 331, 342
The Systems Approach (Churchman), ix, 35,
    170
Systems Dynamics, 191
Systems Engineering, 191
Systems Philosophers, 185
Systems Theory, General, 186
Systems thinking, 190, 251–254. *See also*
    Systems Approach
  critical, 191, 193, 202, 307–311
  hard, 191–192, 202
  rationality of, xii
  soft, 191–193, 202
  Systemic theory of the enterprise v., 191

Taittiriya-Samhita, 178
Tampa incident, 236, 240
Taylor, Frederick, xiii
Taylorist Scientific Management, xiii
Techne, xvii
Technical issues, resolution of, 31
Technology, 177–178

assessment of, 28
DSCIS and, 91–92, 101
expertise v., 91–92
networked societies and, 253
TEI. See Theory of Experimental Inference
Teleconsultations, 83–84, 87, 89–97. *See also* Telemedicine
Telemedicine, 83, 87, 89–90, 92–97. *See also* Teleconsultations
case description, of study, 87
conclusions regarding, 102
data analysis, of study, 89
data collection, of study, 87–89
discussion regarding, 98–101
as DSCIS, 90
further research on, 100–101
practice of, 101
wicked decision problems and, 89–90
Ten Commandments, 158–159
Tenacity, principle of, 132
Terleckyi, N.E., 213
Territory, map v., 308–309
Terrorism, 333–337, 343
development and, 335
war on, 244
"Terrorism: A Systemic View," 333
Text-reality, 76
Thatcher, M., 171
Theoretical wisdom, xiii
Theory
choice of, 132–133
domain of, 282
falsification of, 124
pluralism of, 131–132
selection of, 126
truth of, 120
verification of, 119
Theory of Duty, 11
Theory of Everything, 173
Theory of Experimental Inference (TEI) (Churchman), 2–3, 7
Theory of Social Systems, 210
Theory-ladenness, of observation, 112, 114–115, 123–125, 128, 135
Thick description, 230
Third generation rights, 214
Third Phase Science, 282, 294
Thoreau, H.D., 277
Thought
images of, 314–315, 322, 326
of unthinkable, 312–314, 326–327
Thought and Wisdom (Churchman), xvi–xvii, xix, 166–167

Time-based criteria, for social systems, 209
TIMS. *See* Institute of Management Science
Tolerance, 254
Tomlinson, R., 39
Tools, 37–38, 42, 44
Total Quality Control, 154
Total Quality Management (TQM), 156
TQM. *See* Total Quality Management
Trade-off Analysis, 286
Tragedy of the Commons, 165
Transcendental empiricism, 313
Transcendental Meditation, 151
Transcendental unity of apperception, 113–114
Transcendental wisdom, 177
Transformation-objects, 187, 197
Transformative spaces, 274
Transforming entities, theory of, 216
Transmission, categories and, 52
Transnational borders, 242
Transnational governance, 228
Treatise on Human Nature (Hume), 109
Treaty of Westphalia, 243
Tree of Knowledge (Maturana & Varela), 49
Tree of meaning, 291
Treumann, R.D., 77
Triggering question, 283
Trust, 248–249
Truth
coherence theory of, 114
correspondence of, 116, 120, 132
love of, 24
objective, 173
ontology and, 132–133
probable, 109
as process, 249
productivity and, 76
progress v., 120
science and, 111, 116–117, 119, 120, 124–126, 153
of theories, 120
Truth-epistemology, 12
Tsouvalis, C., 320
Turoff, M., 28–29
Two Dogmas of Empiricism (Mach), 112
Two-way learning, 227–228

Ulrich, W., 193, 307–308, 310
Unanticipated consequences, 338
Uncertainty principle, 111
Underdetermination thesis, 122
Undergraduate education, principles for good practice, 273

Unfolding, of values, 249, 259, 263, 307
United Kingdom, National Health Service of, 311
United Nations Aarhus Convention, 231
United Nations Refugee Convention, 242
United States, 264–265
   aggressive militarism of, 340
   borders of, 334
   escalating authoritarianism of, 340
Universal Declaration of Human Rights, 213–214
Universe, world v., 61
University of California, Berkeley, 272
Unpredictability, 320–321
Utilitarianism, 11

Validity, criteria of, 75–76
Values, 213
   attained system of, 218
   facts v., 5–6
   General Model of, 214–215
   indicators of, 214–215
   projected system of, 218
   reference pattern for, 217–218
   social domains and, 208–209*t*
   unfolding of, 249, 259, 263, 307
   universal, 213–214, 213–217
Values-based approach, xviii
Vanstone, A., 237–239, 241
Varela, F., 48–49, 67, 174
Variable Systems Model (VSM)
   SRISTI, 175–177
   subsytems of, 175–176
Vedanta, 170, 172–180
Vedantic Creation Model (SRISTI), 175–177
Vedantic Management, 179–180
Vedantic mathematics (VM), 178–179
Vedic Mathematics (Shankaracharya), 178

Verifiability Principle, 115, 118
Verificationism, 117, 125, 131
   empiricism and, 108
   science and, 106–111
Vienna Circle, 111–112, 117
Vision, action and, 165
VM. *See* Vedantic mathematics

Waismann, F., 112
War
   ethics and, 159, 166
   rules of, 166
   on terror, 244
Warfield, J., 74
Weak states, 228, 264–266
Weltanshauung, 162–165
Whitehead, A.N., 159
Wicked decision problems, 82–83, 94, 100
   DSCIS and, 89, 98, 102
   before telemedicine, 89
   telemedicine and, 89–90
Will to power, 322
Wisdom, 167
   theoretical v. practical, xiii
   transcendental, 177
Wittgenstein, L., 124, 128, 134
World government, 147
World, universe v., 61
World War II, 145–146, 159, 166
Written word, action v., 142

"X of X," 8, 16–17, 20, 155
X-rays, 95

Yield, 206
Young, V., 234
Yugas, 173–174

Printed in the USA

Wisdom, knowledge, and
 management